WAVELETS AND OPERATORS

ii

Wavelets and Operators

Yves Meyer

Professor, Institut Universitaire de France

Translated by D.H. Salinger
University of Leeds

CAMBRIDGE
UNIVERSITY PRESS

iv

Published by the Press Syndicate of the University of Cambridge
The Pitt Building, Trumpington Street, Cambridge CB2 1RP
40 West 20th Street, New York, NY 10011-4211, USA
10 Stamford Road, Oakleigh, Victoria 3166, Australia

First published in English 1992
Printed in France

Library of Congress cataloguing in publication data available
A catalogue record for this book is available from the British Library

ISBN 0 521 42000 8 hardback

Contents

Contents

"Ce à quoi l'un s'était failli, l'autre est arrivé et ce qui était inconnu à un siècle, le siècle suivant l'a éclairci, et les sciences et les arts ne se jettent pas en moule mais se forment et figurent en les maniant et polissant à plusieurs fois [...] Ce que ma force ne peut découvrir, je ne laisse pas de le sonder et essayer et, en retastant et pétrissant cette nouvelle matière, la remuant et l'eschaufant, j'ouvre à celui qui me suit quelque facilité et la lui rends plus souple et plus maniable. Autant en fera le second au tiers qui est cause que la difficulté ne me doit pas désespérer, ni aussi peu mon impuissance ..."
Montaigne, *Les Essais*, Livre II, Chapitre XII.

"Where someone failed, another has succeeded; what was unknown in one century, the next has discovered; science and the arts do not grind themselves into uniformity, but gain shape and regularity by carving and polishing repeatedly [...] What my own strength has not been able to uncover, I cease not from working at and trying out and, by re-shaping and solidifying this new material, in moulding and heating it, I bequeathe to him who follows some facility and make it the more supple and malleable for him. The second will do the same for the third, which is why difficulty does not make me despair, nor of my own weakness..."

Preface to the English edition

The "Theory of Wavelets" lies on the boundaries between (1) mathematics (2) scientific calculation (3) signal processing and (4) image processing. The aim of the theory is to give a coherent set of concepts, methods and algorithms to deal with the difficulties met in each of these disciplines.

Wavelet analysis has come to have applications in widely differing areas of science, because such analysis gives information (of "time–scale" type) about certain signals, images and operators which is more pertinent than that obtained from standard Fourier analysis (or the "time–frequency" methods that derive from it).

The present book has been written by a mathematician and is intended primarily for mathematicians, without forgetting statisticians and engineers working on signal- and image-processing. Our intention has been to take the greatest care in presenting the various constructions of wavelets, and to describe their utilisation within mathematics. The reader whose interest lies in applications of the "wavelet machine" outside mathematics is invited to turn to the second part of the bibliography, which has been specially prepared for the English edition.

I am very conscious of the honour of being published by the Cambridge University Press and I have admired the quality of David Salinger's translation.

Yves Meyer, Paris, 3rd November 1991.

Introduction

For many years, the sine, cosine and imaginary exponential functions have been the basic functions of analysis. The sequence $(2\pi)^{-1/2}e^{ikx}$, $k = 0, \pm 1, \pm 2, \ldots$ forms an orthonormal basis of the standard space $L^2[0, 2\pi]$; Fourier series are the linear combinations $\sum a_k e^{ikx}$. Their study has been and remains, an unquenchable source of problems and discoveries in mathematical analysis. The problems arise from the absence of a good dictionary for translating the properties of a function into those of its Fourier coefficients. Here is an example of the kind of difficulty that occurs. J.P. Kahane, Y. Katznelson and K. de Leeuw have shown ([150]) that, to get a continuous function $g(x)$ from an arbitrary square-summable function $f(x)$, it is sufficient to increase—or leave unchanged—the moduli of the Fourier coefficients of $f(x)$ and to adjust their phases judiciously. It is thus impossible to predict the properties (size, regularity) of a function solely from knowledge of the order of magnitude of its Fourier coefficients. Indeed it is still difficult if we know the Fourier coefficients explicitly, and many problems are still open.

At the beginning of the 1980s, many scientists were already using "wavelets" as an alternative to traditional Fourier analysis. This alternative gave grounds for hoping for simpler numerical analysis and more robust synthesis of certain transitory phenomena. The "wavelets" of J.S. Liénard or of X. Rodet ([167], [206]) were used for numerical treatment of acoustic signals (words or music) and those of J. Morlet ([124]) for stocking and interpreting seismic signals gathered in the course of oil prospecting expeditions. Among mathematicians, research was just as active: to mention only the most striking, R. Coifman and G. Weiss ([75]) invented the "atoms" and "molecules" which were to form the basic building blocks of various function spaces, the rules of assembly being clearly defined and easy to use. Certain of these atomic decompositions could, moreover, be obtained by making a discrete version of a well-known identity, due to A. Calderón, in which "wavelets" were implicitly involved. That identity was later rediscovered by Morlet and his collaborators .... Lastly, L. Carleson used functions very similar to "wavelets" in order to construct an unconditional basis of the H^1 space of E.M. Stein and G. Weiss.

These separate investigations had such a "family resemblance" that it seemed necessary to gather them together into a coherent theory, mathematically well-founded and at the same time, universally applicable. The **orthonormal wavelet bases**, whose construction is given in the present volume, are a replacement for the empirical "wavelets" of Liénard, Morlet and Rodet.

The same orthonormal wavelet bases give direct access to the "atomic decompositions" of Coifman and Weiss, which are thus—for the first time—related to constructions of **unconditional bases** of the standard spaces of functions and distributions. The wavelet bases are universally applicable: "everything that comes to hand", whether function or distribution, is the sum of a wavelet series and, contrary to what happens with Fourier series, the coefficients of the wavelet series translate the properties of the function or distribution simply, precisely and faithfully.

So we have a new tool at our command, an instrument that lets us perform, without thinking, the delicate constructions that could not formerly be achieved without recourse to lacunary, or random, Fourier series. The exceptional properties of the sums of these special series become the everyday properties of generic sums of wavelet series.

The algorithms for analysis by, and synthesis of, orthogonal wavelet series will, doubtless, play an important role in many different branches of science and technology. Mathematicians, physicists, and engineers who want to know everything about wavelets now have the present volume (Chapters 1–6) of this work at their disposal.

Volumes 2 and 3 are addressed more specifically to an audience of mathematicians. They deal with the operators associated with wavelets. G. Weiss has shown that the study of the operators acting on a space of functions or distributions can become very simple when the elements of the space admit "atomic decompositions". He writes "many problems in analysis have natural formulations as questions of continuity of linear operators defined on spaces of functions or distributions. Such questions can often be answered by rather straightforward techniques if they can first be reduced to the study of the operator on an appropriate class of simple functions which, in some convenient sense, generate the entire space."When these "simple elements" were the functions e^{ikx} of the trigonometric system, the bounded operators T on L^2, which were diagonal with respect to the trigonometric system, did not have any other interesting property (with the exception of translation-invariance, which follows immediately from the definition). It was then necessary to impose quite precise conditions on the eigenvalues of T in order to extend such an operator to other function spaces: the first results in this direction were obtained by J. Marcinkiewicz.

However, the bounded operators which can be diagonalized exactly or approximately, with respect to the wavelet basis, form an algebra **A** of bounded operators on L^2 and the well-known **Calderón-Zygmund real-variable methods** enable the operators of **A** to be extended to other spaces of functions or distributions. The algebra **A**, which extends the pseudo-differential operators in a natural sense, is strictly contained in the set **C** of operators whose study has been recommended by Calderón. Work on these operators should enable us to solve several outstanding classical problems in complex analysis and partial differential equations.

Here is a slightly more precise description of the set **C**, the delicate construction of which we have called "Calderón's programme". After having invented, together with A. Zygmund, what was to become the classical pseudo-differential calculus, Calderón intended to extend the field of application systematically, by weakening, as far as possible, the regularity hypotheses necessary for the algorithms to work.

The fundamental—and unexpected—discovery made by Calderón was the existence of a limit to the search for minimal hypotheses of regularity. There is a "natural boundary" which cannot be transgressed, and the extension of operators to this boundary is precisely the analytic extension of holomorphic functions on certain Banach spaces, as we shall show in Chapter 13.

Chapters 7–9 are devoted to the study and then the construction of the set **C** of operators of Calderón's programme. We call them the **Calderón-Zygmund operators**, although they are very different from the "historical operators" studied by Calderón and Zygmund in the 1950s and 60s.

Just like these "historical operators", those we consider can be defined by singular integrals, in a new sense which we clarify in Chapter 7. To go beyond the context of convolution operators, it becomes indispensable to have a criterion for L^2-continuity, without which the theory collapses like a house of cards. One such criterion is the well-known theorem $T(1)$ due to David and Journé, which we shall prove in Chapter 8. Theorem $T(1)$ replaces the Fourier transform, whose use remains restricted to convolution operators.

Unfortunately, theorem $T(1)$, although giving a necessary and sufficient condition, is not directly applicable to the most interesting operators of the set **C** of Calderón's programme. We do not know why that is. The operators in question have, however, a very special non-linear structure, which, when correctly exploited, allows us to pass from the "local" results given by David and Journé's theorem to the "global" theorems necessary for the functioning of Calderón's programme.

In Volume 3 of this book (Chapters 12–16) and also in Chapter 9 of Volume 2, we have given the most beautiful of the applications of Calderón's programme. First comes the celebrated pseudo-differential calculus, initiated by Calderón, which, at present, has interesting and important applications to non-linear partial differential equations.

Then we pass to complex analysis and the Hardy spaces associated with Lipschitz domains of the complex plane. The object of Chapter 12 is the study of the Cauchy operator on rectifiable curves. We then examine the problem, posed by T. Kato, of determining the domain of the square roots of second order pseudo-differential operators, in the accretive case.

After that, we give an account of the results of B. Dahlberg, D. Jerison, C. Kenig and G. Verchota relative to the Dirichlet and Neumann problems in Lipschitz open sets.

The book ends with a brief presentation of J.M. Bony's paradifferential operators, which serve to analyse the regularity of non-linear partial differential operators.

Wavelets reappear, in a surprising way, as the eigenfunctions of certain paradifferential operators. Correctly handled, they remain present in the study of Hardy spaces and the Cauchy operator on a Lipschitz curve: the operator is "almost diagonal" with respect to a wavelet basis specially designed for complex analysis on that curve (Chapter 11). The construction of wavelet bases is thus sufficiently supple to be adaptable to differing geometric situations: we also obtain "non-orthogonal wavelet bases". At present, there is no universal basis which can simultaneously be used in the analysis of all the operators of Calderón's class **C**.

J.O. Strömberg was the first to construct an orthonormal basis of $L^2(\mathbb{R})$, of the form $2^{j/2}\psi(2^j x - k)$, $j, k \in \mathbb{Z}$, where, for each $m \in \mathbb{N}$, the function $\psi(x)$ was of class C^m and decreased exponentially at infinity.

Subsequent work on orthogonal wavelets has not followed the path discovered by Strömberg. That work is, essentially, due to I. Daubechies, P.G. Lemarié, S. Mallat, and the author. It is given here, with care and with complete proofs.

As far as operators are concerned, the well-known results of Calderón, Zygmund, and Cotlar will be described in Chapter 7, in the new context of the set **C**.

The other names the reader of this work will often encounter are J.M. Bony, G. David, P. Jones, J.L. Journé, C. Kenig, T. Murai, and S. Semmes.

The division into three volumes will allow this work to be read in several ways. As we have already suggested, the reader may wish to go no further than the present volume, which surveys our present knowledge about wavelets. But Volume 2 (Calderón-Zygmund operators) may also be read directly, assuming only the results quoted in the introductions of the first six chapters. Finally, the reader can go straightaway to Chapters 12, 13, 14, 15 or 16 of Volume 3, because each of them forms a coherent account of a subject of independent interest (complex analysis, holomorphic functionals on Banach spaces, Kato theory, elliptic partial differential equations in Lipschitz domains, and, lastly, non-linear partial differential equations). The thread linking these different themes is, quite clearly, the use of wavelets in Calderón's programme in operator theory.

This book has been written at a level appropriate for first-year postgraduates, and we have tested it in France, and in the USA, on various audiences of mathematicians and engineers. To read this book, it is therefore not necessary to have studied the remarkable treatises by E. Stein and G. Weiss ([221]), E. Stein ([217]), or Garcia Cuerva and Rubio de Francia ([115]), not to mention the fundamental text and reference by Zygmund ([239]).

R. Coifman helped me to recognize the importance of Calderón's programme. Since the summer of 1974, our scientific collaboration has been devoted to its realization, and this book has been one of our projects. If our own work no longer appears in its original form in this book, it is because our zeal and enthusiasm have communicated themselves to younger research workers, who have found more elegant solutions to the problems we had been determined to resolve.

1

Fourier series and integrals, filtering and sampling

1 Introduction

Wavelet series provide a simpler and more efficient way to analyse those functions and distributions that have hitherto been studied by means of Fourier series and integrals. But wavelet analysis cannot entirely replace Fourier analysis, indeed, the latter is used in constructing the orthonormal bases of wavelets needed for analysis with wavelet series. To construct the basic wavelets we use what "works well" in Fourier analysis—its algebraic formalism. Once the wavelets have been constructed, they perform incredibly well in situations where Fourier series and integrals involve subtle mathematics or heavy numerical calculations.

The two kinds of analysis are thus complementary rather than competing. What is more, the reader will need to know the rudiments of Fourier analysis in order to proceed!

In the first chapter we restate the classical formulas and theorems of Fourier analysis, but the chapter also serves as an "overture" in which the "themes" of this book—wavelets and operators—make a preliminary appearance.

2 Fourier series

Fourier series are used to analyse periodic functions or distributions. To

begin with, we shall consider the case of one real variable and suppose
the period is 2π.

We start with what works best. Let H be the Hilbert space $L^2[0, 2\pi]$
with the inner product $(f, g) = \int_0^{2\pi} f(x)\bar{g}(x)\, dx$. Then the functions
$(2\pi)^{-1/2}e^{ikx}$, $k \in \mathbb{Z}$, constitute a Hilbert basis of H. With a slight
change of normalization, we define the Fourier coefficients of $f \in H$ by

$$(2.1) \qquad c_k = \frac{1}{2\pi} \int_0^{2\pi} f(x)e^{-ikx}\, dx,$$

and we have

$$(2.2) \qquad f(x) = \sum_{-\infty}^{\infty} c_k e^{ikx},$$

the series converging in H.

The identity (2.2) automatically defines an extension of the function
$f \in L^2[0, 2\pi]$ to the whole real line. This extension is just the 2π-periodic
function whose restriction to $[0, 2\pi]$ is exactly $f(x)$. The interval $[0, 2\pi)$
acts as a fundamental domain for the discrete subgroup $2\pi\mathbb{Z} \subset \mathbb{R}$ and
the 2π-periodic functions which are locally square-summable are canonically identified with functions of $L^2[0, 2\pi]$. The same remark applies to
the spaces $L^p[0, 2\pi]$ which we shall meet later. But it cannot apply to
the space $F \subset \mathcal{D}'(\mathbb{R})$ of 2π-periodic distributions. Recall that $\mathcal{D}(\mathbb{R})$ is
the space of infinitely differentiable functions of compact support and
that $\mathcal{D}'(\mathbb{R})$ is the space of continuous linear functionals on $\mathcal{D}(\mathbb{R})$. A
distribution $S \in \mathcal{D}'(\mathbb{R})$ is 2π-periodic if $\langle S, u \rangle = \langle S, v \rangle$ whenever u and v
belong to $\mathcal{D}(\mathbb{R})$ and satisfy $v(x) = u(x - 2\pi)$. Here we have used $\langle \cdot, \cdot \rangle$ to
denote the bilinear form which implements the duality between distribu-
tions and test functions. 2π-periodic distributions are not characterized
by their restrictions to the open interval $(0, 2\pi)$ because this process can
lose information. One example of such loss involves the "Dirac comb"
defined by

$$(2.3) \qquad S = \sum_{-\infty}^{\infty} \delta_{2k\pi}$$

where δ_a denotes the Dirac measure at a. The restriction of S to the
open interval $(0, 2\pi)$ is zero. Similarly, S cannot sensibly be restricted to
the closed interval $[0, 2\pi]$. It appears, then, that we cannot use the usual
formulas to define the Fourier coefficients of a 2π-periodic distribution.
We get round this difficulty in the following way. Let E be the vector
space of infinitely differentiable 2π-periodic functions. Put E and F in
duality by setting

$$(2.4) \qquad \langle S, f \rangle = \langle S, \phi f \rangle$$

where $S \in F$, $f \in E$ and where $\phi \in \mathcal{D}(\mathbb{R})$ is such that

$$(2.5) \qquad \sum_{-\infty}^{\infty} \phi(x + 2k\pi) = 1.$$

In some sense, such a function ϕ imitates the characteristic function of $[0, 2\pi)$. The left-hand side of (2.4) is defined by the right-hand side, which makes sense because the product ϕf is in $\mathcal{D}(\mathbb{R})$.

It is an exercise to verify the invariance of (2.4) given different choices of ϕ: subtraction of two such choices leads to using the equivalence, for $g \in \mathcal{D}(\mathbb{R})$, of

$$(2.6) \qquad \sum_{-\infty}^{\infty} g(x + 2k\pi) = 0$$

and

$$(2.7) \qquad \exists h \in \mathcal{D}(\mathbb{R}) \qquad \text{such that} \qquad g(x) = h(x + 2\pi) - h(x).$$

As a particular case of (2.4) we shall define the Fourier coefficients c_k, $k \in \mathbb{Z}$, of an arbitrary 2π-periodic distribution S. We put $e_k(x) = e^{ikx}$, $\bar{e}_k(x) = e^{-ikx}$ and

$$(2.8) \qquad c_k = \frac{1}{2\pi}\langle S, \bar{e}_k \rangle.$$

Then, for some integer $m \in \mathbb{N}$ and constant C, $|c_k| \leq C(1 + |k|)^m$ and this property characterizes the Fourier coefficients of a 2π-periodic distribution. Finally

$$(2.9) \qquad S = \sum_{-\infty}^{\infty} c_k e_k$$

and the series on the right-hand side converges in the sense of distributions. This means that, if $u \in \mathcal{D}(\mathbb{R})$ is a test function, then

$$(2.10) \qquad \langle S, \bar{u} \rangle = \sum_{-\infty}^{\infty} c_k \bar{d}_k \qquad \text{where} \qquad d_k = \int_{-\infty}^{\infty} u(x) e^{-ikx}\, dx$$

and the series in (2.10) is absolutely convergent, because the d_k decrease rapidly at infinity.

One can similarly use the duality between E and F, instead of that between $\mathcal{D}(\mathbb{R})$ and $\mathcal{D}'(\mathbb{R})$, to interpret (2.9).

If $f \in E$, that is, if f is infinitely differentiable and 2π-periodic, we denote its Fourier coefficients by d_k, $k \in \mathbb{Z}$ (defined as in (2.1)) and we get

$$(2.11) \qquad \langle S, \bar{f} \rangle = 2\pi \sum_{-\infty}^{\infty} c_k \bar{d}_k \,,$$

which is Plancherel's identity.

We can differentiate equation (2.9), term by term, arbitrarily often, to give the Fourier series of the derivatives of S (in the sense of distributions).

Let us give two classical applications of this rule.

We start with the "saw-tooth" function $s(x)$ which is odd, 2π-periodic and equal to $(\pi-x)/2$ on $[0, 2\pi)$. This function is in $L^2[0, 2\pi]$ and satisfies

$$(2.12) \qquad\qquad s(x) = \sum_1^\infty \frac{1}{n} \sin nx.$$

Before differentiating (2.12) term by term, it is worthwhile drawing the graph of the 2π-periodic function $s(x)$ which highlights the discontinuities of the first kind at each point $2k\pi$, $k \in \mathbb{Z}$.

Now let us differentiate term by term, in the sense of distributions in $\mathcal{D}'(\mathbb{R})$. The derivative of $s(x)$ is the sum of the usual derivative (the constant function $-1/2$) and of πS, where S is the Dirac comb. The series $\sum_1^\infty \cos nx$ appears on the right-hand side and we finally arrive at the result

$$(2.13) \qquad\qquad \sum_{-\infty}^\infty \delta_{2k\pi} = \frac{1}{2\pi} \sum_{-\infty}^\infty e^{ikx},$$

which is the Poisson summation formula.

Here is another amusing application of the same ideas. We start with the even 2π-periodic function $c(x) = \log(|\sin(x/2)|^{-1})$, where log denotes the natural logarithm. Once again, $c(x)$ belongs to $L^2[0, 2\pi]$ and its Fourier series is

$$(2.14). \qquad\qquad \log \frac{1}{|\sin(x/2)|} = \log 2 + \sum_1^\infty \frac{1}{n} \cos nx.$$

This calls for some preliminary remarks which will be developed in Chapter 6. Let $H : L^2[0, 2\pi] \longrightarrow L^2[0, 2\pi]$ denote the operator defined by $H(e^{ikx}) = -i\,\mathrm{sgn}(k)e^{ikx}$ with $H(1) = 0$. In other words, $H(\cos kx) = \sin kx$, for $k \in \mathbb{N}$, and $H(\sin kx) = -\cos kx$, for $k \geq 1$.

The operator H is called the Hilbert transform. It is clearly continuous on $L^2[0, 2\pi]$, but is certainly not continuous on $L^\infty[0, 2\pi]$. Indeed, the function $s(x)$ belongs to $L^\infty[0, 2\pi]$, but the function $c(x)$ is not in L^∞, yet these two functions are related by the identity $(H(s))(x) = -c(x) + \log 2$. We shall discover the explanation of this phenomenon in Chapter 7, when we study the (L^∞, BMO)-continuity of the Calderon-Zygmund operators, of which H is the prototype.

We can make a further remark about (2.14). Just as the derivative of $\log |x|$ is the distribution $\mathrm{PV}\, x^{-1}$, the derivative—in the sense of distributions—of $\log(|\sin(x/2)|^{-1})$ is $\frac{1}{2}\,\mathrm{PV}\cot(x/2)$, a 2π-periodic distribution whose singularities are at $2k\pi$, $k \in \mathbb{Z}$, and are of the same type as those of $\mathrm{PV}\big(1/(x - 2k\pi)\big)$. In fact, in a neighbourhood of $2k\pi$, the difference between the two distributions is an infinitely differentiable function.

On differentiating (2.14) in the sense of distributions we get

$$(2.15) \qquad \mathrm{PV}\cot\frac{x}{2} = \sum_{-\infty}^{\infty} -i\,\mathrm{sgn}(k)e^{ikx} = \sum_{-\infty}^{\infty} H(e^{ikx}).$$

This new identity has a remarkable interpretation: the operator H can be defined as convolution with the distribution $S = (1/2\pi)\,\mathrm{PV}\cot(x/2)$, where the convolution product $S * T$ of two 2π-periodic distributions $S = \sum_{-\infty}^{\infty} c_k e^{ikx}$ and $T = \sum_{-\infty}^{\infty} d_k e^{ikx}$ is defined by

$$S * T = 2\pi \sum_{-\infty}^{\infty} c_k d_k e^{ikx}.$$

This definition extends the usual convolution $f * g$ of two functions in $L^1[0, 2\pi]$. Recall that the convolution product is defined by

$$(2.16) \qquad (f * g)(x) = \int_0^{2\pi} f(x - y)g(y)\,dy$$

where, on the right-hand side, f is extended to the whole of $\mathbb{R}$ by periodicity. The vector space F of 2π-periodic distributions is a topological algebra under the convolution product and the identity of this algebra is the Dirac comb, $\sum_{-\infty}^{\infty} \delta_{2k\pi} = (1/2\pi)\sum_{-\infty}^{\infty} e^{ikx}$.

Finally, for $f \in E$, we write $f(x) = \sum_{-\infty}^{\infty} \alpha_k e^{ikx}$ and get

$$H(f) = \sum_{-\infty}^{\infty} -i\,\mathrm{sgn}\,k\,\alpha_k e^{ikx} = \frac{1}{2\pi}\,\mathrm{PV}\cot\left(\frac{x}{2}\right) * f$$

$$= \frac{1}{2\pi}\lim_{\varepsilon\downarrow 0}\int_{\varepsilon\leq|y|\leq\pi} f(x - y)\cot\frac{y}{2}\,dy.$$

What we have here is the realization of the operator H as a singular integral: this point of view will be developed systematically in the course of Chapter 7.

Fourier analysis in the $L^p[0, 2\pi]$ spaces is subtler than that in E, F and $L^2[0, 2\pi]$.

When $1 < p < \infty$, the partial sums $\sum_{-N}^{N} c_k e^{ikx}$ of the Fourier series of a function $f \in L^p[0, 2\pi]$ converge to f in L^p norm. This result follows from the continuity of the Hilbert transform H on $L^p[0, 2\pi]$ when $1 < p < \infty$. The Hilbert transform is not continuous on L^1 or on the space of continuous 2π-periodic functions on $\mathbb{R}$. As a result, the theorem about norm-convergence of partial sums is no longer true when $p = 1$ or $p = \infty$.

The L^p norm of f cannot be evaluated just from the amplitudes of the

Fourier coefficients of f: the phases of the coefficients play an essential part. Here are two examples.

If $\sum_{-\infty}^{\infty} |c_k|^2 < \infty$, then, for almost all choices of signs $\pm$ (in the sense of independent, centred, Bernoulli random variables), the random Fourier series $\sum_{-\infty}^{\infty} \pm c_k e^{ikx}$ converges to a function that, for $2 \leq p < \infty$, belongs to each L^p space. But the condition $\sum_{-\infty}^{\infty} |c_k|^2 < \infty$ is sufficient for $\sum_{-\infty}^{\infty} c_k e^{ikx}$ to belong to L^p only when $p = 2$. The random choice of signs $\pm$ has let us scale the ramparts of the L^p spaces.

Consider the particular series $\sum_1^{\infty} k^{-\alpha} e^{ikx}$ for $1/2 < \alpha < 1$. Its sum is a 2π-periodic function $f_\alpha(x)$ whose restriction to $[-\pi, \pi]$ is continuous except at the origin. If x tends to 0 from above, $f(x)$ is equivalent to $c(\alpha)x^{-1+\alpha}$, where $c(\alpha)$ is a non-zero complex constant. A necessary and sufficient condition for $f_\alpha(x)$ to belong to $L^p[0, 2\pi]$ is that $p(1 - \alpha) < 1$. But the corresponding random series $\sum_1^{\infty} \pm k^{-\alpha} e^{ikx}$ define (almost everywhere) functions which are continuous (and which belong to the Hölder class of exponent β for $\beta < \alpha - 1/2$). The well-known Brownian motion can be described by such random Fourier series ([149]).

When we want to calculate the L^p norm of a function, knowing each Fourier coefficient gives only the illusion of precision. In calculating every Fourier coefficient, we have taken the analysis too far. We must retrace our steps a little and group the coefficients into so-called dyadic blocks, which we do not break down any further. The dyadic blocks of the Fourier series of f are

$$(2.17) \qquad \Delta_j f(x) = \sum_{2^j \leq |k| < 2^{j+1}} c_k e^{ikx}, \qquad j \in \mathbf{N},$$

and the fundamental result of Littlewood and Paley ([171]) is that, for $1 < p < \infty$, the two norms $\|f\|_p$ and $|c_0| + \|(\sum_0^{\infty} |\Delta_j f(x)|^2)^{1/2}\|_p$ are equivalent. This means that changes of sign, which could not previously be made without changing the L^p status of the functions, become quite harmless when applied to the dyadic blocks $\Delta_j f$. If $\sum_0^{\infty} \Delta_j f$ belongs to L^p then so does $\sum_0^{\infty} \varepsilon_j \Delta_j f$, for every choice of $\varepsilon_j = \pm 1$.

Decomposing a Fourier series into dyadic blocks plays just as essential a role in the analysis of the Hölder spaces C^α. But we then need to define the operators Δ_j more carefully. To do this, we take a function, $\psi(x) \in \mathcal{D}(\mathbf{R})$, which is even, supported by the set $1/2 \leq |x| \leq 3/2$, and such that $1 = \psi(x) + \psi(x/2) + \psi(x/4) + \cdots$, when $|x| \geq 1$. We then change the definition of the Δ_j to

$$(2.18) \qquad \Delta_j f(x) = \sum \psi(k2^{-j}) c_k e^{ikx}.$$

Then, for every $\alpha > 0$, f is in C^α if and only if $\|\Delta_j f\|_\infty = O(2^{-j\alpha})$ as $j \to \infty$. Recall that, for $0 < \alpha < 1$, the Hölder space C^α is composed of those continuous 2π-periodic functions whose moduli of continuity $\omega(h)$ are $O(h^\alpha)$. When $\alpha = 1$, $\|\Delta_j(f)\|_\infty = O(2^{-j})$ means that $f \in \Lambda_*$ and not that $f \in C^1$ (the Zygmund class, Λ_*, is the Banach space of continuous 2π-periodic functions $f(x)$ satisfying $|f(x+y) + f(x-y) - 2f(x)| \le Cy$, for each real x and $0 \le y \le 1$). For $1 < \alpha \le 2$, the C^α spaces, are subspaces of (the usual) C^1 and are defined by the condition $f' \in C^{\alpha-1}$ (where f' is the derivative of f). The definition extends similarly to all $\alpha > 0$.

The partial sums of the Fourier series of a C^α function do not, in general, tend to that function in C^α and are not even uniformly bounded in C^α norm. The same phenomenon occurs for the C^0 norm (the uniform norm) and the L^1 norm.

The remedy lies in the introduction of summability methods. We give Borel's method as an example. Instead of approximating the series $\sum_{-\infty}^{\infty} c_k e^{ikx}$ by its partial sums $\sum_{-N}^{N} c_k e^{ikx}$, we approximate by $\sum_{-\infty}^{\infty} c_k e^{-|k|\varepsilon} e^{ikx}$ and then let ε tend to zero. This amounts to taking the convolution of the function f, whose Fourier coefficients are the c_k, with an approximate identity P_ε. More precisely, we start by extending $f(x)$ to a 2π-periodic function or distribution and put

$$P_\varepsilon(x) = \frac{1}{\varepsilon} P\left(\frac{x}{\varepsilon}\right) \qquad \text{where} \qquad P(x) = \frac{1}{\pi} \frac{1}{1+x^2}.$$

Then, if $f(x)$ is continuous (and 2π-periodic), $f * P_\varepsilon$ converges to $f(x)$ uniformly; the same holds if $f(x)$ is locally integrable.

Many other approximate identities are frequently used and any reader who wishes to study them more closely may refer to [239].

3 Fourier integrals

We start by defining the "naive" Fourier integrals: those of functions f in $L^1(\mathbb{R})$. Here we simply put

$$(3.1) \qquad \hat{f}(\xi) = \int_{-\infty}^{\infty} e^{-i\xi x} f(x) \, dx \qquad \text{for} \qquad -\infty < \xi < \infty.$$

There is no difficulty in showing that $\hat{f}(\xi)$ is a continuous function of ξ and that this function vanishes at infinity.

The convolution product of two functions f and g in $L^1(\mathbb{R})$, gives a third function h in $L^1(\mathbb{R})$ and we get $\hat{h}(\xi) = \hat{f}(\xi)\hat{g}(\xi)$. The Fourier transforms of $L^1(\mathbb{R})$-functions thus form a subalgebra of $C_0(\mathbb{R})$. This subalgebra is called the Wiener algebra, is denoted by $A(\mathbb{R})$ and becomes a Banach algebra if we put $\|\hat{f}\|_A = \|f\|_1$.

Despite its apparent simplicity, the Wiener algebra has not yet given up all its secrets: the interested reader may refer to [148] or to [209].

A property that follows easily from (3.1) is

$$(3.2) \qquad \int \hat{f}(\xi)g(\xi)\,dx = \int f(x)\hat{g}(x)\,dx.$$

Equation (3.2) is a simple consequence of Fubini's theorem applied to $\iint f(x)g(\xi)e^{-ix\xi}\,dx\,d\xi$.

In this book, we shall make systematic use of the following remark:

Lemma 1. Let $f(x)$ be a function in $L^1(\mathbb{R})$. For $T > 0$, let $g(x) = \sum_{-\infty}^{\infty} f(x+kT)$. Then $g(x) \in L^1[0,T]$ and the Fourier coefficients

$$c_k = \frac{1}{T}\int_0^T g(x)\exp\left(\frac{-2\pi ikx}{T}\right)\,dx$$

of g are given by

$$c_k = \frac{1}{T}\hat{f}\left(\frac{2k\pi}{T}\right).$$

The lemma follows immediately from Fubini's theorem and from the fact that $[0,T)$ is a fundamental domain for the subgroup $T\mathbb{Z}$ of $\mathbb{R}$.

As an application of Lemma 1, we give a proof—found in a physics book—of the Fourier inversion formula. Suppose that $f(x)$ and $\hat{f}(\xi)$ are continuous, $O(x^{-2})$ and $O(\xi^{-2})$, respectively, as x and ξ tend to $\pm\infty$. Then we have

$$(3.3) \qquad f(x) = \frac{1}{2\pi}\int_{-\infty}^{\infty} e^{ix\xi}\hat{f}(\xi)\,d\xi.$$

To see this, we apply Lemma 1. The function $g(x)$ is continuous, because the series defining it is uniformly convergent. The Fourier coefficients of $g(x)$ are $O(k^{-2})$, so $g(x)$ is the sum of its Fourier series. We thus get

$$(3.4) \qquad \sum_{-\infty}^{\infty} f(x+kT) = \frac{1}{T}\sum_{-\infty}^{\infty}\hat{f}\left(\frac{2k\pi}{T}\right)\exp\left(i\frac{2k\pi}{T}\right).$$

Having got this far, we fix x and let T tend to infinity, interpreting the right-hand side of (3.4) as a Riemann sum. This gives (3.3).

The physicists' point of view is to consider a function which is $O(x^{-2})$ at infinity as a periodic function of infinite period!

The assumptions of this argument hold in the special case of f belonging to the Schwartz space $\mathcal{S}(\mathbb{R})$. This means that, for each integer $m \geq 1$ and $n \geq 0$, the nth derivative, $f^{(n)}(x)$, of $f(x)$ is $O(x^{-m})$ as $x \to \pm\infty$. Then simple calculations (term by term differentiation and integration by parts) are enough to show that $\hat{f} \in \mathcal{S}(\mathbb{R})$, whenever $f \in \mathcal{S}(\mathbb{R})$.

So (3.3) holds when $f \in \mathcal{S}(\mathbb{R})$. It follows that $f \in \mathcal{S}(\mathbb{R})$ if and only if $\hat{f} \in \mathcal{S}(\mathbb{R})$ and that $\mathcal{F}: \mathcal{S}(\mathbb{R}) \to \mathcal{S}(\mathbb{R})$, where $\mathcal{F}$ denotes the Fourier transform, is a topological isomorphism.

We use (3.3) to define the Fourier transform $\hat{f}$ of a tempered distribution f. We decree that, for every function $g \in \mathcal{S}(\mathbb{R})$, the meaning of $\langle \hat{f}, g \rangle$ should be $\langle f, \hat{g} \rangle$. In other words, $\mathcal{F} : \mathcal{S}'(\mathbb{R}) \to \mathcal{S}'(\mathbb{R})$ is the transpose of the operator $\mathcal{F} : \mathcal{S}(\mathbb{R}) \to \mathcal{S}(\mathbb{R})$.

An important special case is that of distributions S of compact support. In this case we can define the function $\hat{S}(\xi)$ directly by $\hat{S}(\xi) = \langle S, \bar{e}_\xi \rangle$, where $e_\xi = e^{ix\xi}$.

The Paley-Wiener theorem gives a characterization of the functions $\hat{S}(\xi)$ which correspond to distributions S whose supports lie in a fixed interval $[-l, l]$, namely, that $\hat{S}(\xi)$ can be extended to the complex plane as an entire function, $F(z)$ satisfying, for each $\varepsilon > 0$, $|F(z)| \le C(\varepsilon)\exp((l + \varepsilon)|z|)$. Further, $F(\xi) = \hat{S}(\xi)$ grows slowly on the real axis: there is an integer m such that $|F(\xi)| \le C(1 + |\xi|)^m$.

One example of a tempered distribution is given by a 2π-periodic function, f, whose restriction to $[0, 2\pi]$ is in $L^1[0, 2\pi]$. Here the Fourier transform in the sense of distributions, f, is given by

$$(3.5) \qquad \hat{f} = 2\pi \sum_{-\infty}^{\infty} c_k \delta_k \quad \text{where} \quad c_k = \frac{1}{2\pi} \int_0^{2\pi} f(x)e^{-ikx}\,dx$$

and where δ_k is Dirac measure at k.

Let us now give the definition of the Fourier transform, $\mathcal{F}$, on $L^2(\mathbb{R})$. Equations (3.2) and (3.3) imply that $(\hat{f}, \hat{g}) = 2\pi(f, g)$ when f and g belong to $\mathcal{S}(\mathbb{R})$ —we have put $(f, g) = \int_{-\infty}^{\infty} f(x)\bar{g}(x)\,dx$. As a result, $(2\pi)^{-1/2}\mathcal{F} : \mathcal{S}(\mathbb{R}) \to \mathcal{S}(\mathbb{R})$ becomes an isometry when $\mathcal{S}(\mathbb{R})$ is regarded as a subspace of $L^2(\mathbb{R})$. Since $\mathcal{S}(\mathbb{R})$ is dense in $L^2(\mathbb{R})$, the isometry extends to the whole of $L^2(\mathbb{R})$.

For f belonging to $L^2(\mathbb{R})$, the integral $\int_{-\infty}^{\infty} e^{-ix\xi}f(x)\,dx$ will be defined as the limit, in $L^2(\mathbb{R})$ norm, of the truncated integrals $g_T(\xi) = \int_{-T}^{T} e^{-ix\xi}f(x)\,dx$.

Using Carleson's theorem, C. Kenig has shown that the truncated integrals $g_T(\xi)$ converge almost everywhere to $\hat{f}(\xi)$ as $T \to \infty$. We refer the reader to [161].

4. Filtering and sampling

For technological reasons, signals that one wants to analyse have a frequency spectrum limited to a "band" $[-T, T]$. (Recall that mathematical frequencies have a sign which arises from the use of the formulas $\cos \omega t = \frac{1}{2}(e^{i\omega t} + e^{-i\omega t})$ and $\sin \omega t = \frac{1}{2i}(e^{i\omega t} - e^{-i\omega t})$, $\omega > 0$.) So it is essential to make full use of the fact that the frequencies contained in a signal do not exceed T. That is the purpose of this section. To begin with, we study the sampling of such signals. That is, we give a rule which lets us decide whether the measurements $f(k\delta)$, $\delta > 0$, $k \in \mathbb{Z}$, of a signal, are sufficient to recover the signal from those measurements when "Shannon's condition" is satisfied.

After this preamble, we come to the precise mathematical statement. For $T > 0$, let $\mathcal{E}_T$ denote the vector space of distributions $f \in \mathcal{S}'(\mathbb{R})$ whose Fourier transforms S have supports in $[-T, T]$.

The Fourier inversion formula gives $\hat{S}(\xi) = 2\pi f(-\xi)$. Now, $\hat{S}(\xi)$ is the restriction to the real axis of an entire function of exponential type, and so the same holds for $f(\xi)$. In particular, f is infinitely differentiable and thus continuous on the real axis, so the sampling $f(k\delta)$, $k \in \mathbb{Z}$, makes sense.

The answer to the basic problem is given by the following result.

Theorem 1. *If $\delta > \pi T^{-1}$, the sampling $f(k\delta)$, $k \in \mathbb{Z}$, does not determine $f \in \mathcal{E}_T$ uniquely, even under the additional assumption that the sequence $f(k\delta)$ decreases rapidly and that the function f belongs to the Schwartz class $\mathcal{S}(\mathbb{R})$.*

If $\delta = \pi T^{-1}$, the sampling $f(k\delta)$, $k \in \mathbb{Z}$, is sufficient to determine $f \in \mathcal{E}_T$ as long as, in addition, $f(k\delta) \in l^p(\mathbb{Z})$, for $1 < p < \infty$, and $f \in L^p(\mathbb{R})$. If $p = \infty$ and if c_k, $k \in \mathbb{Z}$, is a sequence belonging to $l^\infty(\mathbb{Z})$, then a function $f \in \mathcal{E}_T$ belonging to $L^\infty(\mathbb{R})$, such that $f(k\delta) = c_k$, $k \in \mathbb{Z}$, neither necessarily exists nor is necessarily unique. If c_k, $k \in \mathbb{Z}$, belongs to $l^1(\mathbb{Z})$, a function $f \in L^1(\mathbb{R}) \cap \mathcal{E}_T$, with $f(k\delta) = c_k$, $k \in \mathbb{Z}$, may not exist, but is unique when it does.

Lastly, if $0 < \delta < \pi T^{-1}$, let $\phi \in \mathcal{S}(\mathbb{R})$ be a function whose Fourier transform $\hat{\phi}$ vanishes outside $[-\pi\delta^{-1}, \pi\delta^{-1}]$ and equals 1 on the interval $[-T, T]$. Then, for $f \in \mathcal{E}_T$,

$$(4.1) \qquad f(k\delta) = \int_{-\infty}^{\infty} f(x)\bar{\phi}(x - k\delta)\, dx$$

and

$$(4.2) \qquad f(x) = \delta \sum_{-\infty}^{\infty} f(k\delta)\phi(x - k\delta).$$

The significance of equations (4.1) and (4.2) is that they look as if the functions $\delta^{1/2}\phi(x - k\delta)$, $k \in \mathbb{Z}$, formed an orthogonal basis of the space $V_T = L^2 \cap \mathcal{E}_T$.

But they do nothing of the kind, for three reasons: the functions in question don't lie in V_T but in the slightly larger space $V_{\pi\delta^{-1}}$, they are linearly dependent and they are not mutually orthogonal. In Chapter 2 we shall see how to define an "intermediate space" between V_T and

$V_{\pi\delta^{-1}}$, for which the functions $h^{1/2}\phi(x-kh)$ form an orthonormal basis for a certain value of $h > 0$.

But let us return to Theorem 1. The proof depends on the notion of periodification of a distribution of compact support. Let E_δ denote the linear space of infinitely differentiable periodic functions of period $2\pi\delta^{-1}$: E_1 is the space E of Section 2. The topological dual of E_δ is the space F_δ of $2\pi\delta^{-1}$-periodic distributions—just as the dual of E in Section 2 was F. Now let $S \in \mathcal{D}'(\mathbb{R})$ be a distribution of compact support. We define the periodification $\pi_\delta(S) = \sigma$ by the condition that σ be an element of F_δ and that, for each function $f \in E_\delta$, we have $\langle \sigma, f \rangle = \langle S, f \rangle$.

It amounts to the same thing just to require this condition when $f(x) = e^{ik\delta x}$ and thus to write $\hat{\sigma}(k\delta) = (\delta/2\pi)\hat{S}(k\delta)$, for each $k \in \mathbb{Z}$.

A final way of defining σ is to put $\sigma(x) = (\delta/2\pi)\sum_{-\infty}^{\infty}\hat{S}(k\delta)e^{ik\delta x}$ —the series converges to σ in the sense of distributions.

When S is a function of compact support in $L^1(\mathbb{R})$, the idea of periodification which we have just introduced coincides with that defined by Lemma 1.

Back, once more, to Theorem 1. The idea is to start with the distribution $S = \hat{f}$ with support in $[-T, T]$. Then $\hat{S}(k\delta) = 2\pi f(-k\delta)$, so that the given problem becomes that of defining S knowing only $\sigma = \pi_\delta(S)$.

When the support of S is longer than the period, the phenomenon occurs which specialists in signal processing call "aliasing". That is, "pieces of S" congruent modulo $2\pi\delta^{-1}$ overlap. Continuing this line of argument, we easily prove the first assertion of Theorem 1. Indeed, let $\varepsilon > 0$ be a real number which is small enough for $\pi\delta^{-1} + \varepsilon \leq T$ and $\varepsilon \leq \pi\delta^{-1}$ to hold, and let $\hat{\theta}(\xi)$ be a function in $\mathcal{D}(\mathbb{R})$ with support in $[-\varepsilon, \varepsilon]$. We then put $S = \hat{f} = \hat{\theta}(\xi - \pi\delta^{-1}) - \hat{\theta}(\xi + \pi\delta^{-1})$ or, working with f, $f(x) = -2i\sin(\pi\delta^{-1}x)\theta(x)$. Then $f(x)$ is in $\mathcal{E}_T \cap \mathcal{S}(\mathbb{R})$, but vanishes on $\delta\mathbb{Z}$.

In the same spirit, let us consider the case $T < \pi\delta^{-1}$. Then there is no more aliasing. Indeed, write S_k, $k \in \mathbb{Z}$, for the translates of S by $2k\pi\delta^{-1}$: we get $\sigma = \sum_{-\infty}^{\infty}S_k$, and the supports of the S_k lie in the *disjoint* intervals $[2k\pi\delta^{-1}-T, \ 2k\pi\delta^{-1}+T]$. We thus have $S = S_0 = \hat{\phi}\sigma$, which gives (4.2).

Verification of (4.1) is even easier. We apply Parseval's formula to the right-hand side of (4.1) and get

$$\frac{1}{2\pi}\int_{-\infty}^{\infty}\hat{f}(\xi)e^{ik\delta\xi}\bar{\hat{\phi}}(\xi)\,d\xi = \frac{1}{2\pi}\int_{-\infty}^{\infty}\hat{f}(\xi)e^{ik\delta\xi}\,d\xi = f(k\delta).$$

We are left with the case $T = \pi\delta^{-1}$. We begin by supposing that f

belongs to $V_T = L^2 \cap \mathcal{E}_T$. Then the sequence $\delta f(-k\delta)$, $k \in \mathbb{Z}$, is the sequence of Fourier coefficients of $\hat{f} \in L^2[-\pi\delta^{-1}, \pi\delta^{-1}]$. Naturally we are not distinguishing here between $\hat{f}$ and its periodification of period $2\pi\delta^{-1}$, because we are working in the context of L^2.

We thus write $\hat{f}(\xi) = \delta\chi(\xi)\sum_{-\infty}^{\infty}f(k\delta)e^{-ik\delta\xi}$, where $\chi(\xi)$ is the characteristic function of the interval $[-\pi\delta^{-1}, \pi\delta^{-1}]$. Working back, we get

$$(4.3) \qquad f(x) = \sum_{-\infty}^{\infty}f(k\delta)\frac{\sin(\pi\delta^{-1}(x-k\delta))}{\pi\delta^{-1}(x-k\delta)} .$$

The functions $L_k(x) = \sin(\pi\delta^{-1}(x-k\delta))/(\pi\delta^{-1}(x-k\delta))$ are called sinc functions and the functions $\delta^{-1/2}L_k(x)$ form an orthonormal basis of V_T. This means that the mapping which takes each $f \in V_T$ to the normalized sampling $\delta^{1/2}f(k\delta)$, $k \in \mathbb{Z}$, is an isometric isomorphism.

Now we move to the L^p case, with $1 < p < \infty$. We first show that $f(k\delta)$ belongs to $l^p(\mathbb{Z})$, when $f \in L^p \cap \mathcal{E}_T$, without needing to assume that there is any relationship between $T > 0$ and $\delta > 0$. Indeed, let ϕ be a function in $\mathcal{S}(\mathbb{R})$ whose Fourier transform $\hat{\phi}$ is 1 on the interval $[-T, T]$. Then (4.1) is satisfied. By Hölder's inequality, it follows that

$$|f(k\delta)|^p \leq \left(\int|f(x)|^p|\phi(x-k\delta)|\,dx\right)\|\phi\|_1^{p/q} .$$

It suffices then to sum these inequalities over $k \in \mathbb{Z}$ and to notice that $\sum_{-\infty}^{\infty}|\phi(x-k\delta)| \leq C_0$, because ϕ decreases rapidly.

Conversely, starting with an arbitrary sequence $c_k \in l^p(\mathbb{Z})$, we want to construct a function $f \in L^p \cup \mathcal{E}_T$, with $T = \pi\delta^{-1}$, such that $f(k\delta) = c_k$, $k \in \mathbb{Z}$.

We use (4.3), but have no easy recipe to show that $\sum_{-\infty}^{\infty}c_kL_k(x)$ belongs to $L^p(\mathbb{R})$ when $c_k \in l^p(\mathbb{Z})$. Nonetheless, we start by seeing what an "easy recipe" might be. If we could use (4.2) with ϕ decreasing rapidly, Hölder's inequality would give

$$|f(x)|^p \leq \left(\sum_{-\infty}^{\infty}|\phi(x-k\delta)|\right)^{p/q}\delta^p\sum_{-\infty}^{\infty}|f(k\delta)|^p|\phi(x-k\delta)|.$$

To get our result, we would use the fact that $\|\phi\|_1 = C_1$ was finite and that, once again, $\sum_{-\infty}^{\infty}|\phi(x-k\delta)| \leq C_0$.

Of course, $(\sin x)/x$ is not in $L^1(\mathbb{R})$ and this approach does not work. It has, however, the merit of showing that it is enough to prove

$$\sum_{-\infty}^{\infty}|f(k\delta/2)|^p \leq C\sum_{-\infty}^{\infty}|f(k\delta)|^p$$

for $f \in E_T$, $T = \pi\delta^{-1}$, because then $\delta/2 < \pi T^{-1} = \delta$ and the result is true for that case. Using (4.3), we conclude the proof by means of the following theorem, which will itself be proved in Chapter 7.

Theorem 2. *The matrix* $M = ((k-l)^{-1})_{(k,l)\in\mathbf{Z}^2}$ *(with 0 on the diagonal) defines a bounded linear operator from* $l^p(\mathbf{Z})$ *to* $l^p(\mathbf{Z})$, *when* $1 < p < \infty$.

This operator is called the discrete Hilbert transform, and is a discrete convolution operator whose continuity on $l^2(\mathbf{Z})$ is established as follows. We put $y_j = \sum'(j-k)^{-1}x_k$ where $\sum'$ denotes that the term $j = k$ is omitted from the sum. We then have

$$\sum_{-\infty}^{\infty} y_j e^{ijx} = \sum_{-\infty}^{\infty}\sideset{}{'}\sum \frac{1}{j-k} x_k e^{i(j-k)x} e^{ikx} = m(x)f(x)$$

where

$$m(x) = 2i \sum_{1}^{\infty} \frac{\sin jx}{j} \qquad \text{and} \qquad f(x) = \sum_{-\infty}^{\infty} x_k e^{ikx}.$$

Up to a factor of $2i$, $m(x)$ is the saw-tooth function, so we have $\|m\|_{\infty} = \pi$. The continuity of H on $l^2(\mathbf{Z})$ is equivalent to the continuity on $L^2[0, 2\pi]$ of the operator defined by pointwise multiplication by $m(x)$, which is clear because $m(x) \in L^{\infty}[0, 2\pi]$.

Of course, this approach does not allow us to establish the continuity of H on $l^p(\mathbf{Z})$, for $1 < p < \infty$, when $p \neq 2$. For that we shall have to call on the general theorems of Chapter 7.

If we were to use the matrix $(|j-k|^{-1})$ instead of $((j-k)^{-1})$, the corresponding operator would not be continuous on $l^2(\mathbf{Z})$. In fact, the same calculation as above would give

$$m(x) = 2 \sum_{1}^{\infty} \frac{\cos jx}{j} = -2\log 2 + 2\log \frac{1}{|\sin(x/2)|}$$

which does not, this time, belong to $L^{\infty}[0, 2\pi]$.

In applying Theorem 2, we use the matrix

$$\left((-1)^k \frac{\sin(\pi l/2)}{(l-2k)} \right),$$

where $l \in 2\mathbf{Z} + 1$—a sampling at the points $l\delta/2$ of the signal obtained from its sampling at $k\delta$, $k \in \mathbf{Z}$. The discrete Hilbert transform is restricted to sequences in $2\mathbf{Z}$ and the results are calculated for $l \in 2\mathbf{Z}+1$: the multiplicative factor $(-1)^k \sin(\pi l/2)$ has no effect on the estimation and can thus be ignored.

It is clear that the results above no longer hold when $p = 1$ or $p = \infty$. Indeed, if a matrix $m(j,k)$ is bounded on $l^1(\mathbf{Z})$, there is necessarily a constant C such that $\sum_j |m(j,k)| \leq C$, uniformly in k (and conversely). This condition is not satisfied by the matrix giving the values of the signal at the points $j\delta/2$ in terms of its values at $k\delta$.

Equally, the matrices bounded on $l^\infty(\mathbb{Z})$ are characterized by the condition $\sum_k |m(j,k)| \leq C$, where, this time, the inequality is satisfied uniformly in j. Once again, the operator we are studying does not satisfy the condition.

We end this section with the classical inequalities of S. Bernstein, which we give in a slightly imprecise version: for $1 \leq p \leq \infty$ and $f \in \mathcal{E}_T \cap L^1$, we have $\|f'\|_p \leq CT\|f\|_p$. In fact the constant C is 1. To prove the inequality, we go back to the proof of (4.1). Denoting by ϕ a function in the Schwarz class $\mathcal{S}(\mathbb{R})$ such that $\hat{\phi}(\xi) = 1$ on $[-1,1]$, we set $\phi_T(x) = T\phi(Tx)$. Then the Fourier transform of ϕ_T is $\hat{\phi}_T(\xi) = \hat{\phi}(T^{-1}\xi)$ and, for each function $f \in \mathcal{E}_T$, we get $f = f * \phi_T$. This identity gives $f' = f * \phi_T'$ and thus $\|f'\|_p \leq \|f\|_p \|\phi_T'\|_1$, for $1 \leq p \leq \infty$. But $\phi_T'(x) = T^2\phi'(Tx)$, so Bernstein's inequality is established with $C = \|\phi'\|_1$.

We refer the reader to [148] for a proof that $C = 1$, and to [2] for an explanation of the connection between Bernstein's theorem and the design of amplifiers.

Another inequality due to Bernstein concerns the calculation of the L^q norm of a function f from its L^p norm, when $1 \leq p \leq q \leq \infty$. We use the Hausdorff-Young inequality

(4.4) $$\|f * g\|_w \leq \|f\|_u \|g\|_v,$$

where $1 \leq u, v, w \leq \infty$ and $1 + 1/w = 1/u + 1/v$.

We apply this inequality with $w = q$ and $u = p$, which leads to $1/v = 1 + 1/q - 1/p$. Taking $g = \phi_T$, we get

(4.5) $$\|f\|_q \leq CT^{\frac{1}{p}-\frac{1}{q}}\|f\|_p.$$

To get n-dimensional versions, we suppose that f belongs to $L^p(\mathbb{R}^n)$ and that the Fourier transform of f has support in the ball $|\xi| \leq R$. Then we have

$$\|\nabla f\|_p \leq R\|f\|_p \qquad \text{when} \qquad 1 \leq p \leq \infty$$

and

$$\|f\|_q \leq CR^{n(\frac{1}{p}-\frac{1}{q})}\|f\|_p \qquad \text{when} \qquad 1 \leq p \leq q \leq \infty.$$

5 "Wavelets" in the work of Lusin and Calderon

The first appearence of "wavelets" can be detected in the work of Lusin in the 1930s. Indeed, if the well-known characterization of Hardy's $\mathbb{H}^p$ spaces in terms of Lusin's area function is written in the language of atomic decomposition due to Coifman and Weiss, then we can see the prototypes of "wavelets" appearing.

Let us be a little more precise. We let P denote the open upper half-plane, given by $z = x + iy$, for $y > 0$. Then a function $f(x + iy)$ belongs to $\mathbb{H}^p(\mathbb{R})$, if it is holomorphic in the half-plane P and if

$$(5.1) \qquad \sup_{y>0} \left(\int_{-\infty}^{\infty} |f(x + iy)|^p \, dx \right)^{1/p} < \infty.$$

We shall restrict p to the range $1 \le p < \infty$: the left-hand side of (5.1) is then the $\mathbb{H}^p$ norm of f.

Lusin's work was about the analysis and synthesis of functions in $\mathbb{H}^p$, using the integral representations $f(z) = \iint_P (z - \bar{\zeta})^{-2} \alpha(\zeta) \, du \, dv$, where $\zeta = u + iv$ and where $\alpha(\zeta)$ is a measurable function on P which is subject to certain growth conditions on the boundary.

First of all, observe that each function $(z - \bar{\zeta})^{-2}$ clearly belongs to $\mathbb{H}^p$ (when $\zeta \in P$).

Synthesis is obtained by the following "assembly rule". Starting with an arbitrary measureable function $\alpha(\zeta)$, we form the *quadratic functional*

$$(5.2) \qquad A(x) = \left(\iint_{\Gamma(x)} |\alpha(u + iv)|^2 v^{-2} \, du \, dv \right)^{1/2},$$

where $\Gamma(x) = \{(u, v) \in \mathbb{R}^2 : v > |u - x|\}$.

Then, if $1 < p < \infty$ and $f(z) = \iint_P (z - \bar{\zeta})^{-2} \alpha(\zeta) \, du \, dv$, we get

$$(5.3) \qquad \|f\|_p \le C(p) \|A\|_p.$$

The left-hand side of (5.3) is $\left(\int_{-\infty}^{\infty} |f(x)|^p \, dx \right)^{1/p}$, which is also the norm of f in $\mathbb{H}^p$ as defined by (5.1). The term on the right-hand side is $\left(\int_{-\infty}^{\infty} (A(x))^p \, dx \right)^{1/p}$.

This estimate can be far too imprecise. If $f(z) = (z + i)^{-2}$, for example, then the Dirac measure at the point i is a natural choice for $\alpha(\zeta)$, and the right-hand side of (5.3) is infinite.

This paradox arises because the representation

$$(5.4) \qquad f(z) = \iint_P (z - \bar{\zeta})^{-2} \alpha(\zeta) \, du \, dv$$

is not unique.

The space of those measurable functions $\alpha(\zeta)$ on the upper half-plane which satisfy the condition $A(x) \in L^p(\mathbb{R})$ is studied in [71], where the space is called a *tent space*.

Of all possible representations (5.4) of a function $f(z)$ of $\mathbb{H}^p$, there is one which is singled out by the following three properties:

  (a) the coefficients $\alpha(\zeta)$ of f are calculated from f as if the functions $(z - \bar{\zeta})^{-2}$ were an orthonormal basis of $\mathbb{H}^2$;

(b) (5.3) is a norm equivalence for the coefficients $\alpha(\zeta)$ calculated in this way;

(c) every function $f \in \mathbb{H}^p$ admits a representation of this kind.

We shall verify these properties using the language of "coherent states in quantum mechanics" ([125]). To this end, we start with the basic function $\psi(t) = (t+i)^{-2}$, a function notable for its regularity, its localization and its oscillatory nature: $\int_{-\infty}^{\infty} \psi(t)\,dt = 0$. This is why A. Grossmann and J. Morlet describe $\psi(t)$ as a "wavelet"—a "wavelet" oscillates for "a little while" like a wave, but is then localized by damping.

We now put $\psi_{(a,b)}(t) = a^{-1/2}\psi((t-b)/a)$, $a > 0$, $b \in \mathbb{R}$: this set is the orbit of ψ under the irreducible unitary action of the affine group (acting on $\mathbb{H}^2(\mathbb{R})$). *Then we can use the collection of functions* $\psi_{(a,b)} \in \mathbb{H}^2(\mathbb{R})$ *as if it formed an orthonormal basis of* $\mathbb{H}^2$. That is to say, for f in $\mathbb{H}^2$, we start by calculating the "wavelet coefficients" of f as $W(a,b) = (f, \psi_{(a,b)})$. Then we recover $f(x)$ by superposing the "wavelets", $\psi_{(a,b)}(x)$, modified by the corresponding coefficients. So we get

$$(5.5) \qquad f(x) = \frac{1}{\pi^2} \int_0^\infty \int_{-\infty}^\infty W(a,b)\psi_{(a,b)}(x)\,db\frac{da}{a^2}$$

for every function $f \in \mathbb{H}^2$.

The explicit expression for the coefficients $W(a,b)$ is

$$(5.6) \qquad W(a,b) = 2\pi i a^{3/2} f'(b + ia).$$

With the notation of (5.4), we thus have $\alpha(\zeta) = (2i/\pi)vf'(u + iv)$ and the norm equivalence given by (5.3) is the well-known characterisation of $\mathbb{H}^p$ by Lusin's area function.

It was only in 1965 that Calderón proved (5.3) for $p = 1$. This discovery was the starting point for significant work both in the theory of operators on $\mathbb{H}^1$ ([109],[217]) and on the duality between $\mathbb{H}^1$ and BMO. We shall come back to these matters in Chapters 5 and 6.

Looking for the historical origins of "wavelets" leads us to the well-known identity of Calderón, which we now recall. We begin with a function $\psi(x) \in L^1(\mathbb{R}^n)$ whose integral over $\mathbb{R}^n$ is zero and whose Fourier transform $\hat{\psi}(\xi)$, $\xi \in \mathbb{R}^n$, satisfies the following condition:

$$(5.7) \qquad \int_0^\infty |\hat{\psi}(t\xi)|^2 \frac{dt}{t} = 1 \text{ for every } \xi \neq 0.$$

For example, if $\psi(x)$ is sufficiently regular and localized, if its integral over $\mathbb{R}^n$ is zero, and if it is a radial function, then there is a constant $c > 0$ such that $c\psi(x)$ satisfies (5.7). We then put $\psi_t(x) = t^{-n}\psi(x/t)$ and $\psi_{(u,t)}(x) = t^{-n}\psi((x-u)/t)$. Following Grossman and Morlet ([90],

[124]) we define the "wavelet coefficients" of $f \in L^2(\mathbb{R}^n)$ by

(5.8) $$\alpha(u,t) = (f, \psi_{(u,t)})$$

and we can construct f from the coefficients by

(5.9) $$f(x) = \int_0^\infty \int_{\mathbb{R}^n} \alpha(u,t)\psi_{(u,t)}(x)\, du\, \frac{dt}{t}.$$

An equivalent formulation of (5.9) is

(5.10) $$I = \int_0^\infty Q_t Q_t^\star \frac{dt}{t},$$

where $Q_t(f) = f * \psi_t$ and where $Q_t^\star$ is the adjoint of Q_t. Calderón's identity has a natural application to the analysis of classical function spaces by means of conditions involving the modulus of the gradient of a harmonic extension. This can be found in [217]. In fact, if we denote by P_t the Poisson semi-group and if $P_t f(x) = F(x,t)$ is the function which is harmonic in the half-space $t > 0$, satisfies $F(x,0) = f(x) \in L^2(\mathbb{R}^n)$, and tends to zero at infinity, then $Q_t = -t(\partial/\partial t)P_t$.

In the formulas (5.8) and (5.9), it is as if the functions $\psi_{(u,t)}$, $u \in \mathbb{R}^n$, $t > 0$ formed an orthonormal basis of $L^2(\mathbb{R}^n)$ —the coefficients of the decomposition of f with respect to this basis are given by (5.8) and the expression for f in terms of those coefficients is provided by (5.9).

Is it possible to replace the redundant set of functions $\psi_{(u,t)}$, $u \in \mathbb{R}^n$, $t > 0$, by an actual orthonormal basis ψ_λ, $\lambda \in \Lambda$, whose construction conforms to the same algebraic rules? In Chapters 2 and 3 we shall see that this is, in fact, the case. The work of Lusin, Calderón, Zygmund, Stein and their school assumes a new significance, in that the orthonormal basis ψ_λ, $\lambda \in \Lambda$, is a *universal unconditional basis* for all the classical spaces of functions or distributions, apart from those spaces which are constructed from L^1 and L^∞ and which, therefore, cannot have any unconditional bases at all.

2

Multiresolution approximations of $L^2(\mathbb{R}^n)$

1 Introduction

Orthonormal wavelet bases have only been available for the last few years, but similar constructions were previously used in mathematics, theoretical physics and signal processing.

It is interesting and surprising that all these related constructions saw the light of day almost simultaneously (in fact, during the 1980s) in constructive field theory, in the geometry of Banach spaces and even in the processing of signals obtained during prospecting trips of the Elf-Aquitaine group (Morlet's work on reflection seismology).

Research workers in the various specialities were hoping to find practical algorithms for decomposing arbitrary functions into sums of special functions which would combine the advantages of the trigonometric and the Haar systems. These systems stand at two extremes, in the following sense: the functions of the trigonometric system are exactly localized by frequency, that is, in the Fourier variable, but have no precise localization in space. On the other hand, the functions of the Haar system (whose definition we recall in Chapter 3) are perfectly localized in space (the x variable) but are badly localized in the Fourier variable (the ξ variable). This is due to two defects of the functions of the Haar system: their lack of regularity and their lack of oscillation.

R. Balian has given the following justification for trying to find Hilbert bases which are simultaneously well-localized in space and in the Fourier variable. "In the theory of communications, it is appropriate to represent an oscillatory signal as the superposition of elementary wavelets possessing both a well-defined frequency and a localization in time. Indeed, the pertinent information is often carried simultaneously by the frequency emitted and by the temporal structure of the signal—the example of music is characteristic. The representation of a signal as a function of time does not bring out the frequencies in play, whereas on the other hand, the Fourier representation conceals the moment of emission and the duration of each of the components of the signal. An adequate representation ought to combine the advantages of the two complementary descriptions; it ought also to be in a discrete form, which is better suited to the theory of communications."

To achieve the joint localization (in the Fourier and space variables) which we want to impose on our orthonormal wavelet basis, it seems natural first to prescribe a localization in the Fourier variable and then to follow this by a localization in the space variable. In order not to annul the first, the second localization must respect Heisenberg's Uncertainty Principle.

We want the orthonormal wavelet basis (which will be a basis of the standard space L^2) to be a tool for the analysis of most of the spaces of functions or distributions in general use. But such spaces are often easily characterized by means of the Littlewood-Paley decomposition, and this suggests using that decomposition to obtain the Fourier variable (or frequency) localization which we require.

The Littlewood-Paley decomposition of a 2π-periodic function was described in Chapter 1. It can be written as

$$(1.1) \qquad f = c_0 + \sum_0^\infty \Delta_j(f),$$

and corresponds to a frequency localization using an exponential scale given by the dyadic intervals $\frac{1}{2}2^j \leq |k| \leq \frac{3}{2}2^j$ (the octave scale). We don't yet have any localization in space. To get that, we want to take the dyadic blocks $\Delta_j(f)$ and to divide them into "small packets" using a partition of unity (with respect to the space variable) to get quantities which are localized in both the Fourier and space variables. Heisenberg's Uncertainty Principle says that we cannot get such "small packets" if their size is less than 2^{-j}. This leads us to write $\Delta_j(f)$ as the sum $\sum \Delta_j(f)\phi(2^j x - k)$, where ϕ, belonging to the Schwartz class $\mathcal{S}(\mathbb{R})$, satisfies $\sum_{-\infty}^\infty \phi(x - k) \equiv 1$. On requiring, in addition, the Fourier transform of ϕ to be zero outside the interval $[-1/4, 1/4]$, the attempted localization in space should not unduly distort the frequency localization obtained by the Littlewood-Paley decomposition.

This heuristic approach does not give us the "wavelet miracle" that allows $\alpha(j,k)2^{j/2}\psi(2^j x - k)$ to appear instead of $\Delta_j(f)\phi(2^j x - k)$, where $\alpha(j,k)$ is a numerical coefficient and where ψ is a function in $\mathcal{S}(\mathbb{R})$ whose integral is zero—indeed we could require all the moments of ψ to vanish—and such that the functions $2^{j/2}\psi(2^j x - k)$, $j,k \in \mathbb{Z}$, form an orthonormal basis of $L^2(\mathbb{R})$. (We should remark that the function $\psi(x)$ is not the same as the function giving the operators Δ_j in Chapter 1.)

At least, the heuristic approach has the merit of associating wavelet series with Littlewood-Paley decompositions, which gives an explanation of the role played by the octave scale in wavelet series. This lets us anticipate the effectiveness—inherited from the Littlewood-Paley decomposition—of wavelet series in the analysis of classical spaces of functions or distributions.

However, a "wavelet miracle" will never take place if we start with the traditional Littlewood-Paley decomposition. What is needed is to replace that approach by a simpler, more robust version: a *multiresolution approximation*. The idea of a multiresolution approximation enables us to combine analysis in the space variable with analysis in the Fourier variable while satisfying Heisenberg's Uncertainty Principle. To be more precise, it is a question of approximating a general function f by a sequence of simple functions f_j. The simplicity of the f_j lies in that the f_j are sufficiently explicit and regular to be completely determined when they are sampled on the lattice $\Gamma_j = 2^{-j}\mathbb{Z}^n$ in $\mathbb{R}^n$. The regularity of the functions f_j corresponds to their Fourier transforms $\hat{f}_j$ being essentially supported by the associated balls $C2^j$. The relation between the sampling step 2^{-j} and the radius of the corresponding ball $C2^j$ is roughly that resulting from Heisenberg's Uncertainty Principle, or from Shannon's rule, which was cited in Chapter 1.

Most examples of multiresolution approximations come from nested sequences of splines associated with refinements of lattices. So the theory which we shall develop in Chapter 3 will contain, as a particular case, the classical results of approximation by spline functions associated with finer and finer lattices.

But there are other important examples of multiresolution approximations which cannot be seen as part of the theory of splines.

The purpose of this second chapter is to familiarize ourselves with the notion of a multiresolution approximation. We shall see that r-regular multiresolution approximations enable us to analyse every space of functions or distributions whose regularity is bounded by r.

The construction of wavelets from a multiresolution approximation will await Chapter 3.

2 Multiresolution approximations: definition and examples

We give the main definition immediately and comment on it afterwards.

Definition 1. *A multiresolution approximation of $L^2(\mathbb{R}^n)$ is, by definition, an increasing sequence V_j, $j \in \mathbb{Z}$, of closed linear subspaces of $L^2(\mathbb{R}^n)$ with the following properties:*

$$(2.1) \qquad \bigcap_{-\infty}^{\infty} V_j = \{0\}, \qquad \bigcup_{-\infty}^{\infty} V_j \text{ is dense in } L^2(\mathbb{R}^n);$$

for all $f \in L^2(\mathbb{R}^n)$ and all $j \in \mathbb{Z}$,

$$(2.2) \qquad f(x) \in V_j \iff f(2x) \in V_{j+1};$$

for all $f \in L^2(\mathbb{R}^n)$ and all $k \in \mathbb{Z}^n$

$$(2.3) \qquad f(x) \in V_0 \iff f(x-k) \in V_0;$$

(2.4) *there exists a function, $g(x) \in V_0$, such that the sequence*

$g(x-k)$, $k \in \mathbb{Z}^n$, *is a Riesz basis of the space V_0.*

Recall that if H is a Hilbert space and if $e_0, e_1, \ldots, e_k, \ldots$ is a sequence of elements of H, the sequence is a *Riesz basis* of H if there exist constants $C' > C > 0$ such that, for every sequence of scalars $\alpha_0, \alpha_1 \ldots$, we have

$$(2.5) \qquad C(\sum |\alpha_k|^2)^{1/2} \leq \|\sum \alpha_k e_k\| \leq C'(\sum |\alpha_k|^2)^{1/2}$$

and the vector space of finite sums $\sum \alpha_k e_k$ (on which (2.5) is tested) is dense in H. Another way to express these two conditions is to say that there is an isomorphism $T : l^2(\mathbb{N}) \to H$, not necessarily isometric, such that $T(\varepsilon_k) = e_k$, where ε_k denotes the sequence which is zero except for the kth element, which is 1.

Definition 2. *A multiresolution approximation, V_j, $j \in \mathbb{Z}$, is called r-regular ($r \in \mathbb{N}$) if the function $g(x)$ of (2.4) can be chosen in such a way that*

$$(2.6) \qquad |\partial^\alpha g(x)| \leq C_m(1+|x|)^{-m}$$

for each integer $m \in \mathbb{N}$ and for every multi-index $\alpha = (\alpha_1, \ldots, \alpha_n)$ satisfying $|\alpha| \leq r$. (Here, $\partial^\alpha = (\partial/\partial x_1)^{\alpha_1} \cdots (\partial/\partial x_n)^{\alpha_n}$ and $|\alpha| = \alpha_1 + \cdots + \alpha_n$.)

Our *first example* of a multiresolution approximation is given, in dimension 1, by the nested spaces of splines of order r: the nodes of the functions $f \in V_j$ being precisely the points $k2^{-j}$, $k \in \mathbb{Z}$.

Recall the definition of V_0. We start with an integer $r \in \mathbb{N}$ and denote by V_0 the subspace of $L^2(\mathbb{R})$ consisting of the functions in C^{r-1} (this condition is void if $r = 0$) whose restrictions to each interval $[k, k + 1)$, $k \in \mathbb{Z}$, coincide with polynomials of degree less than or equal to r. V_j is then defined by (2.2) and properties (2.1), (2.3) and (2.4) can be verified immediately. A possible choice for $g(x)$ is given by the convolution product $\chi * \cdots * \chi$ ($r + 1$ factors), where χ is the characteristic function of the interval $[0, 1]$. When $r = 0$, the functions $\chi(x - k)$, $k \in \mathbb{Z}$ form a Riesz basis of V_0. When $r \geq 1$, the functions $g(x - k)$, $k \in \mathbb{Z}$, form a Riesz basis of V_0.

When $r = 0$, this is just the familiar situation of dyadic martingales, and the theory of multiresolution approximations appears as a variant— in the direction of set theory rather than functional analysis—of the theory of martingales.

In our *second example*, Fourier analysis plays its part. We take V_0 to be the closed subspace of $L^2(\mathbb{R})$ composed of those functions f whose Fourier transforms $\hat{f}$ are supported by the interval $[-\pi, \pi]$. The function $g(x)$ is $(\sin \pi x)/\pi x$, and the functions $g(x - k)$, $k \in \mathbb{Z}$, form an orthonormal basis of V_0. In addition, the coefficients of the decomposition of $f \in V_0$, with respect to this basis, are given by the sampling values $f(k)$, $k \in \mathbb{Z}$. This example does not lead to an r-regular multiresolution approximation because the function $g(x)$ does not decrease rapidly enough. In fact, it is necessary to verify that no possible choice of $g(x)$ can satisfy (2.6)—this very simple task is left to the reader.

Let V_j, $j \in \mathbb{Z}$, be any multiresolution approximation of $L^2(\mathbb{R}^n)$. We consider the nested sequence of lattices $\Gamma_j = 2^{-j}\mathbb{Z}^n$ in $\mathbb{R}^n$ and define the *sampling operator* $T_j : V_j \to l^2(\Gamma_j)$ in the following way. If $j = 0$, then $T_0(\sum_{k \in \mathbb{Z}^n} \alpha_k g(x - k))$ is the sequence $(\alpha_k)_{k \in \mathbb{Z}^n}$. If $j > 0$, we require the definition of T_j to be compatible with (2.2), that is, T_j applied to the function $g(2^j x - k)$ is the sequence in $l^2(\Gamma_j)$ which is 1 at $2^{-j}k$ and 0 elsewhere. This construction makes T_j depend on the choice of function $g(x)$ of (2.4).

In our second example, the operator T_j applied to V_j gives the restriction (in the usual sense) of f to $\Gamma_j = 2^{-j}\mathbb{Z}$. In every case, the operator T_j is an isomorphism between V_j and $l^2(\Gamma_j)$.

Our *third example* is an improvement of the second. We start with a function $\theta(\xi)$, of the real variable ξ, belonging to $\mathcal{D}(\mathbb{R})$, which is even, equals 1 on $[-2\pi/3, 2\pi/3]$ and is 0 outside $[-4\pi/3, 4\pi/3]$. We suppose in addition that $\theta(\xi) \in [0, 1]$, for all $\xi \in \mathbb{R}$, and that $\theta^2(\xi) + \theta^2(2\pi - \xi) = 1$ when $0 \leq \xi \leq 2\pi$.

Let g denote the function in $\mathcal{S}(\mathbb{R})$ whose Fourier transform is $\theta(\xi)$. Then we can verify immediately that the sequence $g(x - k)$, $k \in \mathbb{Z}$, is the orthonormal basis of a closed subspace of $L^2(\mathbb{R})$ which we call

V_0. In fact, applying the Fourier transform, $\mathcal{F}V_0$ is the vector space of products $m(\xi)\theta(\xi)$, where $m(\xi)$ is 2π-periodic and $m(\xi)$ restricted to $[0, 2\pi]$ belongs to $L^2[0, 2\pi]$. If $g_1(\xi) = m_1(\xi)\theta(\xi)$ and $g_2(\xi) = m_2(\xi)\theta(\xi)$, we get

$$(g_1, g_2) = \int_{-\infty}^{\infty} g_1(\xi)\bar{g}_2(\xi)\,d\xi$$
$$= \int_{-\infty}^{\infty} m_1(\xi)\bar{m}_2(\xi)\theta^2(\xi)\,d\xi = \int_0^{2\pi} m_1(\xi)\bar{m}_2(\xi)\,d\xi,$$

since $\sum_{-\infty}^{\infty} \theta^2(\xi + 2k\pi) = 1$. This shows why $g(x - k)$, $k \in \mathbb{Z}$, is an orthonormal basis of V_0.

V_1 is then defined by (2.2). We need to show that V_0 is contained in V_1, which is not obvious because we have no geometric description of the functions in V_0 (unlike the case of the spline functions in the first example). But $\mathcal{F}V_1$ is the set of functions $m_1(\xi)\theta(\xi/2)$, where $m_1(\xi)$ is 4π-periodic. We make the fundamental observation that $\theta(\xi) = \lambda(\xi)\theta(\xi/2)$, where we can require $\lambda(\xi)$ to be 4π-periodic and to be in $C^\infty(\mathbb{R})$ (taking advantage of the indeterminacy of $0/0$). So the functions $m(\xi)\theta(\xi)$ of $\mathcal{F}V_0$ belong to $\mathcal{F}V_1$.

The other properties of a multiresolution approximation can be verified without difficulty in this example.

We now move to examples in higher dimensions.

Our *fourth example* shows that we can always use a multiresolution approximation V_j, $j \in \mathbb{Z}$, of $L^2(\mathbb{R})$ to construct a multiresolution approximation $V_j \hat{\otimes} V_j$ of $L^2(\mathbb{R}^2)$.

As in the previous examples, we shall confine ourselves to defining $V_0 \hat{\otimes} V_0$. A Riesz basis of this closed subspace of $L^2(\mathbb{R}^2)$ will consist of the functions $g(x - k)g(y - l)$, where $(k, l) \in \mathbb{Z}^2$, and so $V_0 \hat{\otimes} V_0$ will consist of the (infinite) sums

$$\sum_{-\infty}^{\infty}\sum_{-\infty}^{\infty} \alpha(k, l)g(x - k)g(y - l) \quad \text{where} \quad \sum_{-\infty}^{\infty}\sum_{-\infty}^{\infty} |\alpha(k, l)|^2 < \infty.$$

The other properties of a multiresolution approximation follow immediately.

We finish this section by going back to the first example (splines of order r) which we examine via the Fourier transform: this will be useful for the calculations of the next chapter.

If f belongs to V_0, then f is in $L^2(\mathbb{R})$ and

$$\left(\frac{d}{dx}\right)^{r+1} f = \sum_{-\infty}^{\infty} c_k \delta_k,$$

where δ_k is Dirac measure at k and where $c_k \in l^2(\mathbb{Z})$. Moreover, these two properties characterize V_0. Passing to the Fourier transforms, for each $f \in V_0$ we get $\xi^{r+1}\hat{f}(\xi) = q(\xi)$, where $q(\xi)$ is a 2π-periodic function in $L^2[0, 2\pi]$. Naturally, $\hat{f}(\xi) \in L^2(\mathbb{R})$, so the function $m(\xi) = q(\xi)(1 - e^{i\xi})^{-r-1}$ has the same properties (of 2π-periodicity and of being in $L^2[0, 2\pi]$) as $q(\xi)$. In fact, up to a constant multiple, $m(\xi)$ is equivalent to $\hat{f}(\xi)$ in a neighbourhood of 0.

Finally, taking the Fourier transform, $\mathcal{F}V_0$ is the set of products $m(\xi)\left((1 - e^{-i\xi})/(i\xi)\right)^{r+1}$, where $m(\xi)$ is 2π-periodic and belongs to $L^2[0, 2\pi]$. In other words, the function $\left((1 - e^{-i\xi})/(i\xi)\right)^{r+1}$ acts like the function $\theta(\xi)$ of the third example. Going back, we see that $(1 - e^{-i\xi})/(i\xi)$ is the Fourier transform of the characteristic function of the interval $[0, 1]$, a function that we denote by χ. Then $g = \chi \star \cdots \star \chi$ $(r + 1$ factors) has $\left((1 - e^{-i\xi})/(i\xi)\right)^{r+1}$ as its Fourier transform and $g(x - k)$, $k \in \mathbb{Z}$, is a Riesz basis of V_0.

We now define a *Lagrangian spline* to be a function $h \in V_0$ such that $h(0) = 1$ and $h(k) = 0$ for $k \in \mathbb{Z}$, $k \neq 0$. We can calculate such a function using the following lemma, which follows from Lemma 1 of Chapter 1.

Lemma 1. *Let $f(x)$ be a function in $L^1(\mathbb{R})$. f satisfies the conditions $\hat{f}(0) = 1$ and $\hat{f}(k) = 0$ for $k \in \mathbb{Z}$, $k \neq 0$, if and only if*

$$\sum_{-\infty}^{\infty} f(x + 2k\pi) = \frac{1}{2\pi}.$$

In our case

$$f(\xi) = \hat{h}(\xi) = m(\xi)\left(\frac{1 - e^{-i\xi}}{i\xi}\right)^{r+1} \in L^1(\mathbb{R}),$$

if $r \geq 1$, which we suppose to be the case. We want to have $\sum_{-\infty}^{\infty} f(\xi + 2k\pi) = 1/2\pi$, so we need to calculate the sum of the series $\sum_{-\infty}^{\infty} (\xi + 2k\pi)^{-r-1}$.

If r is even, the sum of the series is zero for $\xi = \pi$. The sum is, in fact, the analytic 2π-periodic function obtained by taking the rth derivative of the function $\cot \xi/2$. So we cannot construct Lagrangian splines.

However, if r is odd, the function $(1 - e^{-i\xi})^{r+1} \sum_{-\infty}^{\infty} (\xi + 2k\pi)^{-r-1}$ is 2π-periodic, real-analytic and non-zero on the real axis. So there is a unique function $m_0(\xi)$ such that $\hat{h}(\xi) = m_0(\xi)\left((1 - e^{-i\xi})/(i\xi)\right)^{r+1}$ defines a Lagrangian spline h. Further, $m_0(\xi)$ is real-analytic and nowhere zero. It follows that the functions $h(x - k)$, $k \in \mathbb{Z}$, form a Riesz basis of V_0. Finally, for the multiresolution approximations formed by spline functions of odd order, the sampling operators $T_j : V_j \to l^2(\Gamma_j)$ can be defined as the usual restrictions of the functions in V_j to the lattices $2^{-j}\mathbb{Z}$. This property does not hold for splines of even order.

We finish this section by fixing some notation.

Definition 3. *We say that V_j, $j \in \mathbb{Z}$, is a multiresolution approximation by splines of order r, if the V_j are constructed as in the first example. We say that V_j, $j \in \mathbb{Z}$, is a multiresolution approximation of Littlewood-Paley type, if the V_j are constructed as in the third example.*

We shall presently see the reason for the second part of this definition.

We shall show that an r-regular multiresolution approximation allows us to approximate functions or distributions f, whose regularity is bounded by r, by functions $f_j \in V_j$. Moreover, many spaces of functions or distributions have simple characterizations in terms of the differences $f_{j+1} - f_j$. The operator E_j which takes f to f_j will, in fact, be the orthogonal projection of $L^2(\mathbb{R}^n)$ onto V_j, an operator which extends to the usual spaces of functions or distributions, as long as the index of regularity is bounded by r.

In order to construct the operator E_j, it is convenient to have a simple orthogonal basis of V_j at our disposal. That is what we shall obtain in the next section.

3 Riesz bases and orthonormal bases

We start by recalling the two classical constructions of an orthonormal basis of a Hilbert space H from a Riesz basis e_k, $k \in \mathbb{N}$, of H.

The first uses the Gram matrix $G = (g(j, k))_{j,k \in \mathbb{N}}$ of the Riesz basis: $g(j, k) = \langle e_j, e_k \rangle$, where $\langle \cdot , \cdot \rangle$ is the inner product on the Hilbert space H.

Lemma 2. *There exist two constants $c_2 \geq c_1 > 0$ such that, for every sequence of scalars ξ_k, $k \in \mathbb{N}$, we have*

$$(3.1) \qquad c_1 \sum_0^\infty |\xi_k|^2 \leq \sum_0^\infty \sum_0^\infty g(j, k) \xi_j \bar{\xi}_k \leq c_2 \sum_0^\infty |\xi_k|^2 .$$

In other words, G is positive-definite. The lemma follows immediately from the definition of Riesz basis because the middle term of (3.1) is just $\| \sum_0^\infty \xi_k e_k \|^2$.

Using the symbolic calculus on positive-definite matrices, we can form the matrix $G^{-1/2}$. We denote its entries by $\gamma(j, k)$, $j, k \in \mathbb{N}$.

Lemma 3. *The vectors $f_j = \sum_0^\infty \gamma(j, k) e_k$ form an orthonormal basis of the Hilbert space, H.*

Indeed, if we let ε_k, $k \in \mathbb{N}$, denote an arbitrary orthonormal basis of

H and let $S : H \to H$ denote the isomorphism defined by $S(\varepsilon_k) = e_k$, $k \in \mathbf{N}$, then $g(j,k) = \langle S(\varepsilon_j), S(\varepsilon_k) \rangle = \langle S^{\star}S(\varepsilon_j), \varepsilon_k \rangle$. That is, the $g(j,k)$ are the entries of the matrix reprsenting the operator $S^{\star}S$ with respect to the basis $\{\varepsilon_j\}$. It follows that the $\gamma(j,k)$ are the entries of the matrix of $(S^{\star}S)^{-1/2}$ with respect to the same basis. So $\sum_0^{\infty} \gamma(j,k)\varepsilon_k = (S^{\star}S)^{-1/2}\varepsilon_j$ and

$$f_j = \sum_0^{\infty} \gamma(j,k)e_k = S(\sum_0^{\infty} \gamma(j,k)\varepsilon_k) = S(S^{\star}S)^{-1/2}\varepsilon_j.$$

Lemma 3 thus follows from the observation that $S(S^{\star}S)^{-1/2}$ is unitary.

To proceed to the second algorithm—which gives the same orthonormal basis—we use the following remark.

Lemma 4. *If S is an isomorphism of a Hilbert space H onto itself, then* $S(S^{\star}S)^{-1/2} = (S^{\star}S)^{-1/2}S$.

Indeed

$$B^{-1/2} = \frac{1}{\pi} \int_0^{\infty} (B + \lambda)^{-1}\lambda^{-1/2}\,d\lambda$$

for every operator B such that $c_1 I \le B \le c_2 I$, where $c_2 \ge c_1 > 0$ and where I is the identity operator. We apply this in the case that $B = S^{\star}S$. To conclude, it is enough to observe that, for each $\lambda \ge 0$,

$$S(S^{\star}S + \lambda)^{-1}S^{-1} = (SS^{\star} + \lambda)^{-1}.$$

We can now describe the second algorithm. We start with the operator defined (formally) by $T(x) = \sum_0^{\infty} \langle x, e_k \rangle e_k$. To prove that T exists we write, as before, $e_k = S(\varepsilon_k)$, where $S : H \to H$ is an isomorphism and ε_k, $k \in \mathbf{N}$, forms an orthonormal basis of H. Then $\langle x, e_k \rangle e_k = S(\langle S^{\star}(x), \varepsilon_k \rangle \varepsilon_k)$ and this remark leads immediately to $T = SS^{\star}$. Then the symbolic calculus on (bounded) positive self-adjoint operators produces $T^{-1/2}$ and, by Lemma 4, we get $T^{-1/2}(e_k) = f_k$.

In comparison with the usual Gram-Schmidt process, this algorithm has the advantage of transforming a Riesz sequence of the form $g(x - k)$, $k \in \mathbf{Z}^n$, into an orthonormal basis with the same structure. On the other hand, we could forget this preamble and use "bare hands" to prove the following theorem, which gives the canonical orthonormal basis $\phi(x - k)$, $k \in \mathbf{Z}^n$, of V_0.

The notation is that of Chapter 1: in particular, $\hat{f}(\xi)$ will always denote the Fourier transform of f.

Theorem 1. *Let V_j, $j \in \mathbf{Z}$, be a multiresolution approximation of $L^2(\mathbf{R}^n)$. Then there exist two constants, $c_2 \ge c_1 > 0$, such that, for* almost all $\xi \in \mathbb{R}^n$, we have

$$(3.2) \qquad c_1 \leq \left(\sum_{k \in \mathbb{Z}^n} |\hat{g}(\xi + 2k\pi)|^2 \right)^{1/2} \leq c_2 \, .$$

Further, if $\phi \in L^2(\mathbb{R}^n)$ is defined by

$$(3.3) \qquad \hat{\phi}(\xi) = \hat{g}(\xi) \left(\sum_{k \in \mathbb{Z}^n} |\hat{g}(\xi + 2k\pi)|^2 \right)^{-1/2} ,$$

then $\phi(x - k)$, $k \in \mathbb{Z}^n$, is an orthonormal basis of V_0.

Finally, let $f \in V_0$ be a function such that the sequence $f(x - k)$, $k \in \mathbb{Z}^n$, is orthonormal. Then the sequence is an orthonormal basis of V_0 and we have $\hat{f}(\xi) = \theta(\xi)\hat{\phi}(\xi)$, where $\theta(\xi) \in C^\infty(\mathbb{R}^n)$, $|\theta(\xi)| = 1$ almost everywhere, and $\theta(\xi + 2k\pi) = \theta(\xi)$, for each $k \in \mathbb{Z}^n$.

We start by proving (3.2). Using (2.4) and (2.5), we get

$$C \left(\sum |\alpha_k|^2 \right)^{1/2} \leq \| \sum \alpha_k g(x - k) \|_2 \leq C' \left(\sum |\alpha_k|^2 \right)^{1/2} .$$

We now work only with finite sums $\sum \alpha_k g(x - k)$. By putting $m(\xi) = \sum \alpha_k e^{-ik \cdot \xi}$ and applying Plancherel's theorem, we get

$$(3.4) \qquad C\|m\|_2 \leq \left(\int_{\mathbb{R}^n} |m(\xi)|^2 |\hat{g}(\xi)|^2 \, d\xi \right)^{1/2} \leq C'\|m\|_2 \, ,$$

where $\|m\|_2 = \left(\int_{Q_0} |m(\xi)|^2 \, d\xi \right)^{1/2}$ and $Q_0 = [0, 2\pi)^n$. Next we put $\omega(\xi) = \sum_{k \in \mathbb{Z}^n} |\hat{g}(\xi + 2k\pi)|^2$, so that $\omega(\xi) \in L^1(Q_0)$, since $\hat{g} \in L^2(\mathbb{R}^n)$. The double inequality (3.4) then becomes

$$(3.5) \qquad C\|m\|_2 \leq \left(\int_{Q_0} |m(\xi)|^2 \omega(\xi) \, d\xi \right)^{1/2} \leq C'\|m\|_2 \, ,$$

where both $|m(\xi)|^2$ and $\omega(\xi)$ are 2π-periodic.

We finally choose $m(\xi)$ appropriately. Put $m(\xi) = m_N(\xi - \xi_0)$, where $m_N(\xi) = (2\pi N)^{-1/2} \sum_{0 \leq k_j < N} e^{ik \cdot \xi}$. Then $|m_N(\xi)|^2 = K_N(\xi)$ is the Fejer kernel and (3.5) takes the simpler form

$$C \leq (K_N * \omega)^{1/2}(\xi_0) \leq C' \, ,$$

where $*$ denotes the convolution product of periodic functions. Then $(K_N * \omega)(\xi_0) \to \omega(\xi_0)$ almost everywhere , as $N \to \infty$, and (3.2) follows.

Let us move to the proof of the second assertion of Theorem 1. We start by observing that

$$(3.6) \qquad \mathcal{F}V_0 = \{ m(\xi)\hat{g}(\xi) : m \in L^2(Q_0) \},$$

where $\mathcal{F}$ denotes the Fourier transform. We then get

Lemma 5. *The operator* $U : V_0 \rightarrow L^2(Q_0, |Q_0|^{-1} \, d\xi)$ *which maps*

$f(x) = \sum \alpha_k g(x - k)$ to $m(\xi)(\omega(\xi))^{1/2}$, where $m(\xi) = \sum \alpha_k e^{-ik\cdot\xi}$, is an isometric isomorphism.

Indeed, if the sum defining $f(x)$ is finite, we have

$$\|f\|_2 = \frac{1}{(2\pi)^{n/2}} \left(\int |\hat{f}(\xi)|^2 \, d\xi \right)^{1/2} = \frac{1}{|Q_0|^{1/2}} \left(\int |m(\xi)|^2 |\hat{g}(\xi)|^2 \, d\xi \right)^{1/2}$$

$$= \frac{1}{|Q_0|^{1/2}} \left(\int_{Q_0} |m(\xi)|^2 \omega(\xi) \, d\xi \right)^{1/2}.$$

Extension to the whole of V_0 follows by continuity. Finally, to show that U is surjective, we note that, because $0 < c_1^2 \leq \omega(\xi) \leq c_2^2$, every function $q(x)$ in $L^2(Q_0, |Q_0|^{-1} \, d\xi)$ is of the form $m(\xi)(\omega(\xi))^{1/2}$, where $m(\xi) = \sum \alpha_k e^{-ik\cdot\xi}$ and $\sum |\alpha_k|^2 < \infty$, so that the preimage $f(x) \in V_0$ can be constructed.

The isometry U has the property that $U\tau_k = \chi_k U$, where τ_k is translation by k and χ_k is multiplication by $e^{-ik\cdot\xi}$.

In order to construct the orthonormal basis $\phi(x - k)$, $k \in \mathbf{Z}^n$, of V_0, it is therefore enough to construct its image by U, which leads us to look for an orthonormal basis of $L^2(Q_0, |Q_0|^{-1} \, d\xi)$ of the form $\chi_k U(\phi)$. The simplest choice is to set $U(\phi) = 1$, so that $m(\xi) = (\omega(\xi))^{-1/2}$ and thus $\hat{\phi}(\xi) = (\omega(\xi))^{-1/2}\hat{g}(\xi)$, as required.

The last part of Theorem 1 is established by the same method.

Now we can show that the algorithm described at the beginning of the section would have led to the same result. Indeed, that algorithm has the following virtue. If it produces the vectors f_k when applied to the vectors e_k, then it will produce the vectors $U(f_k)$ when applied to the vectors $U(e_k)$, whenever U is an isometry. Returning to our particular situation, the e_k are the functions $g(x - k)$ and the $U(e_k)$ are the functions $\chi_k(\omega(\xi))^{1/2}$, which it will be convenient to orthonormalize in $L^2(Q_0, |Q_0|^{-1} \, d\xi)$. The operator which we have called T in the preamble is then defined by

$$T(f)(\xi) = \sum_{k\in\mathbf{Z}^n} c_k e^{-ik\cdot\xi} (\omega(\xi))^{1/2}$$

where

$$c_k = \frac{1}{|Q_0|} \int_{Q_0} f(\xi)(\omega(\xi))^{1/2} e^{ik\cdot\xi} \, d\xi.$$

Quite simply, all we have done is to write the Fourier series of $\omega^{1/2} f$ and multiply by $\omega^{1/2}$. Finally, $T(f) = \omega f$ and $T^{-1/2}(\chi_k \omega^{1/2}) = \chi_k$, as claimed.

4 Regularity of the function ϕ

Suppose now that the multiresolution approximation is r-regular. Then we have the following fundamental result.

Theorem 2. *Let V_j, $j \in \mathbb{Z}$, be an r-regular multiresolution approximation of $L^2(\mathbb{R}^n)$. Then the function $\phi \in V_0$, defined by $\hat{\phi}(\xi) = \hat{g}(\xi)(\sum |\hat{g}(\xi + 2k\pi)|^2)^{-1/2}$, satisfies the estimates*

(4.1) $$|\partial^\alpha \phi(x)| \le C_m (1 + |x|)^{-m},$$

for every $\alpha \in \mathbb{N}^n$ such that $|\alpha| \le r$ and for every $m \in \mathbb{N}$.

To prove this result, let us recall some simple properties of the Sobolev spaces $H^m(\mathbb{R}^n)$.

For $m \in \mathbb{N}$, a function f is in $H^m(\mathbb{R}^n)$ if and only if $f \in L^2(\mathbb{R}^n)$ and all its derivatives (in the sense of distributions) $\partial^\alpha f$ of order $|\alpha| \le m$ belong to $L^2(\mathbb{R}^n)$ as well. The norm of f in $H^m(\mathbb{R}^n)$ is $(\sum_{|\alpha| \le m} \|\partial^\alpha f\|_2^2)^{1/2}$.

Let $\chi \in \mathcal{D}(\mathbb{R}^n)$ be a C^∞-function of compact support such that $\sum_{k \in \mathbb{Z}^n} |\chi(\xi - k)| \ge 1$. We can use χ to reduce the "global" H^m norm to H^m norms "localized" about the points $k \in \mathbb{Z}^n$.

Lemma 6. *If $f \in H^m(\mathbb{R}^n)$, the sequence $\omega(k) = \|f(\xi)\chi(\xi - k)\|_{H^m}$ is in $l^2(\mathbb{Z}^n)$. Conversely, if $f \in L^2(\mathbb{R}^n)$ and if $\omega(k) \in l^2(\mathbb{Z}^n)$, then $f \in H^m(\mathbb{R}^n)$.*

The proof follows immediately from applying Leibniz's formula to the product $f(\xi)\chi(\xi - k)$ and from observing that the L^2 norm can be recovered from its localizations via an l^2 sum.

Lemma 7. *If $|g(x)| \le C_m(1 + |x|)^{-m}$, for each integer $m \in \mathbb{N}$, then $\sum |\hat{g}(\xi + 2k\pi)|^2)^{1/2}$ is a C^∞-function.*

Indeed, $\int |g(x)|^2 (1 + |x|^2)^m \, dx < \infty$ for each positive integer $m \in \mathbb{N}$. So $\hat{g}(\xi)$ belongs to $H^m(\mathbb{R}^n)$ for each m.

Now we use Sobolev's embedding theorem, $H^m \subset C^{m-n/2}$, without defining the Hölder spaces C^r too precisely. It will be enough for our purposes to use the weaker inclusion $H^m \subset C^{m-n/2-\varepsilon}$, $\varepsilon > 0$. Put $s = m - n/2 - \varepsilon$. By Lemma 6, we have $\sum \|\hat{g}(\xi)\chi(\xi - k)\|_{C^s}^2 < \infty$ for every $s \in \mathbb{N}$.

This means that the series $\sum |\hat{g}(\xi + 2k\pi)|^2$ and its derivatives up to order s converge uniformly on compacta. Since s is arbitrary, Lemma 7 has been proved.

We are now in a position to prove Theorem 2. We know that $(\sum |\hat{g}(\xi + 2k\pi)|^2)^{1/2} \ge c > 0$. It follows that the inverse of this function is infinitely differentiable as well and can be written as $\sum \alpha_k e^{ik\cdot\xi}$, where the coefficients α_k decrease rapidly. So $\hat{\phi}(\xi) = (\sum \alpha_k e^{ik\cdot\xi})\hat{g}(\xi)$, which gives

$\phi(x) = \sum \alpha_k g(x + k)$. The regularity and the estimates of $\phi(x)$ follow immediately.

The remarks which follow serve as motivation for the calculations of Sections 6 and 10.

Firstly, note that, by a simple change of scale, we can check that the functions $2^{nj/2}\phi(2^j x - k)$, $k \in \mathbb{Z}^n$, form an orthonormal basis of V_j.

Next, the orthogonal projection E_j of $L^2(\mathbb{R}^n)$ onto V_j is given by

$$(4.2) \qquad E_j f(x) = \sum_{k \in \mathbb{Z}^n} \alpha(j, k)\phi(2^j x - k),$$

where

$$(4.3) \qquad \alpha(j, k) = 2^{nj} \int f(y)\bar{\phi}(2^j y - k)\, dy.$$

In Section 10 we shall show that ϕ can always be chosen so that its integral is 1. As a result, $\sum \phi(x - k) \equiv 1$ on $\mathbb{R}^n$.

Now, each of the operations (4.2) and (4.3) have a remarkable significance for numerical analysis, a point of view that will be developed in Section 11. The integrals of (4.3) are, in fact, the mean values of f at the points $k2^{-j}$, $k \in \mathbb{Z}^n$, on the scale 2^{-j}. *These mean values enable us to sample $f \in L^2(\mathbb{R}^n)$ on the lattice $\Gamma_j = 2^{-j}\mathbb{Z}^n$.*

The identity (4.2) serves to *extrapolate* from the sampling. The function f_j is constructed to be the simplest possible function which has the same sampling on Γ_j as f. This last condition is guaranteed precisely by the fact that the functions $\phi(x - k)$, $k \in \mathbb{Z}^n$, form an orthonormal sequence.

The sampling operator T_j, defined by (4.3), is not the same as that introduced after Definition 1.

5 Bernstein's inequalities

The properties of the function ϕ, which we have just established, enable us to extend the multiresolution approximation V_j of $L^2(\mathbb{R}^n)$ to other function spaces.

Let us try to replace $L^2(\mathbb{R}^n)$ by $L^p(\mathbb{R}^n)$, for an exponent $p \in [1, \infty]$. This is made possible by the following lemma.

Lemma 8. *Let ϕ be the function of Theorem 2. Then there are two constants $c_2 > c_1 > 0$ such that, for each $p \in [1, \infty]$ and for each finite sum $f(x) = \sum \alpha(k)\phi(x - k)$, we have*

$$(5.1) \qquad c_1\|f\|_p \le \left(\sum |\alpha(k)|^p \right)^{1/p} \le c_2\|f\|_p.$$

Let us start at the extremes. When $p = \infty$, then

$$|f(x)| \le \sum |\alpha(k)||\phi(x-k)| \le \sup_k |\alpha(k)|C(\phi),$$

where

$$C(\phi) = \sup_{x \in \mathbb{R}^n} \sum |\phi(x-k)|.$$

In the opposite direction we get $\alpha(k) = \int f(x)\bar{\phi}(x-k)\,dx$, so $|\alpha(k)| \le \|f\|_\infty \|\phi\|_1$.

The case $p = 1$ is similar and is left to the reader.

For the general case, let q be the conjugate exponent of p, so that $1/p + 1/q = 1$. We write $|\phi(x-k)| = |\phi(x-k)|^{1/p}|\phi(x-k)|^{1/q}$, which gives

$$|f(x)| \le \sum |\alpha(k)||\phi(x-k)|$$
$$\le \left(\sum |\alpha(k)|^p |\phi(x-k)|\right)^{1/p} \left(\sum |\phi(x-k)|\right)^{1/q}.$$

As above, $C(\phi)$ is finite and the left-hand inequality of (5.1) follows immediately. To prove the right-hand inequality, we start with the identity $\alpha(k) = \int f(x)\bar{\phi}(x-k)\,dx$. This gives

$$|\alpha(k)| \le \|\phi\|_1 \left(\int |f(x)|^p |\phi(x-k)|\,dx\right)^{1/p},$$

which, in turn, gives the right-hand inequality of (5.1).

This lemma leads us to define $V_0(p)$ as the intersection $V_0 \cap L^p(\mathbb{R}^n)$, for $1 \le p \le 2$, and as the completion of V_0 in the $L^p(\mathbb{R}^n)$ norm, when $2 \le p < \infty$.

By the lemma, $f \in V_0(p)$ if and only if

$$f(x) = \sum \alpha(k)\phi(x-k) \quad \text{where} \quad \alpha(k) \in l^p(\mathbb{Z}^n).$$

We define $V_0(\infty)$ as the vector space whose elements f can be written $f(x) = \lim f_m(x)$, where the limit is uniform on compact subsets and where $f_m \in V_0$ and $\sup_{m \ge 0} \|f_m\|_\infty < \infty$. In other words, $f \in V_0(\infty)$ if and only if

$$f(x) = \sum \alpha(k)\phi(x-k) \quad \text{where} \quad \alpha(k) \in l^\infty(\mathbb{Z}^n).$$

Finally, we define $V_j(p)$ by a simple change of scale:

$$f(x) \in V_0(p) \iff f(2^j x) \in V_j(p)$$

and we think of $V_j(p)$ as imbedded in $L^p(\mathbb{R}^n)$.

We shall show that the $V_j(p)$ approximate to $L^p(\mathbb{R}^n)$ in the following sense: if E_j is the orthogonal projection of $L^2(\mathbb{R}^n)$ onto V_j, then, for every function $f \in L^p(\mathbb{R}^n)$, $E_j(f)$ converges to f, both in $L^p(\mathbb{R}^n)$ norm and almost everywhere. When $p = \infty$, $E_j(f)$ converges to f in the $\sigma(L^\infty, L^1)$-topology defined by the duality between L^∞ and L^1.

We can now state and prove the inequalities of "Bernstein type".

Theorem 3. *Let V_j, $j \in \mathbb{Z}$, be an r-regular multiresolution approxima-tion of $L^2(\mathbb{R}^n)$. Then there exists a constant C such that, for $1 \leq p \leq \infty$, $j \in \mathbb{Z}$, $f \in V_j(p)$ and $|\alpha| \leq r$, we have*

$$(5.2) \qquad \|\partial^\alpha f\|_p \leq C 2^{|\alpha| j} \|f\|_p .$$

To establish (5.2), we first reduce to the case $j = 0$ by a simple change of variable. We write $f = \sum \xi(k)\phi(x - k)$ and get $|\partial^\alpha f(x)| \leq \sum |\xi(k)||\partial^\alpha \phi(x-k)|$. A repetition of the argument which proved the left-hand inequality of Lemma 8 gives $\|\partial^\alpha f\|_p \leq C \left(\sum |\xi(k)|^p\right)^{1/p}$, for which a bound can be found using the right-hand inequality of the self-same lemma.

Let us recall what the classical inequalities of S. Bernstein are. For $1 \leq p \leq \infty$, let f be a function in $L^p(\mathbb{R}^n)$ whose Fourier transform $\hat{f}$ has support contained in the ball $|\xi| \leq R$.

Then f is infinitely differentiable and, for every $\alpha \in \mathbb{N}^n$, we get

$$(5.3) \qquad \|\partial^\alpha f\|_p \leq R^{|\alpha|} \|f\|_p .$$

By an obvious change of variable we can reduce to the case $R = 1$. Then (5.3) is proved by induction on $|\alpha|$ and it only remains to check the inequality

$$(5.4) \qquad \left\|\frac{\partial f}{\partial x_1}\right\|_p \leq \|f\|_p ,$$

when the support of the Fourier transform of f is contained in the unit ball.

To prove (5.4) it is enough to establish the corresponding one-dimensional inequality and then apply Fubini's theorem.

So we need to prove that

$$(5.5) \qquad \|f'\|_p \leq \|f\|_p$$

is satisfied, when $1 \leq p \leq \infty$ and the Fourier transform of f has support in $[-1, 1]$.

The idea is to verify that, for such a function, $f'(x) = (f * \mu)(x)$, where μ is a measure of total mass 1. To prove this identity, we let $\sigma(\xi)$ denote the periodic function of period 4 which is odd and equal to ξ on $[-1, 1]$. The measure μ is a sum of point masses placed at $(2k + 1)\pi/2$ corresponding to the Fourier coefficients of $i\sigma(\xi)$. That $\|\mu\| = 1$ follows immediately ([148]).

6 A remarkable identity satisfied by the operator E_j

Let V_j, $j \in \mathbb{Z}$, be an r-regular multiresolution approximation of $L^2(\mathbb{R}^n)$.

We intend to show that this multiresolution approximation can be used in function spaces other than the initial $L^2(\mathbb{R}^n)$. For that, however, we need to improve our understanding of the operators E_j, the orthogonal projections of $L^2(\mathbb{R}^n)$ onto the subspaces V_j. In each of the cases where V_j is described explicitly—examples 1 and 3 of Section 2—the kernel of E_j is easily calculated. Indeed, we have

$$E_j(x,y) = 2^{nj}E(2^jx, 2^jy) \quad \text{and} \quad E(x,y) = \sum \phi(x-k)\bar{\phi}(y-k).$$

Then we calculate ϕ, using the algorithm of Section 4. From that we derive the exact expression for $E(x,y)$ and the conclusion of Theorem 4 below can be checked directly.

The reader who is only interested in the multiresolution approximations of examples 1 and 3 may thus omit the proof of Theorem 4 and retain only the conclusion—that $E_j(P) = P$, for every $j \in \mathbb{Z}$ and for every polynomial P of degree less than or equal to r.

Throughout this section, V_j, $j \in \mathbb{Z}$, will be an r-regular multiresolution approximation of $L^2(\mathbb{R}^n)$ on which no additional conditions are imposed.

Since the functions $\phi(x-k)$, $k \in \mathbb{Z}^n$, form an orthonormal basis of V_0, the orthogonal projection operator E_0 of $L^2(\mathbb{R}^n)$ onto V_0 is given by

$$(6.1) \qquad E_0 f(x) = \int E(x,y)f(y)\,dy\,,$$

where $E(x,y) = \sum_{k \in \mathbb{Z}^n} \phi(x-k)\bar{\phi}(y-k)$.

We immediately deduce the following properties:

$$(6.2) \qquad |\partial_x^\alpha \partial_y^\beta E(x,y)| \leq C_m (1 + |x-y|)^{-m}\,,$$

for every $m \in \mathbb{N}$, $|\alpha| \leq r$, and $|\beta| \leq r$, and

$$(6.3) \qquad E(x+k, y+k) = E(x,y) \qquad \text{for} \qquad k \in \mathbb{Z}^n.$$

Moreover, the kernel of the operator E_j is $2^{nj}E(2^jx, 2^jy)$, as can be seen by a simple change of variable.

We know that, as $j \to +\infty$, $E_j(f) \to f$ in $L^2(\mathbb{R}^n)$ norm, for each function f in $L^2(\mathbb{R}^n)$. In order to establish a similar result for many other spaces of functions or distributions, we shall prove the following fundamental result.

Theorem 4. *With the above notation,*

$$(6.4) \qquad \int E(x,y)y^\alpha\,dy = x^\alpha\,,$$

for every multi-index $\alpha \in \mathbb{N}^n$ such that $|\alpha| \leq r$.

In the proof of (6.4) we shall use the inclusion $V_0 \subset V_j$, $j \in \mathbb{N}$ (for the first time in this chapter). At the level of orthogonal projection operators this becomes $E_0 = E_j E_0$ and, at the level of the kernels of such operators,

(6.5) $$E(x,y) = 2^{nj} \int E(2^j x, 2^j u) E(u,y) du \,.$$

We begin the proof with the case $r = 0$.

The function $E(x,y)$ belongs to $L^\infty(\mathbb{R}^n \times \mathbb{R}^n)$ and satisfies the inequality $|E(x,y)| \leq C_m (1 + |x - y|)^{-m}$ for every $m \in \mathbb{N}$. We form the function

$$m_0(x) = \int E(x,y) \, dy,$$

which is $\mathbb{Z}^n$-periodic because of (6.3).

Next, we let j tend to $+\infty$ on the right-hand side of (6.5) and use the following result.

Lemma 9. *For every $y \in \mathbb{R}^n$ and for almost all $x \in \mathbb{R}^n$,*

(6.6) $$\lim_{j\uparrow+\infty} \left\{ 2^{nj} \int E(2^j x, 2^j u) E(u,y) \, du - m_0(2^j x) E(x,y) \right\} = 0 \,.$$

Let us assume this result and show how to conclude the case $r = 0$. We deduce—using (6.5)—that $\lim_{j\uparrow\infty} (1 - m_0(2^j x)) E(x,y) = 0$ for every $y \in \mathbb{R}^n$ and almost all $x \in \mathbb{R}^n$. We can multiply this identity by $f \in V_0$, integrate and then apply Lebesgue's dominated convergence theorem. We get $\lim_{j\uparrow+\infty} m_0(2^j x) f(x) = f(x)$, for every $f \in V_0$. By a simple change of variable, the same conclusion applies to $f \in V_{j_0}$, $j_0 \geq 0$.

Convergence takes place in $L^2(\mathbb{R}^n)$, because m_0 lies in $L^\infty(\mathbb{R}^n)$. Finally, by the denseness of $\bigcup V_j$, $\lim_{j\uparrow\infty} m_0(2^j x) f(x) = f(x)$ for every function $f \in L^2(\mathbb{R}^n)$. In particular, if $f(x)$ is the characteristic function of the unit cube $[0,1)^n$, we get $\int_0^1 \cdots \int_0^1 |m_0(2^j x) - 1|^2 \, dx \to 0$. But, because of the periodicity of m_0, this integral is independent of $j \in \mathbb{N}$ and we get $m_0(x) = 1$, as required.

We still have to prove Lemma 9. The proof turns on the idea of a Lebesgue point of a function $f \in L^1_{\text{loc}}(\mathbb{R}^n)$. We say that x_0 is a *Lebesgue point* of f if

$$\lim_{r\downarrow 0} \frac{1}{r^n} \int_{|y|\leq r} |f(x_0 + y) - f(x_0)| \, dy = 0 \,.$$

It is known that almost all $x_0 \in \mathbb{R}^n$ are Lebesgue points of $f \in L^1_{\text{loc}}(\mathbb{R}^n)$.

Lemma 10. *If $f \in L^\infty(\mathbb{R}^n)$ and if $x_0 \in \mathbb{R}^n$ is a Lebesgue point of f, then*

(6.7) $$\lim_{\varepsilon\downarrow 0} \int |f(x_0 + \varepsilon y) - f(x_0)|(1 + |y|)^{-n-1} \, dy = 0 \,.$$

To see this, put $J(r) = r^{-n} \int_{|y|\leq r} |f(x_0 + y) - f(x_0)| \, dy$. After integration by parts, the integral whose limit we are seeking becomes

$$c_n \int_0^\infty J(\varepsilon r) \frac{r^n}{(1+r)^{n+2}} \, dr \, .$$

Since $J(r) \le 2\omega(n)\|f\|_\infty$, where $\omega(n)$ is the volume of the unit ball of $\mathbb{R}^n$, we can apply Lebesgue's dominated convergence theorem to conclude the proof of Lemma 10.

Returning to Lemma 9, we have

$$2^{nj} \int E(2^j x, 2^j u) E(u,y) \, du = \int E(2^j x, 2^j x + v) E(x + 2^{-j} v, y) \, dv$$

$$= \int K_j(x,v) f(x + 2^{-j} v) \, dv \, ,$$

where we have put $f(x) = E(x,y)$.

The difference that we need to estimate can then be written more simply as

$$\int K_j(x,v) \big(f(x + 2^{-j} v) - f(x) \big) \, dv$$

which—by (6.2)—is less than

$$C \int |f(x + 2^{-j} v) - f(x)| (1 + |v|)^{-n-1} \, dv \, .$$

Using (6.7), we conclude the proof.

For $r = 0$, we could have used a simpler method. But the proof which we have just presented extends, without much change, to the case $r > 1$, which we shall deal with now.

Since $\int E(x,y) \, dy = 1$ and from (6.2), we see that $2^{nj} E(2^j x, 2^j y)$ is an approximation to the identity, that is, to Dirac measure at x. This means that the function $2^{nj} E(2^j x, 2^j u)$ enforces a localization of u about x on the scale 2^{-j}. This remark suggests that we expand $E(u,y)$ as a Taylor series in powers of $u - x$ in order to estimate $2^{nj} \int E(2^j x, 2^j u) E(u,y) \, du$.

So we write

$$E(u,y) = \sum_{0 \le |\alpha| \le r} \frac{(u-x)^\alpha}{\alpha!} \partial_x^\alpha E(x,y) + R(u,x,y) \, ,$$

where

$$R(u,x,y) = \sum_{|\alpha|=r} \frac{(u-x)^\alpha}{\alpha!} S_\alpha(u,x,y)$$

and finally

$$S_\alpha(u,x,y) = r \int_0^1 \partial_x^\alpha E(tu + (1-t)x, y)(1-t)^{r-1} \, dt \, .$$

To estimate our integral, we make the change of variable $u = x + 2^{-j} v$, so that $S_\alpha(u, x, y)$ becomes

$$S_\alpha^{(j)}(v, x, y) = r \int_0^1 \partial_x^\alpha E(x + t2^{-j}v, y)(1 - t)^{r-1} \, dt \, .$$

We then put $\alpha! m_\alpha(x) = \int E(x, u)(u - x)^\alpha \, du$, in order to integrate the leading terms, observing that this function is continuous and $\mathbf{Z}^n$-periodic. So we get

$$(6.8) \quad 2^{nj} \int E(2^j x, 2^j u) E(u, y) \, dy = \sum_{0 \le \alpha < r} 2^{-j|\alpha|} m_\alpha(2^j x) \partial_x^\alpha E(x, y) +$$

$$+ r2^{-jr} \sum_{|\alpha|=r} \iint_0^1 E(2^j x, 2^j x + v) \partial_x^\alpha E(x + t2^{-j}v, y)(1-t)^{r-1} \, dt \, dv$$

$$= E(x, y) \, .$$

Let us leave the first term of (6.8) in order to exploit the equality of the second term with $E(x, y)$ systematically.

We shall show that $m_\alpha(x) = 0$, by induction on $|\alpha| = s \in [1, r]$. Let us indicate what happens for $s = 1$. The term $E(x, y)$ cancels, because $m_0(x) = 1$. After this first simplification, we get

$$(6.9) \quad \sum_{1 \le |\alpha| \le r-1} 2^{-j|\alpha|} m_\alpha(2^j x) \partial_x^\alpha E(x, y) + O(2^{-jr}) = 0 \, ,$$

where O is uniform in x and y. This lets us multiply (6.9) by 2^j and then pass to the limit. So we have

$$\lim_{j \to +\infty} \sum_{|\alpha|=1} m_\alpha(2^j x) \partial_x^\alpha E(x, y) = 0 \, .$$

We then repeat our analysis of the case $r = 0$, that is, we multiply by $f \in V_0$ and integrate. We thus get

$$\lim_{j \to +\infty} \sum_{|\alpha|=1} m_\alpha(2^j x) \partial^\alpha f = 0 \, .$$

Now, for almost all x in $\mathbf{R}^n$, the sequence $2^j x$ is equidistributed modulo $\mathbf{Z}^n$. In particular, for all $u \in \mathbf{R}^n$, we can find, for almost all $x \in \mathbf{R}^n$, a sequence of values of j such that $m_\alpha(2^j x)$ tends to $m_\alpha(u)$. So we have $\sum_{|\alpha|=1} m_\alpha(u) \partial^\alpha f = 0$ for every $u \in \mathbf{R}^n$ and for every function $f \in V_0$. By a simple change of scale, the same identity can be obtained for $f \in V_j$, $j \ge 0$. By the denseness of $\bigcup V_j$, we have $\sum_{|\alpha|=1} m_\alpha(u) \partial^\alpha f = 0$, in the sense of distributions, for every $u \in \mathbf{R}^n$ and for every $f \in L^2(\mathbf{R}^n)$. Hence all the m_α are identically zero.

If we do not want to use equidistribution modulo $\mathbf{Z}^n$, we argue as follows. Let $g(x)$ be a continuous $\mathbf{Z}^n$-periodic function. We multiply the equation

(6.10) $$\lim_{j \to +\infty} \sum_{|\alpha|=1} m_\alpha(2^j x)\partial^\alpha f = 0, \qquad \text{for} \qquad f \in V_0$$

by the product $g(2^j x)h(x)$, where $h \in \mathcal{D}(\mathbb{R}^n)$, and integrate. Then we use the following lemma, whose proof we delay until we have concluded the proof of Theorem 4.

Lemma 11. *If $u(x) \in L^\infty(\mathbb{R}^n)$ is $\mathbb{Z}^n$-periodic and if $v(x) \in L^1(\mathbb{R}^n)$, then*

(6.11) $$\lim_{N \to \infty} \int_{\mathbb{R}^n} u(Nx)v(x)\,dx = \left(\int u \right) \int_{\mathbb{R}^n} v\,dx\,,$$

where

$$\int u = \int_0^1 \cdots \int_0^1 u(x_1, \ldots, x_n)\,dx_1 \cdots dx_n\,.$$

Returning to (6.10), it follows that, after integrating and passing to the limit, $\sum_{|\alpha|=1} \left(\int gm_\alpha \right) \langle \partial^\alpha f, h \rangle = 0$. That is, on setting $c_\alpha = \int gm_\alpha$, the vector field $\sum_{|\alpha|=1} c_\alpha \partial^\alpha$ annihilates all the functions of V_0. Again, by a simple change of variable, this vector field annihilates all the functions in V_j, $j \in \mathbb{N}$, and, by the denseness of $\bigcup V_j$, it annihilates—in the sense of distributions—all functions $f \in L^2(\mathbb{R}^n)$. Thus $c_\alpha = 0$ for every function g, which certainly implies that $m_\alpha = 0$ when $|\alpha| = 1$.

No modification of this proof is needed as long as $s = |\alpha| < r$. We do get $m_\alpha(x) = 0$ for $|\alpha| < r$ and we are left with the case $|\alpha| = r$. Here, the identity (6.8) reduces to

$$\sum_{|\alpha|=r} \iint_0^1 E(2^j x, 2^j x + v)\partial_x^\alpha E(x + t2^{-j}v, y)(1-t)^{r-1}\,dt\,dv = 0$$

We next use the fact that $\partial_x^\alpha E(x, y) \in L^\infty(dx)$, for each fixed y. Almost every $x \in \mathbb{R}^n$ is a Lebesgue point of this function. If x is, indeed, a Lebesgue point, the integral can be replaced by the corresponding integral wherein $\partial_x^\alpha E(x + t2^{-j}v, y)$ is replaced by $\partial_x^\alpha E(x, y)$.

Then the limit of the sum

$$S_j(x, y) = \sum_{|\alpha|=r} m_\alpha(2^j x)\partial_x^\alpha E(x, y)$$

equals 0 for all $y \in \mathbb{R}^n$ and almost all $x \in \mathbb{R}^n$. The functions $S_j(x, y)$ satisfy the uniform estimates $|S_j(x, y)| \le C_m \left(1 + |x - y|\right)^{-m}$. On multiplying by $f \in V_0$ and integrating, we find that

$$\sum_{|\alpha|=r} m_\alpha(2^j x)\partial^\alpha f$$

converges to 0 almost everywhere and also in $L^2(\mathbb{R}^n)$, because we can apply Lebesgue's dominated convergence theorem.

We conclude, at last, that, as in the case $s < r$, all the coefficients m_α vanish when $|\alpha| = r$.

The proof of the theorem is then a simple exercise. We have shown that $\int E(x, y)(x - y)^\alpha \, dy = 0$ for $1 \leq |\alpha| \leq r$ and that $\int E(x, y) \, dy = 1$. It can immediately be deduced that $\int E(x, y) y^\alpha \, dy = x^\alpha$, when $|\alpha| \leq r$. Indeed, we write

$$\frac{y^\alpha}{\alpha!} = \sum_{\beta + \gamma = \alpha} \frac{(y - x)^\beta}{\beta!} \frac{x^\gamma}{\gamma!}$$

and, on integrating with respect to y, all the right-hand side terms are 0, except when $\beta = 0$ and $\gamma = \alpha$.

We still have to prove Lemma 11. On decomposing $\mathbb{R}^n$ as $Q_0 + \mathbb{Z}^n$, where Q_0 is the unit cube $[0, 1)^n$, $v(x)$ can be replaced by $w(x) = \sum_{k \in \mathbb{Z}^n} v(x + k)$. This reduces the proof to the case when u and w are periodic and the integral is taken over the n-dimensional torus $\mathbb{R}^n / \mathbb{Z}^n$. We then exploit the denseness of the trigonometric polynomials as a linear subspace of $L^1(\mathbb{R}^n / \mathbb{Z}^n)$. By linearity, we reduce to considering $w(x) = e^{2\pi i l \cdot x}$. Finally,

$$\int_{\mathbb{R}^n / \mathbb{Z}^n} u(Nx) e^{2\pi i l \cdot x} \, dx = 0 \,,$$

when $l \neq 0$ and N is large enough for l not to be in $N\mathbb{Z}^n$. When $l = 0$, the integral equals that of u.

Theorem 4 can be paraphrased by extending the domain of definition of the operators E_j to the functions in $L^2(\mathbb{R}^n, (1 + |x|)^{-m} \, dx)$, for some integer $m \in \mathbb{N}$. This extension does not create any problems, because the kernels of the operators E_j decrease rapidly when $|y - x|$ tends to infinity. The image space is no longer V_j, but will be complete in the norm of $L^2(\mathbb{R}^n, (1 + |x|)^{-m} \, dx)$, exactly as in Section 5.

Then we have the following remarkable identity:

Corollary. Let V_j be an r-regular multiresolution approximation of $L^2(\mathbb{R}^n)$ and let $E_j : L^2(\mathbb{R}^n) \to V_j$ be the orthogonal projection onto V_j. Then

(6.12) $$E_j(P) = P \,,$$

for every polynomial P of degree less than or equal to r.

The theorem which we present below is an easy but important consequence of Theorem 4. The significance of Theorem 5 is that the operators E_j behave like convolution operators. The latter commute exactly with the partial differentiation operators ∂^α, whereas the operators E_j commute approximately with the operators ∂^α as long as $|\alpha| \leq r$. This approximate commutativity is what we are about to describe.

We fix a function $g \in \mathcal{D}(\mathbb{R}^n)$ whose integral equals 1, we let g_j denote the function $2^{nj}g(2^j x)$, and then we let G_j denote the operation of convolution by g_j.

We intend to compare $\partial^\alpha E_j$ with $G_j \partial^\alpha$, for every multi-index α satisfying $|\alpha| \leq r$.

Theorem 5. *Let V_j, $j \in \mathbf{Z}$, be an r-regular multiresolution approximation of $L^2(\mathbb{R}^n)$. Then, for every multi-index α satisfying $|\alpha| \leq r$, there exist functions $R^{(\alpha,\beta)} \in L^\infty(\mathbb{R}^n \times \mathbb{R}^n)$, indexed by the multi-indices β such that $|\beta| = |\alpha|$, satisfying*

$$(6.13) \qquad |R^{(\alpha,\beta)}(x,y)| \leq C_m (1 + |x - y|)^{-m},$$

for every integer $m \in \mathbf{N}$. Those functions also satisfy

$$(6.14) \qquad \int R^{(\alpha,\beta)}(x,y)\,dy = 0$$

identically in x and they define, for every $j \in \mathbf{Z}$, operators $\mathcal{R}_j^{(\alpha,\beta)}$ by

$$(6.15) \qquad \mathcal{R}_j^{(\alpha,\beta)} f(x) = 2^{nj} \int R^{(\alpha,\beta)}(2^j x, 2^j y) f(y)\,dy,$$

which are such that

$$(6.16) \qquad \partial^\alpha E_j = G_j \partial^\alpha + \sum_{|\beta|=|\alpha|} \mathcal{R}_j^{(\alpha,\beta)} \partial^\beta,$$

where we have put $\partial^\alpha = (\partial/\partial x_1)^{\alpha_1} \ldots (\partial/\partial x_n)^{\alpha_n}$.

Once we have established (6.16), we shall show, by an analogous argument, that $E_{j+1} - E_j = 2^{-jr} \sum_{|\beta|=r} \mathcal{R}_j^{(\beta)} \partial^\beta$, where each $\mathcal{R}_j^{(\beta)}$ is defined by the kernel $2^{nj} R^{(\beta)}(2^j x, 2^j y)$ and the $R^{(\beta)}$ satisfy similar properties to those of the $R^{(\alpha,\beta)}$.

Clearly, it will be enough to prove (6.16) when $j = 0$. A simple change of variable then gives the general case.

For each fixed x, we thus consider the function $\partial_x^\alpha E(x,y)$ of the variable y. By Theorem 4, we have

$$\int \partial_x^\alpha E(x,y) y^\beta \, dy = 0 \qquad \text{if } |\beta| \leq |\alpha| \text{ and } \beta \neq \alpha,$$

whereas

$$\int \partial_x^\alpha E(x,y) y^\alpha \, dy = \alpha! \,.$$

Now, the kernel $g(x-y)$ of the operator $G = G_0$ shares these properties with $E(x,y)$. This leads us to form the difference $R^\alpha(x,y) = \partial_x^\alpha E(x,y) - \partial_x^\alpha g(x - y)$, for which $\int R^\alpha(x,y) y^\beta \, dy = 0$ when $|\beta| \leq |\alpha|$. Finally, we put $f^\alpha(x,y) = R^\alpha(x, x + y)$ and consider it as a function of y for fixed x. This function belongs to the space $\mathcal{S}_r(\mathbb{R}^n)$, which we are about to define.

We write $S_r(\mathbb{R}^n)$ for the vector space of functions f which satisfy the inequality $|\partial^\alpha f(x)| \leq C_m(1 + |x|)^{-m}$, for every $m \in \mathbb{N}$ and every multi-index $\alpha \in \mathbb{N}^n$ such that $|\alpha| \leq r$. We have the following lemma.

Lemma 12. *Let* $0 \leq s \leq r$. *If* f, *belonging to* $S_r(\mathbb{R}^n)$, *is such that* $\int f(x)x^\alpha \, dx = 0$, *for each* $\alpha \in \mathbb{N}^n$ *satisfying* $|\alpha| \leq s$, *then* $f(x) = \sum_{|\alpha|=s} \partial^\alpha f_\alpha(x)$, *where* $f_\alpha \in S_r(\mathbb{R}^n)$ *and* $\int f_\alpha(x) \, dx = 0$, *for every* $\alpha \in \mathbb{N}^n$ *such that* $|\alpha| = s$.

This result will be proved below. We now apply it to $f^\alpha(x, y)$, regarded as a function of y depending on the parameter x. So we have

$$\partial^\alpha_x E(x, y) = (-1)^{|\alpha|}\partial^\alpha_y g(x - y) + \sum_{|\beta|=|\alpha|} \partial^\beta_y R^{(\alpha,\beta)}(x, y),$$

where $|R^{(\alpha,\beta)}(x, y)| \leq C_m(1 + |x - y|)^{-m}$, for every $m \in \mathbb{N}$. Further, $\int R^{(\alpha,\beta)}(x, y) dy = 0$, for each x.

To conclude, it is enough to notice that, if $S(x, y)$ is a sufficiently regular function of y defining an operator S, then integration by parts shows that the kernel of the operator $S\partial^\beta$ is exactly $(-1)^{|\beta|}\partial^\beta_y S(x, y)$.

Let us go back and prove Lemma 12. The proof is by induction on the dimension n. If $n = 1$, the result is elementary, f_s being the primitive of order s of f which tends to zero at infinity.

For $n \geq 2$, we shall restrict ourselves to dealing with the case $s = 1$ — to simplify the presentation — and let the reader take care of the general case.

Let $x = (x', x_n)$, where $x' \in \mathbb{R}^{n-1}$ and set $g(x') = \int f(x', x_n) \, dx_n$ and $h(x') = \int f(x', x_n)x_n \, dx_n$. By hypothesis, we have $\int g(x') \, dx' = \int x_j g(x') \, dx' = 0$ when $1 \leq j \leq n - 1$. Further, g belongs to $S_r(\mathbb{R}^{n-1})$. By the induction hypothesis, we have

$$g(x') = \sum_1^{n-1} \frac{\partial}{\partial x_j} g_j(x') \qquad \text{where} \qquad g_j \in S_r(\mathbb{R}^{n-1})$$

and $\int g_j(x') \, dx' = 0$. Moreover, $\int h(x') \, dx' = 0$.

We denote by ϕ an even function in $\mathcal{D}(\mathbb{R})$ whose integral is 1 and we set $\psi = -\phi'$. So $\int \psi(x_n) \, dx_n = 0$ and $\int x_n\psi(x_n) \, dx_n = 1$.

We define the auxiliary function $r(x)$ by

$$r(x) = f(x) - g(x')\phi(x_n) - h(x')\psi(x_n).$$

By the definitions of g and h, we get

$$\int r(x', x_n) \, dx_n = 0, \qquad \int r(x', x_n)x_n \, dx_n = 0$$

and this gives

$$r(x', x_n) = \frac{\partial}{\partial x_n} g_n(x', x_n),$$

where $g_n(x', x_n)$ belongs to $\mathcal{S}_r(\mathbb{R}^n)$, just like $r(x', x_n)$. We have finally obtained

$$f(x) = \sum_{j=1}^{n-1} \frac{\partial}{\partial x_j}(g_j(x')\phi(x_n)) + \frac{\partial}{\partial x_n}(g_n(x) - h(x')\phi(x_n)),$$

as claimed.

It is worth noting that the algorithm used depends linearly and continuously on f. Thus, if f is a continuous function of certain parameters, the same will be true for the f_α.

Returning to Theorem 5, this means that if V_j, $j \in \mathbb{Z}$, is a multiresolution approximation of $L^2(\mathbb{R}^n)$ such that the corresponding function ϕ is of class C^r, then the functions $R^{(\alpha, \beta)}$ are continuous in all the variables.

7 Effectiveness of a multiresolution approximation

How well do the projections of a multiresolution approximation V_j, $j \in \mathbb{Z}$, converge to a given function? We shall show that the answer depends on the regularity r of the functions in V_0.

We start by studying the approximation $E_j(f)$ of f when f belongs to the Sobolev space $H^s(\mathbb{R}^n)$, where $E_j : L^2(\mathbb{R}^n) \to V_j$ denotes the orthogonal projection operator, as before.

Let us recall the definition of the Sobolev spaces. If $s = 0$, $H^s(\mathbb{R}^n)$ coincides with $L^2(\mathbb{R}^n)$. If $s \in \mathbb{N}$, a function f belongs to $H^s(\mathbb{R}^n)$ if it lies in $L^2(\mathbb{R}^n)$ and if all its derivatives $\partial^\alpha f$, $|\alpha| \leq s$, taken in the sense of distributions, belong to $L^2(\mathbb{R}^n)$. It comes to the same thing to say that $\int (1 + |\xi|^2)^s |\hat{f}(\xi)|^2 \, d\xi$ is finite, and it is this latter definition we use when s, $0 \leq s \leq r$, is not necessarily an integer.

The definition of the spaces H^s, $s < 0$, is based on the idea that, for $s = -\sigma$, $\sigma > 0$, H^σ is a Hilbert space of test functions and H^s represents the corresponding dual space of distributions—the duality being the natural duality between $\mathcal{S}(\mathbb{R}^n)$ and $\mathcal{S}'(\mathbb{R}^n)$. On applying Parseval's identity, we immediately find that, for $s < 0$, a tempered distribution S belongs to $H^s(\mathbb{R}^n)$ if and only if its Fourier transform $\hat{S}$ belongs to $L^2_{\text{loc}}(\mathbb{R}^n)$ and satisfies $\int |\hat{S}(\xi)|^2 (1 + |\xi|^2)^s \, d\xi < \infty$.

The next theorem is the first of a series describing the effectiveness of the functional approximation by the operators E_j.

Theorem 6. *Let $r \in \mathbb{N}$ and let V_j, $j \in \mathbb{Z}$, be an r-regular multiresolution approximation of $L^2(\mathbb{R}^n)$. If f belongs to the Sobolev space $H^s(\mathbb{R}^n)$ and if $-r \leq s \leq r$, then the $E_j(f)$ converge to f in $H^s(\mathbb{R}^n)$ norm.*

To prove this result, we use the following very simple lemma.

Lemma 13. Throughout this lemma, $K(x,y)$ denotes a function in $L^\infty(\mathbb{R}^n \times \mathbb{R}^n)$ satisfying $|K(x,y)| \le C(1 + |x-y|)^{-n-1}$ and, for $1 \le p \le \infty$ and $\lambda > 0$, $T_\lambda : L^p(\mathbb{R}^n) \to L^p(\mathbb{R}^n)$ denotes the operator whose kernel is $\lambda^n K(\lambda x, \lambda y)$. Explicitly,

$$T_\lambda f(x) = \lambda^n \int K(\lambda x, \lambda y) f(y) \, dy \,.$$

If $\int K(x,y)\, dy = 0$, identically in x, and $1 \le p < \infty$, then

$$\lim_{\lambda \uparrow \infty} \|T_\lambda(f)\|_p = 0 \qquad \text{for } f \in L^p.$$

If $\int K(x,y)\, dy = 1$, identically in x, and $1 \le p < \infty$, then

$$\lim_{\lambda \uparrow \infty} \|T_\lambda(f) - f\|_p = 0 \qquad \text{for } f \in L^p.$$

When $p = \infty$, we suppose, in addition, that f is uniformly continuous and that $K(x,y)$ is a continuous function of x for each fixed y. Then, if $\int K(x,y)\, dy = 1$ and $f \in L^\infty(\mathbb{R}^n)$, we have

$$\lim_{\lambda \uparrow \infty} \|T_\lambda(f) - f\|_\infty = 0 \,.$$

The proof of Lemma 13 is classical. We first note that the norm of the operator $T_\lambda : L^p(\mathbb{R}^n) \to L^p(\mathbb{R}^n)$ is a finite constant, independent of λ. When $1 \le p < \infty$, we can reduce to a dense subset of $L^p(\mathbb{R}^n)$. So, if f is a continuous function of compact support, then $T_\lambda f(x)$ is $O(|x|^{-n-1})$ as $|x| \to \infty$, uniformly in $\lambda \ge 1$. Further, $T_\lambda f(x) = \lambda^n \int K(\lambda x, \lambda y) f(y)\, dy = \lambda^n \int K(\lambda x, \lambda y)\big(f(y) - f(x)\big)\, dy$ which is less than $C \int (1 + |y|)^{-n-1} \omega(\lambda^{-1}|y|)\, dy = \varepsilon(\lambda)$, where ω denotes the modulus of continuity of f: $\omega(h) = \sup_{|x-y| \le h} |f(y) - f(x)|$. An immediate application of Lebesgue's dominated convergence theorem supplies $\lim_{\lambda \uparrow \infty} \varepsilon(\lambda) = 0$. The functions $T_\lambda f(x)$ thus converge uniformly to 0 and are uniformly $O(|x|^{-n-1})$ as $|x| \to \infty$. Hence they converge to 0 in $L^p(\mathbb{R}^n)$ norm.

The second case of the lemma, where $\int K(x,y)\, dy = 1$, is a corollary of the first and of classical results on approximate identities. Indeed, we let g denote a continuous function of compact support whose integral is 1 and we replace $K(x,y)$ by $R(x,y) = K(x,y) - g(x-y)$, to which we apply the first case.

The assertion which brings the lemma to a close follows from the calculations above.

Let us return to Theorem 6. At first we suppose that $s = r$. From the definition of the norm on H^s, it follows that we have to prove that $\|\partial^\alpha E_j(f) - \partial^\alpha f\|_2 \to 0$ as $j \to +\infty$, for $|\alpha| \le r$, knowing that f belongs to $H^s(\mathbb{R}^n)$.

Applying Theorem 5 gives

$$\partial^\alpha E_j = G_j \partial^\alpha + \sum_{|\beta|=|\alpha|} \mathcal{R}_j^{(\alpha,\beta)} \partial^\beta$$

so all we need do is use Lemma 13.

If s is not an integer and $0 < s < r$, we first remark that the operators $E_j : H^s \to H^s$ are uniformly bounded. This is due to the fact that the property holds in the cases $s = 0$ and $s = r$ (by virtue of the preceding argument).

We get the intermediate case by the following classical proposition.

Proposition 1. *Every continuous linear operator $T : L^2 \to L^2$, whose restriction to the Sobolev space H^r is a continuous linear operator of H^r to itself, is also continuous as an operator of H^s to H^s, if $0 < s < r$.*

If the norms of T as an operator on L^2 and H^r are denoted by C_0 and C_1 respectively, then the norm of T as an operator on H^s, $0 \le s \le r$, is not greater than $\max(C_0, C_1)$.

We assume this result and continue the proof of Theorem 6. $H^r(\mathbb{R}^n)$ is a dense subspace of $H^s(\mathbb{R}^n)$. Since, for every f in $H^r(\mathbb{R}^n)$, $E_j(f)$ converges to f in H^r norm, we can conclude by a classical "dense subset" argument.

To prove Theorem 6 when $-r \le s \le 0$, we must first define E_j on $H^s(\mathbb{R}^n)$. Observe that $L^2(\mathbb{R}^n)$ is dense in $H^s(\mathbb{R}^n)$. In what follows, we write $(u, v) = \int_{\mathbb{R}^n} u(x)\bar{v}(x)\,dx$ for $u(x)$ and $v(x)$ in $L^2(\mathbb{R}^n)$.

If f and g belong to $L^2(\mathbb{R}^n)$, then $\big(E_j(f), g\big) = \big(f, E_j(g)\big)$. Now we suppose that f lies in $H^s(\mathbb{R}^n)$, where $-r \le s \le 0$ and that f_m is a sequence of functions in $L^2(\mathbb{R}^n)$ converging to f in $H^s(\mathbb{R}^n)$ norm. Then, for $t = -s$ and $g \in H^t(\mathbb{R}^n)$, we have

$$\big(E_j(f_m), g\big) = \big(f_m, E_j(g)\big) \to \big(f, E_j(g)\big)$$

as $m \to \infty$. So we define $E_j(f) \in H^s(\mathbb{R}^n)$ by $\big(E_j(f), g\big) = \big(f, E_j(g)\big)$. Because $E_j^2 = E_j$, we get $\big(E_j(f), E_j(g)\big) = \big(E_j(f), g\big)$. These remarks will be useful in what follows.

The convergence of the sequence $E_j(f)$ to f, for $f \in H^s(\mathbb{R}^n)$ is then immediate. The operators $E_j : H^s(\mathbb{R}^n) \to H^s(\mathbb{R}^n)$ are uniformly bounded, as can be seen by duality. Further $L^2(\mathbb{R}^n)$ is dense in $H^s(\mathbb{R}^n)$, and $E_j(f)$ converges to f when f is in $L^2(\mathbb{R}^n)$.

The standard argument "with 2ε" over a dense set then allows us to conclude that $\lim_{j\uparrow\infty} E_j(f) = f$, for every distribution f belonging to $H^s(\mathbb{R}^n)$ with $-r \le s < 0$.

We still need to prove Proposition 1. Taking Fourier transforms, we see that it is enough to prove the following result:

Proposition 2. Let $\omega(x) > 0$ be a function which is continuous on $\mathbb{R}^n$ and let $T : L^2(\mathbb{R}^n, dx) \to L^2(\mathbb{R}^n, dx)$ be a continuous linear operator which is also continuous from $L^2(\mathbb{R}^n, \omega(x)\,dx)$ to $L^2(\mathbb{R}^n, \omega(x)\,dx)$. Then, for $0 \le s \le 1$, T is also continuous from $L^2(\mathbb{R}^n, \omega^s(x)\,dx)$ to itself.

We prove this by complex analysis. Consider two continuous functions f and g of compact support, satisfying $\|f\|_2 \le 1$ and $\|g\|_2 \le 1$. For a complex number z such that $0 \le \operatorname{Re} z \le 1$, we take the entire function

$$(7.1) \qquad F(z) = (T(f\omega^{-z/2}), g\omega^{z/2}),$$

where the inner product is the $L^2(\mathbb{R}^n)$ inner product.

If $\operatorname{Re} z = 0$, then $\|f\omega^{-z/2}\|_2 = \|f\|_2 \le 1$ and similarly for $\|g\omega^{z/2}\|_2$. The continuity of T on $L^2(\mathbb{R}^n)$ gives $|F(z)| \le C_0$, the operator norm of $T : L^2(\mathbb{R}^n) \to L^2(\mathbb{R}^n)$.

If $\operatorname{Re} z = 1$, then $\int |f\omega^{-z/2}|^2 \omega\,dx = \|f\|_2^2 \le 1$. $\int |g\omega^{z/2}|^2 \omega^{-1}\,dx = \|g\|_2^2 \le 1$. The continuity of T on $L^2(\mathbb{R}^n, \omega\,dx)$ gives $|F(z)| \le C_1$, where C_1 is the operator norm on $L^2(\mathbb{R}^n, \omega\,dx)$.

Finally, we apply the Maximum Principle to the entire function

$$e^{\varepsilon z^2} F(z)$$

which, for $\varepsilon > 0$, is continuous on the strip $0 \le \operatorname{Re} z \le 1$, holomorphic on the interior and zero at infinity. This gives, for $0 < \operatorname{Re} z < 1$,

$$(7.2) \qquad |e^{\varepsilon z^2} F(z)| \le \max(C_0, e^{\varepsilon} C_1).$$

Now we let ε tend to 0 to conclude that $|F(z)| \le \max(C_0, C_1)$ when $0 \le s = \operatorname{Re} z \le 1$. But a moment's thought shows that this inequality means precisely that T is continuous on the intermediate spaces $L^2(\mathbb{R}^n, \omega^s\,dx)$, $0 \le s \le 1$.

We can generalize all the above to the $W^{s,p}$ spaces, $-r \le s \le r$, $1 < p < \infty$, which are defined as follows. If $0 \le s \le r$ and s is an integer, then f is in $W^{s,p}$ if and only if f and all its derivatives $\partial^\alpha f$, with $|\alpha| \le s$, belong to $L^p(\mathbb{R}^n)$. It comes to the same thing ([217] Chapter V, Theorem 3) to require that $(I - \Delta)^{s/2} f$ belongs to $L^p(\mathbb{R}^n)$, and it is the latter condition that enables us to define $W^{s,p}$ for arbitrary real s. It can then be shown, as in the case of Sobolev spaces, that, for each function or distribution f in $W^{s,p}$, $E_j(f)$ converges to f in $W^{s,p}$ norm.

The extreme case $p = \infty$ has to be examined separately, as usual. We first extend the definition of the $W^{m,p}$-spaces, $m \in \mathbb{N}$, by defining $W^{m,\infty}$ via the condition that $\partial^\alpha f$ belongs to $L^\infty(\mathbb{R}^n)$ for all multi-indices α of order $|\alpha| \le m$.

Then, for $f \in W^{m,\infty}$, the sequence $E_j(f)$ is uniformly bounded in

$W^{m,\infty}$ and converges weakly to f. That is, $\int E_j(f)g\,dx \to \int fg\,dx$ for each $g \in \mathcal{S}(\mathbb{R}^n)$ (as long as $0 \le m \le r$).

We can also replace each $W^{m,\infty}$ by the space C^m, where C^m is defined as usual except for the addition of conditions at infinity. We define C^m by the condition that all the derivatives $\partial^\alpha f$, $|\alpha| \le m$ are bounded and uniformly continuous on $\mathbb{R}^n$. Then, for every function f in $C^m(\mathbb{R}^n)$, the $E_j(f)$ converge to f in C^m norm, as long as $0 \le m < r$. Extension to $m = r$ requires the function ϕ (which is needed to calculate the kernel of E_j) to belong to C^m as well (and to decrease rapidly at infinity together with all its derivatives of order $|\alpha| \le r$).

Finally, this definition of C^m can be replaced by the usual definition of C^m, where no assumption is made on behaviour at infinity, as long as ϕ is in C^m and has compact support. In Chapter 3, Section 8, we shall see that this can occur for certain multiresolution approximations.

8 The operators $D_j = E_{j+1} - E_j$, $j \in \mathbb{Z}$

Let V_j, $j \in \mathbb{Z}$, be an r-regular multiresolution approximation of $L^2(\mathbb{R}^n)$ and let W_j denote the orthogonal complement of V_j in V_{j+1}. The orthogonal projection of $L^2(\mathbb{R}^n)$ onto W_j is precisely $D_j = E_{j+1} - E_j$. Then $f(x) \in W_0$ is equivalent to $f(2^j x) \in W_j$. Moreover,

$$(8.1) \qquad L^2(\mathbb{R}^n) = \bigoplus_{-\infty}^{\infty} W_j,$$

because the union of the V_j is dense in $L^2(\mathbb{R}^n)$ and the intersection of the V_j reduces to $\{0\}$.

It is therefore possible to decompose each $f \in L^2(\mathbb{R}^n)$ into an orthogonal series

$$(8.2) \qquad f = \sum_{-\infty}^{\infty} D_j(f) = \sum_{-\infty}^{\infty} d_j$$

and we intend to verify that the classical norms in spaces of functions or distributions are easily calculated via this decomposition.

The first thing to do is to extend the decomposition (8.1) to $L^p(\mathbb{R}^n)$, $1 \le p \le \infty$. This necessitates extending or restricting W_j to the L^p spaces.

If $1 \le p \le 2$, $W_j(p)$ is defined as the intersection of W_j with $L^p(\mathbb{R}^n)$, which is compatible with the definition of $V_j(p)$. We get $V_{j+1}(p) = V_j(p) + W_j(p)$ once again, where the sum is still direct, but no longer orthogonal. This results from the operators E_j and D_j, which project $V_{j+1}(p)$ onto $V_j(p)$ and $W_j(p)$, being continuous on L^p when $1 \le p \le 2$.

Thus we have $L^p(\mathbb{R}^n) = +_{-\infty}^{\infty} W_j(p)$ where the sum is direct but not orthogonal. This means that every function $f \in L^p(\mathbb{R}^n)$ can be written uniquely as

(8.3) $$f = \sum_{-\infty}^{\infty} d_j \qquad \text{where} \qquad d_j \in W_j(p)$$

and where the partial sums of the series (8.3) converge to f, in the sense of the L^p norm, for $1 < p \le 2$. Once again, the proof is clear and follows from the uniform continuity of the operators E_j on $L^p(\mathbb{R}^n)$ when $1 \le p \le 2$.

In the case where $2 \le p < \infty$, $W_j(p)$ is defined as the completion of W_j in $L^p(\mathbb{R}^n)$ and, lastly, if $p = \infty$, $W_j(p)$ is the completion of W_j in $L^\infty(\mathbb{R}^n)$ with the $\sigma(L^\infty, L^1)$-topology.

This enables us to extend the range of (8.3) to $1 < p < \infty$. The extreme cases $p = 1$ and $p = \infty$ will be examined in Chapter 3, Section 10 and in Chapters 5 and 6.

The following result completes and sharpens Bernstein's inequalities.

Theorem 7. Let V_j, $j \in \mathbb{Z}$, be an r-regular multiresolution approximation of $L^2(\mathbb{R}^n)$. Then there exist two constants, $C_2 \ge C_1 > 0$ such that, for every integer $s \in \mathbb{N}$ with $0 \le s \le r$, for every $p \in [1, +\infty]$, and for every function $f \in W_0(p)$,

(8.4) $$C_1 \|f\|_p \le \sum_{|\alpha|=s} \|\partial^\alpha f\|_p \le C_2 \|f\|_p.$$

If s is not an integer and if $0 < s < r$, then we still have

(8.5) $$C_1 \|f\|_p \le \|\Lambda^s f\|_p \le C_2 \|f\|_p,$$

where $\Lambda = (-\Delta)^{1/2}$.

The right-hand inequality of (8.4) is none other than Bernstein's inequality. To prove the left-hand inequality, we let $D(x, y)$ denote the kernel of the projection $D_0 = E_1 - E_0$. Then it is clear that $f(x) = \int D(x, y) f(y) \, dy$ when f belongs to W_0. We then use the cancellation properties of the function $D(x, \cdot)$ resulting from Theorem 4. On the one hand, $|D(x, y)| \le C_m (1 + |x - y|)^{-m}$, for every $m \in \mathbb{N}$. On the other, $D(x, y) = 2^n E(2x, 2y) - E(x, y)$, where $E(x, y)$ is the kernel of E_0, which gives $\int D(x, y) y^\alpha = 0$ when $|\alpha| \le r$. We then apply Lemma 12 to get $D(x, y) = \sum_{|\beta|=s} \partial_y^\beta D_\beta(x, y)$, where the functions D_β also satisfy $|D_\beta(x, y)| \le C_m' (1 + |x - y|)^{-m}$, for every $m \in \mathbb{N}$.

Thus, $f(x) = \int D(x, y) f(y) \, dy = (-1)^s \sum_{|\beta|=s} \int D_\beta(x, y) \partial^\beta f(y) \, dy$. Because the kernels $D_\beta(x, y)$ define bounded operators on $L^p(\mathbb{R}^n)$, for $1 \le p \le \infty$, we get the left-hand inequality of (8.4).

The adaptation of this argument to the case where s is not an integer is fairly routine, but we shall describe it for the reader's convenience. Let m be the integer such that $m < s < m+1$. We use the two decompositions $D_0 = \sum_{|\beta|=m} D_\beta \partial^\beta = \sum_{|\beta|=m+1} D_\beta \partial^\beta$ which, with the help of a little trick, will lead to $D_0 = G\Lambda^s$, where G is an operator which is bounded on all the $L^p(\mathbb{R}^n)$, $1 \le p \le \infty$. The trick is to write $I = U + V$, where U is the convolution operator corresponding to a multiplier $u(\xi) \in \mathcal{D}(\mathbb{R}^n)$, which is radial, equals 1 if $|\xi| \le 1/2$ and equals 0 if $|\xi| \ge 1$.

From this we deduce that

$$D_0 = D_0 U + D_0 V = \sum_{|\beta|=m+1} D_\beta \partial^\beta \Lambda^{-s} U \Lambda^s + \sum_{|\beta|=m} D_\beta \partial^\beta \Lambda^{-s} V \Lambda^s = G\Lambda^s.$$

To show that G is continuous on all the $L^p(\mathbb{R}^n)$, it is enough to examine the operators $\partial^\beta \Lambda^{-s} U$ and $\partial^\beta \Lambda^{-s} V$, where $|\beta| = m + 1$ in the first case and m in the second. They are the convolution operators associated with the multipliers $\xi^\beta |\xi|^{-s} u(\xi)$ and $\xi^\beta |\xi|^{-s}(1 - u(\xi))$ (we have omitted powers of i). These two functions are the Fourier transforms of integrable functions, which concludes the proof.

Corollary. *If $1 \le p \le \infty$ and if f belongs to $W_j(p)$, then*

$$(8.6) \qquad C_1 2^{js} \|f\|_p \le \sum_{|\alpha|=s} \|\partial^\alpha f\|_p \le C_2 2^{js} \|f\|_p,$$

if s is an integer and $0 \le s \le r$.

Similarly, in the general case when $0 \le s \le r$ and s is a real number,

$$(8.7) \qquad C_1 2^{js} \|f\|_p \le \|\Lambda^s f\|_p \le C_2 2^{js} \|f\|_p .$$

To verify this, it is enough to make the change of variable $x \to 2^j x$ in (8.4) and (8.5).

We observe that, for $j \in \mathbb{N}$, Λ^s can be replaced by $(1 - \Delta)^{s/2}$ in (8.7). Clearly,

$$\|\Lambda^s f\|_p \le C(s,n) \|(I - \Delta)^{s/2}\|_p \qquad \text{if } 1 \le p \le \infty.$$

We see this by comparing the Fourier transforms of $\Lambda^s f$ and of $(I - \Delta)^{s/2} f$. This gives the left-hand side of the inequality we are trying to establish.

The right-hand side comes from the inequality

$$\|(I - \Delta)^{s/2} f\|_p \le C'(s,n) \|(I + \Lambda^s) f\|_p \qquad \text{if } 1 \le p \le \infty.$$

Proposition 3. *If $-r \le s \le r$, $1 \le p \le \infty$, and $f \in W_j(p)$ for $j \in \mathbb{N}$, then*

$$(8.8) \qquad C_1' 2^{js} \|f\|_p \le \|(I - \Delta)^{s/2} f\|_p \le C_2' 2^{js} \|f\|_p ,$$

where $C_2' > C_1' > 0$ are two constants.

To prove this result, we may restrict ourselves to the case where $-r \le$

$s < 0$, since the case $0 \leq s \leq r$ has already been dealt with. We make the change of variable $x \mapsto 2^j x$ and find we must verify that

$$(8.9) \qquad C_1' \|f\|_p \leq \|(\varepsilon I - \Delta)^{s/2} f\|_p \leq C_2' \|f\|_p ,$$

for $f \in W_0(p)$ and $\varepsilon = 4^{-j}$.

To establish the right-hand side, we re-use the identity $I = U + V$. This lets us decompose $(\varepsilon I - \Delta)^{s/2}$ into $(\varepsilon I - \Delta)^{s/2} U + (\varepsilon I - \Delta)^{s/2} V$. The second term is a convolution operator and corresponds to the multiplier $(\varepsilon + |\xi|^2)^{s/2}(1 - u(\xi))$. Since $u(\xi) = 1$ when $|\xi| \leq 1/2$, the singularity of the first factor at the origin is cancelled by the second. The operator $(\varepsilon I - \Delta)^{s/2} V$ is thus uniformly bounded on $L^p(\mathbb{R}^n)$ when $1 \leq p \leq \infty$.

To study the action of the first term on $W_0(p)$, we again use the *reproducing kernel* $D(x, y)$ of W_0. Then $f(x) = \int D(x, y) f(y) \, dy$ for $f \in W_0$. The operator D_0 is self-adjoint and the identity $D_0 = \sum_{|\beta|=r} D_\beta \partial^\beta$, which we have already used, leads—after a slight change of notation—to $D_0 = \sum_{|\beta|=r} \partial^\beta D_\beta$, where the D_β are uniformly bounded on $L^p(\mathbb{R}^n)$ for $1 \leq p \leq \infty$.

We thus have

$$(\varepsilon I - \Delta)^{s/2} U f = \sum_{|\beta|=r} \left((\varepsilon I - \Delta)^{s/2} U \partial^\beta \right) D_\beta f$$

and $(\varepsilon I - \Delta)^{s/2} U \partial^\beta : L^p(\mathbb{R}^n) \to L^p(\mathbb{R}^n)$ is uniformly bounded on $L^p(\mathbb{R}^n)$ when $|\beta| = r > |s|$ and $0 \leq \varepsilon \leq 1$. Here again, the property is a result of calculating the multiplier associated with the convolution operator. The multiplier is $(\varepsilon + |\xi|^2)^{s/2} u(\xi) \xi^\beta$ (up to a multiple of i) and coincides with the Fourier transform of an integrable function.

We have finished the proof of the right-hand inequality of (8.9). The left-hand inequality is proved by setting $t = -s$ and writing $f = E_0(f) = E_0(\varepsilon I - \Delta)^{t/2}(\varepsilon I - \Delta)^{s/2} f$. We then observe that the operators $E_0(\varepsilon I - \Delta)^{t/2}$ are uniformly bounded on $L^p(\mathbb{R}^n)$ when $0 \leq p \leq \infty$. This is because the adjoints $(\varepsilon I - \Delta)^{t/2} E_0$ are uniformly bounded on $L^p(\mathbb{R}^n)$, by virtue of Bernstein's inequalities.

We now come to the characterization of the Sobolev space $H^s(\mathbb{R}^n)$ by the decomposition (8.2).

Theorem 8. *If f belongs to $H^{-r}(\mathbb{R}^n)$, if V_j, $j \in \mathbb{Z}$, is an r-regular multiresolution approximation of $L^2(\mathbb{R}^n)$, and if $-r < s < r$, then f belongs to the Sobolev space $H^s(\mathbb{R}^n)$ if and only if $E_0(f) \in L^2(\mathbb{R}^n)$ and $\|D_j f\|_2 = \varepsilon_j 2^{-js}$, for all $j \in \mathbb{N}$, where $\varepsilon_j \in l^2(\mathbb{N})$.*

Further, the H^s norm of f is equivalent to the sum of the L^2 norm of $E_0(f)$ and of the $l^2(\mathbb{N})$ norm of the sequence ε_j.

We shall first show that, if the functions $d_j \in W_j$ are such that $\|d_j\|_2 = \varepsilon_j 2^{-js}$ with $\varepsilon_j \in l^2(\mathbb{N})$, then $\sum_0^\infty d_j$ belongs to $H^s(\mathbb{R}^n)$. We already know that V_0 is contained in $H^r(\mathbb{R}^n)$ and it will follow that $u + \sum_0^\infty d_j \in H^s$ if $u \in V_0$, $d_j \in W_j$ and $\|d_j\|_2 \leq \varepsilon_j 2^{-js}$. We shall then show the converse.

To deal with $\sum_0^\infty d_j$, we let $\varepsilon > 0$ denote a number which is small enough for the inequality $|s| + 2\varepsilon \leq r$ to hold. We use the Hilbert space structure of $H^s(\mathbb{R}^n)$ defined by the inner product $\left((I - \Delta)^{s/2}u, (I - \Delta)^{s/2}v\right) = \left((I-\Delta)^{s/2+\varepsilon}u, (I-\Delta)^{s/2-\varepsilon}v\right)$, where $(\cdot, \cdot)$ denotes the inner product on $L^2(\mathbb{R}^n)$. Denoting by $\|\cdot\|_s$ the Hilbert norm on H^s, we get

$$(8.10) \qquad |((I - \Delta)^{s/2}u, (I - \Delta)^{s/2}v)| \leq \|u\|_{s+2\varepsilon}\|v\|_{s-2\varepsilon}.$$

This leads to

$$\left\|\sum_0^\infty d_j\right\|_s^2 \leq 2\sum_{j<k}\sum \|d_j\|_{s+2\varepsilon}\|d_k\|_{s-2\varepsilon} + \sum_j \|d_j\|_s^2.$$

Proposition 3 then gives $\|d_j\|_\sigma \simeq 2^{j\sigma}2^{-js}\varepsilon_j$, for $-r \leq \sigma \leq r$. So we have $\|d_j\|_{s+2\varepsilon}\|d_k\|_{s-2\varepsilon} \leq C\varepsilon_j\varepsilon_k 4^{\varepsilon(j-k)}$. It all boils down to showing that $\sum\sum_{j<k}\varepsilon_j\varepsilon_k 4^{-\varepsilon|j-k|} + \sum\varepsilon_j^2$ converges. By hypothesis, the second sum is finite and the first is dealt with by remarking that the convolution of an l^1-function with an l^2-function lies in l^2.

The second assertion of Theorem 8 is that the conditions $E_0(f) \in L^2(\mathbb{R}^n)$ and $\|D_j f\|_2 \leq \varepsilon_j 2^{-js}$ are necessary for the function f to belong to H^s. To prove it we use the sufficiency of the conditions (in the case $t = -s$) and a very simple remark on duality.

Let f be a distribution in H^{-r} and let g be a test function such that $D_j(g) = 0$ for large enough j. It is then clear that

$$(8.11) \qquad (f, g) = \left(E_0(f), E_0(g)\right) + \sum_0^\infty \left(D_j(f), D_j(g)\right),$$

where $(u, v) = \int u(x)\bar{v}(x)\, dx$.

We suppose that $f \in H^s$ is given. We choose g by specifying that $E_0(g) = \alpha E_0(f)$, $\alpha > 0$, and then $D_j(g) = \varepsilon_j D_j(f)$, where $\varepsilon_j > 0$, for $0 \leq j \leq N$, and $\varepsilon_j = 0$, for $j > N$.

Lastly, we choose $\alpha > 0$ and $\varepsilon_j > 0$, so that what we have already proved shows us that $\|g\|_{H^t} \leq C$ where C is a constant. For this, it is enough that $\alpha\|E_0(f)\|_2 \leq 1$ and that $\sum_0^N \varepsilon_j^2 4^{-js}\|D_j(f)\|_2^2 \leq 1$. Under these conditions, we get $|(f, g)| \leq C\|f\|_{H^s}$ and thus $\alpha\|E_0(f)\|_2^2 + \sum_0^\infty \|D_j\|_2^2 \leq C\|f\|_{H^s}$. On choosing α and the ε_j appropriately, we get the desired result.

Now that Theorem 8 has been completely proved, we draw the following consequence.

Corollary. *If $-r < s < r$, if f belongs to $H^s(\mathbb{R}^n)$, and if g lies in*

$H^{-r}(\mathbb{R}^n)$, then (8.11) holds and the series on the right-hand side converges absolutely.

Indeed, we have $\|D_j(f)\|_2 \le \varepsilon_j 2^{-js}$ and $\|D_j g\|_2 \le \eta_j 2^{js}$, so that $|(D_j(f), D_j(g))| \le \varepsilon_j \eta_j$ and the series given by these terms converges.

9 Besov spaces

We shall restrict ourselves to the inhomogeneous Besov spaces $B_p^{s,q}$ and start by considering the case where the index of regularity s is strictly positive.

The first definition of Besov spaces avoids Fourier analysis and is based, instead, on the idea of best approximation.

Let $1 \le p, q \le \infty$, let $s > 0$ and let m be an integer such that $m > s$ (the exact choice of m does not matter but the conditions which follow are the easier to verify if m is small).

Then f belongs to the inhomogeneous Besov space $B_p^{s,q}$ if, on the one hand, $f \in L^p(\mathbb{R}^n)$ and if, on the other, there exists a sequence of positive numbers $\varepsilon_j \in l^q(\mathbb{N})$ and a sequence of functions $f_j \in L^p(\mathbb{R}^n)$ such that

$$(9.1) \qquad \|f - f_j\|_p \le \varepsilon_j 2^{-js} \qquad \text{for} \qquad j \in \mathbb{N}$$

and

$$(9.2) \qquad \|\partial^\alpha f_j\|_p \le \varepsilon_j 2^{(m-s)j},$$

for every multi-index $\alpha \in \mathbb{N}^n$ such that $|\alpha| = m$ (the derivatives are taken in the sense of distributions).

Now let us put $g_j = f_{j+1} - f_j$. We get

$$(9.3) \qquad f = f_0 + g_0 + g_1 + \cdots$$

where

$$\|\partial^\alpha g_j\|_p \le \varepsilon_j 2^{(m-s)j} \qquad \text{when} \qquad |\alpha| = m$$

and

$$\|g_j\|_p \le \varepsilon_j 2^{-js}.$$

In the opposite direction, if $0 < s < m$, the decomposition (9.3) and the properties of the g_j imply (9.1) and (9.2).

A second possible definition lies in providing the explicit algorithm leading to the functions $g_j(x)$. To this end, we start with an r-regular multiresolution approximation V_j, $j \in \mathbb{Z}$, of $L^2(\mathbb{R}^n)$, let $E_j : L^2(\mathbb{R}^n) \to V_j$ denote the orthogonal projection, and then let D_j denote the operator $E_{j+1} - E_j$. So D_j is the orthogonal projection on the orthogonal complement W_j of V_j in V_{j+1}. In these circumstances we have

Proposition 4. *Suppose that $0 < s < r$ and that f is a function in $L^p(\mathbb{R}^n)$. Then f belongs to the inhomogeneous Besov space $B_p^{s,q}$ if and only if the sequence $2^{js}\|D_j(f)\|_p$ belongs to $l^q(\mathbb{N})$. Moreover, the norm of f in $B_p^{s,q}$ is equivalent to the sum of the l^q norm of the sequence and $\|E_0(f)\|_p$.*

In one direction, Proposition 4 is straightforward. In fact, Bernstein's inequalities give an estimation of the L^p norms of $\partial^\alpha(D_j f)$ for $|\alpha| \leq r$ and (9.3) drops out.

In the opposite direction, we use the definition of $B_p^{s,q}$ given by (9.3) to get

$$D_j(f) = D_j(f_0) + \sum_{k=0}^{\infty} D_j(g_k).$$

If $j \leq k$, we simply majorize $\|D_j(g_k)\|_p$ by $C\|g_k\|_p$, using the fact that the operators $D_j : L^p \to L^p$ are uniformly bounded when $1 \leq p \leq \infty$.

If $j > k$, we apply the remark following Theorem 5 to decompose D_j as $2^{-jr}\sum_{|\alpha|=r} \mathcal{R}_j^\alpha \partial^\alpha$. This gives $C\varepsilon_k 2^{(r-s)k}2^{-rj}$ as an upper bound for $\|D_j(g_k)\|_p$. Finally,

$$\|D_j(f)\|_p \leq C2^{-rj} \sum_{0}^{j} \varepsilon_k 2^{(r-s)k} + C \sum_{j+1}^{\infty} \varepsilon_k 2^{-sk} \leq C'\tilde{\varepsilon}_j 2^{-sj},$$

where $\tilde{\varepsilon}_j = \sum \varepsilon_k 2^{-(r-s)|j-k|}$ is the convolution of an l^1 sequence with an l^q sequence, and is thus itself in l^q.

We have completely proved Proposition 4.

By analogy with the special case of Sobolev spaces, we can extend the definition of Besov spaces to negative values of s and to $s = 0$. Letting r denote an integer such that $r > |s|$, we write $f \in B_p^{s,q}$ if $E_0(f)$ belongs to $L^p(\mathbb{R}^n)$ and if

$$\|D_j f\|_p \leq \varepsilon_j 2^{-js} \qquad \text{where} \qquad \varepsilon_j \in l^q(\mathbb{N}).$$

Then, if $s' = -s$, $s' > 0$, $1/p + 1/p' = 1$ and $1/q + 1/q' = 1$, we immediately see that the duality between the distributions $f \in B_p^{s,q}$ and the test functions $g \in B_{p'}^{s',q'}$ is given by

$$(f,g) = (E_0(f), E_0(g)) + \sum_{0}^{\infty}(D_j(f), D_j(g)).$$

If $p > 1$ and $q > 1$, $B_p^{s,q}$ is the dual of $B_{p'}^{s',q'}$, and its definition is independent of the choice of multiresolution approximation. For $p = 1$ or $q = 1$ the fact that the definition of $B_p^{s,q}$ is also intrinsic can be verified directly.

A noteworthy special case of this duality is that of the Hölder space C^r, $r > 0$.

Let us at first suppose that $0 < r < 1$ and define the (inhomogeneous) Hölder space C^r as the set of bounded continuous functions $f : \mathbb{R}^n \to \mathbb{C}$ such that $\sup_{x \neq y} |f(y) - f(x)|/|y - x|^r < \infty$. The norm of f in C^r is the sum of this upper bound and the $L^\infty(\mathbb{R}^n)$ norm of f. Then C^r is a Banach space.

Ascoli's theorem tells us that, if f_j is a bounded sequence of functions in C^r, then there is a subsequence $f_{j(k)}$ which converges uniformly on $\mathbb{R}^n$ to a function f in C^r. This property is very similar to the following classical result: if E is a separable Banach space, the unit ball of the dual E^* of E is a compact metric space in the topology $\sigma(E^*, E)$.

These two results, Ascoli's theorem and the weak* compactness of the unit ball of the dual of a separable Banach space, are related by the fact that the Hölder space C^r is the same as the Besov space $B_\infty^{r,\infty}$ which is the dual E^* of the separable Banach space $E = B_1^{-r,1}$.

We shall verify these assertions when $0 < r < 1$. Then we shall extend the definition of the C^r spaces to $r \geq 1$ in such a way that the identity $C^r = B_\infty^{r,\infty}$ is preserved. This will lead us to substitute the Zygmund class Λ_* for the space C^1.

But let us start by verifying that

(9.4) $C^r = B_\infty^{r,\infty} = (B_1^{-r,1})^*.$

To begin with, we take $f \in C^r$. Then

$$D_j f(x) = 2^{nj} \int D(2^j x, 2^j y) f(y) \, dy$$

$$= 2^{nj} \int D(2^j x, 2^j y)(f(y) - f(x)) \, dy \, .$$

The required upper bound of $\|D_j f\|_\infty$ follows immediately, because $|f(y) - f(x)| \leq C|y - x|^r$ and $|D(x, y)| \leq C|x - y|^{-n-1}$.

Conversely, we suppose that $f = g_0 + f_0 + f_1 + \cdots$, where $\|f_j\|_\infty \leq C2^{-jr}$ and $\|\partial f_j / \partial x_k\|_\infty \leq C2^{j(1-r)}$, for $1 \leq k \leq n$. We also suppose that g_0 and its gradient belong to $L^\infty(\mathbb{R}^n)$. Then f does belong to C^r. To show that, we estimate $|f(y) - f(x)|$ for $|y - x| < 1$. We define the integer $N \geq 1$ by $2^{-N} \leq |y - x| < 2.2^{-N}$. We denote the gradient by ∇ and majorize $|f_j(y) - f_j(x)|$ by $|y - x| \|\nabla f_j\|_\infty$, if $j \leq N$, and by $2\|f_j\|_\infty$, otherwise. Clearly, $|g_0(y) - g_0(x)| \leq C_0|y - x|$ and putting this together leads to the upper bound

$$C|y - x| \sum_0^N 2^{j(1-r)} + C \sum_{N+1}^\infty 2^{-jr} \leq C'|y - x|^r$$

for $|f(y) - f(x)|$.

We now describe the space $B_1^{-r,1}$ explicitly. A distribution h belongs to this space if and only if

$$h = u_0 + \sum_0^\infty v_j \qquad \text{where} \qquad u_0 \in V_0 \cap L^1 \quad \text{and} \quad v_j \in W_j \cap L^1$$

with $\|u_0\|_1 + \sum_0^\infty 2^{-jr}\|v_j\|_1 < \infty$. The latter sum is, by definition, the norm of h in $B_1^{-r,1}$. The duality between $f \in C^r$ and $h \in B_1^{-r,1}$ is clearly given by

$$(9.5) \qquad (f,h) = (g_0, u_0) + \sum_0^\infty (f_j, v_j)$$

and this series converges, since $|(f_j, v_j)| \le \|f_j\|_\infty \|v_j\|_1 \le C2^{-jr}\|v_j\|_1$.

To verify that C^r is indeed the dual of $B_1^{-r,1}$ it thus is enough to show that every function $f \in W_j(\infty)$ satisfies $\|f\|_\infty \le C \sup |(f,h)|$, where the supremum is taken over $h \in W_j$ satisfying $\|h\|_1 \le 1$. To see this, we let h be a function in $L^1(\mathbb{R}^n)$ such that $\|h\|_1 \le 1$ and $\|f\|_\infty \le 2|(f,h)|$. Since $f \in W_j(\infty)$, we get $(f,h) = (f, D_j h) = (f, h_j)$. Now $D_j : L^1 \to L^1$ is a bounded operator and the norm of D_j is independent of j. So we have $\|h_j\|_1 \le C$ and $\|f\|_\infty \le 2|(f, h_j)|$, which gives the required inequality— with $2C$ instead of C.

We shall extend the duality between $B_1^{-r,1}$ and C^r, by using the correct definition of the C^r spaces for $r \ge 1$.

For $r = 1$, we reject the usual C^1 space for the Zygmund class Λ_*, which we now define. We say that $f \in \Lambda_*$ if the function f is continuous on $\mathbb{R}^n$ and there is a constant C such that $|f(x+y)+f(x-y)-2f(x)| \le C|y|$, for every $x \in \mathbb{R}^n$ and all $y \in \mathbb{R}^n$ with $|y| \le 1$. We further impose the condition $f \in L^\infty(\mathbb{R}^n)$, which deals with the case $|y| > 1$. The norm of f in Λ_* is the sum of $\|f\|_\infty$ and the infimum of the constants C which can appear in the inequality defining Λ_*.

For $m < s \le m+1$, we write $f \in C^s(\mathbb{R}^n)$, if f is a C^m-function all of whose derivatives $\partial^\alpha f$, of order $|\alpha| = m$, belong to the spaces C^{s-m} defined above. Again, we further require that f belongs to $L^\infty(\mathbb{R}^n)$.

With this notation, we have

Proposition 5. *Let V_j, $j \in \mathbb{Z}$, be an r-regular multiresolution approximation, where $r > s$. Then $f \in C^s$ if and only if $E_0(f) \in L^\infty$ and if $\|D_j(f)\|_\infty \le C2^{-js}$ for some constant C.*

We shall resume the study of inhomogeneous Hölder spaces in Chapter 6 and shall therefore restrict ourselves, for now, to looking at the main difficulty presented by the proof of Proposition 5. It is a matter of showing that $f \in C^s$ implies $\|D_j(f)\|_\infty \le C2^{-js}$ when $s \ge 1$ is an integer. The difficulty arises because the C^s spaces are not the usual ones, but come from the Zygmund class $\Lambda_* \dots$

We first show how to reduce to the case $s = 1$. To this end, we put $s = m + 1$ and apply the identity

$$(9.6) \qquad D_j = 2^{-js} \sum_{|\alpha|=s} D_j^\alpha \partial^\alpha$$

which we have used before.

The operator D_j^α is defined by the kernel $2^{nj} D_\alpha(2^j x, 2^j y)$, where $|D_\alpha(x, y)| \leq C_N (1 + |x - y|)^{-N}$, for every $N \in \mathbb{N}$, and $\int D_\alpha(x, y)\, dy = \int D_\alpha(x, y) y_j\, dy = 0$ for $1 \leq j \leq n$.

We thus need deal only with the case $s = 1$, that is, the case where f is in the Zygmund class Λ_*. We finish the proof by means of the following lemma, which is not obvious and will be proved in Chapter 6.

Lemma 14. *If $g \in L^1(\mathbb{R}^n)$ satisfies, in addition,*
$$|g(x)| \leq C(1 + |x - x_0|)^{-n-2}$$

$$(9.7)$$

$$\int g(x)\, dx = 0 \qquad \text{and} \qquad \int x_k g(x)\, dx = 0,$$

for $1 \leq k \leq n$, then every function f in the Zygmund class Λ_* satisfies the inequality

$$\left| \int f(x) g(Nx)\, dx \right| \leq C' \|f\|_{\Lambda_*} N^{-n-1},$$

where $N \geq 1$ and where C' depends only on C and n, but not on x_0.

10. The operators E_j and pseudo-differential operators

Recall that a pseudo-differential operator T is defined by means of a symbol $\sigma(x, \xi)$ which is a function of $x \in \mathbb{R}^n$ and $\xi \in \mathbb{R}^n$ (sometimes restricted to $\xi \neq 0$) and by the formal rule

$$(10.1) \qquad T(e^{ix \cdot \xi}) = \sigma(x, \xi) e^{ix \cdot \xi},$$

which is reminiscent of amplitude modulation in radio detection.

If the symbol $\sigma(x, \xi)$ does not depend on x, the operator T is diagonalized with respect to the "basis of characters", which is a basis only in the periodic case when $\xi \in \mathbb{Z}^n$.

If the symbol $\sigma(x, \xi)$ has certain regularity properties with respect to the variable x, the operator T will be "almost diagonal" and we can expect it to have good functional analytic properties. Remember that the Hörmander classes $S_{\rho, \delta}^0$ are defined by

$$(10.2) \qquad |\partial_\xi^\alpha \partial_x^\beta \sigma(x, \xi)| \leq C(\alpha, \beta)(1 + |\xi|)^{\rho|\beta| - \delta|\alpha|}.$$

We clearly cannot let multiplication by the symbol $\sigma(x, \xi)$ completely destroy the oscillations of the function $e^{ix\cdot\xi}$. To avoid this, we exclude $\rho = \delta = 1$ in (10.2).

In most cases, the relation between the operator T and the symbol $\sigma(x,\xi)$ is linearized. Using the Fourier transform, we write every function f in the class $S(\mathbb{R}^n)$ as a superposition of functions $e^{ix\cdot\xi}$,

$$(10.3) \qquad f(x) = \frac{1}{(2\pi)^n} \int e^{ix\cdot\xi} \hat{f}(\xi)\, d\xi$$

and, by linearity, we get

$$(10.4) \qquad Tf(x) = \frac{1}{(2\pi)^n} \int e^{ix\cdot\xi} \sigma(x,\xi) \hat{f}(\xi)\, d\xi.$$

The algorithm of (10.4) makes sense when $\sigma(x,\xi) \in L^\infty(\mathbb{R}^n \times \mathbb{R}^n)$, which will be the case in the example to follow.

Let $V_j, j \in \mathbb{Z}$, be an r-regular multiresolution approximation of $L^2(\mathbb{R}^n)$. We let ϕ denote the function given by the algorithm of Section 4 and we write $E_j : L^2(\mathbb{R}^n) \to V_j$ for the orthogonal projection operator.

We then have

Proposition 6. *The symbol of the orthogonal projection E_j of $L^2(\mathbb{R}^n)$ onto V_j is $\sigma(2^j x, 2^{-j}\xi)$, where*

$$(10.5) \qquad \sigma(x,\xi) = \sum_{k\in\mathbb{Z}^n} e^{2\pi i k\cdot x} \hat{\phi}(\xi + 2k\pi)\bar{\hat{\phi}}(\xi).$$

Before proving this result, we shall make some straightforward remarks about the regularity of the symbol.

We first observe that $\sigma(x,\xi)$ is a $\mathbb{Z}^n$-periodic function in the variable x. Here's why. Let $\sigma(x,\xi)$ be a symbol and T the corresponding operator. Then, for every $a \in \mathbb{R}^n$, $\sigma(x-a,\xi)$ is the symbol of $R_a T R_a^{-1}$, where $R_a f(x) = f(x-a)$. The operator E_0 commutes with the translations R_k, $k \in \mathbb{Z}^n$, so the symbol $\sigma(x,\xi)$ of E_0 is $\mathbb{Z}^n$-periodic in x.

The series appearing on the right-hand side of (10.5) is the Fourier series of the function $x \mapsto \sigma(x,\xi)$. One can easily calculate the sum, since $\hat{\phi}(\xi + 2k\pi)$ is the sequence, over $k \in \mathbb{Z}^n$, of the Fourier coefficients of the periodic version of the function $x \mapsto e^{-i\xi\cdot x}\phi(x)$ (this remark is a consequence of Lemma 1 in Chapter 1).

So we have

$$(10.6) \qquad \sigma(x,\xi) = \left(\sum_{l\in\mathbb{Z}^n} e^{-i\xi\cdot(x+l)}\phi(x+l) \right) \bar{\hat{\phi}}(\xi).$$

The properties of the symbol can be read off from this formula, and, taking account of the rapid decrease of the derivatives of ϕ of order less than or equal to r, we get

$$(10.7) \qquad |\partial_\xi^\alpha \partial_x^\beta \sigma(x,\xi)| \leq C(\alpha),$$

for every $\alpha \in \mathbf{N}^n$ and every $\beta \in \mathbf{N}^n$ of order $|\beta| \leq r$.

We shall now use the identity $E_0(e^{ix \cdot \xi}) = \sigma(x, \xi)e^{ix \cdot \xi}$ to prove (10.6). Since the kernel of E_0 is $\sum_{k \in \mathbf{Z}^n} \phi(x - k)\bar{\phi}(y - k)$, it follows that

$$E_0(e^{ix \cdot \xi}) = \sum_{k \in \mathbf{Z}^n} \phi(x - k) \int \bar{\phi}(y - k)e^{iy \cdot \xi} \, dy$$

$$= \sum_{k \in \mathbf{Z}^n} \phi(x - k)e^{ik \cdot \xi}\hat{\bar{\phi}}(\xi) = \sigma(x, \xi)e^{ix \cdot \xi}.$$

This gives (10.6).

We return to our remarkable identity, $E_0(x^\alpha) = x^\alpha$, $|\alpha| \leq r$, whose consequences include particular properties of the symbol $\sigma(x, \xi)$, which we shall state explicitly.

We calculate $E_0(x^\alpha)$ by using (10.4). This leads to approximating x^α by functions of the Schwartz class. So we consider, for each $\varepsilon > 0$, the function $f_\varepsilon(x) = x^\alpha e^{-\varepsilon|x|^2}$ whose Fourier transform is $i^{|\alpha|}\partial_\xi^\alpha g_\varepsilon(\xi)$, where

$$g_\varepsilon(\xi) = \left(\frac{\pi}{\varepsilon}\right)^{n/2} e^{-|\xi|^2/4\varepsilon}.$$

Then

$$f_\varepsilon(x) = \frac{i^{|\alpha|}}{(2\pi)^n} \int e^{ix \cdot \xi}\sigma(x, \xi)\partial_\xi^\alpha g_\varepsilon(\xi) \, d\xi,$$

which we integrate by parts. The symbol $\sigma(x, \xi)$ is infinitely differentiable in ξ and all its derivatives are bounded. So this gives

$$(10.8) \qquad E_0 f_\varepsilon(x) = \frac{(-i)^{|\alpha|}}{(2\pi)^n} \int \partial_\xi^\alpha \{e^{ix \cdot \xi}\sigma(x, \xi)\}g_\varepsilon(\xi) \, d\xi.$$

Finally, $E_0(x^\alpha) = \lim_{\varepsilon \downarrow 0} E_0(f_\varepsilon)(x)$. Indeed, the kernel $E(x, y)$ of E_0 is $O(|x - y|^{-m})$ at infinity, for every m, so Lebesgue's dominated convergence theorem applies. On passing to the limit in (10.8), it follows that

$$(10.9) \qquad x^\alpha = (-i)^{|\alpha|}\partial_\xi^\alpha \{e^{ix \cdot \xi}\sigma(x, \xi)\}_{\xi=0}.$$

This identity holds for all $\alpha \in \mathbf{N}^n$ such that $|\alpha| \leq r$. We thus have $\sigma(x, 0) = 1$ and $\partial_\xi^\alpha \sigma(x, \xi) = 0$ at $\xi = 0$ for $1 \leq |\alpha| \leq r$.

Now, (10.5) gives $\sigma(x, 0) = \sum_{k \in \mathbf{Z}^n} e^{2\pi i k \cdot x}\hat{\phi}(2k\pi)\hat{\bar{\phi}}(0) = 1$. This implies that $\hat{\phi}(2k\pi) = 0$ for $k \neq 0$. We continue, and, step by step, get $(\partial^\alpha \hat{\phi})(2k\pi) = 0$, for $k \neq 0$ and $|\alpha| \leq r$.

We return to the identity $1 = \sum_{k \in \mathbf{Z}^n} |\hat{\phi}(\xi + 2k\pi)|^2$.

We know that this series and all its derivatives are uniformly convergent on compacta. We have just established that $\hat{\phi}$ and all its derivatives up to order r vanish at $2k\pi$, $k \in \mathbf{Z}^n$, $k \neq 0$. Hence we have

Proposition 7. *With the above notation, $|\hat{\phi}(\xi)|^2 = 1 + O(|\xi|^{2r+2})$ as $|\xi|$ tends to 0.*

This remarkable result will enable us to prove, in Chapter 3, that the moments $\int x^\alpha \psi_\lambda(x)\, dx$ of the wavelets constructed from the multiresolution approximation provided by the V_j are zero for $|\alpha| \le r$.

Finally, we can adjust the function ϕ in a way that will give us the more precise information that $\hat\phi(\xi) = 1 + O(|\xi|^{2r+2})$, while preserving the other properties of ϕ. That is, the function ϕ, thus corrected, will satisfy

$$\int \phi(x)\, dx = 1 \qquad \text{and} \qquad \int x^\alpha \phi(x)\, dx = 0\,,$$

for $1 \le |\alpha| \le 2r + 2$.

The other properties of ϕ which are preserved are the fact that $\phi(x-k)$, $k \in \mathbb{Z}^n$, is an orthonormal basis of V_0 and that the partial derivatives $\partial^\alpha \phi$, of order $|\alpha| \le r$, decrease rapidly.

There is, however, a situation where we should not change ϕ: if, in Proposition 7, ϕ has compact support, this property will disappear during the course of the manipulations about to be made.

We start by multiplying $\hat\phi$ by a constant of modulus 1 to get $\hat\phi(0) = 1$. Then we let $\alpha(\xi)$ denote a continuous choice of argument of $\hat\phi(\xi)$, defined on a neighbourhood of 0. We thus have $\hat\phi(\xi) = |\hat\phi(\xi)|e^{i\alpha(\xi)}$ for $|\xi| < \delta$, where $\alpha(0) = 0$ and $\alpha(\xi)$ is infinitely differentiable on $|\xi| < \delta$. Then we extend $\alpha(\xi)$ to the whole of $\mathbb{R}^n$ as a real, infinitely differentiable, 2π-periodic function. Calling this function $\beta(\xi)$, we replace $\hat\phi(\xi)$ by $e^{-i\beta(\xi)}\hat\phi(\xi)$ and get $\hat\phi(\xi) = |\hat\phi(\xi)|$ in a neighbourhood of 0, which is what we wanted.

11 Multiresolution approximations and finite elements

Let us recall the definitions of the usual "restriction" and "extension" operators used in finite element methods in numerical analysis. We shall then realize that there is total agreement with our theory of r-regular multiresolution approximations if the function g, which occurs in Definition 1, has compact support instead of just decreasing rapidly.

We consider a sampling step $h > 0$ and, following finite element methods, we shall replace the *continuous* space $\mathbb{R}^n$ by the *discrete* lattice $h\mathbb{Z}^n$, which we denote by Γ. We want to imitate this geometric approximation by defining a functional approximation of the various function spaces by spaces of sequences indexed by Γ.

The first algorithm we construct is called the "restriction" and transforms functions or distributions defined on $\mathbb{R}^n$ into numerical sequences indexed by Γ. Then follows the "extension" which, starting from a sequence over Γ, provides the simplest possible extrapolation of the sequence.

The words "restriction" and "extension" do not have their usual meaning here. In fact, we intend to "restrict" functions $f \in L^2(\mathbb{R}^n)$ and, indeed, distributions which do not have any restriction to Γ in the usual sense (or even in the sense of Sobolev's trace theorems).

What is more surprising is that the "extension" is also not what might be expected. Starting with a sequence $c(\gamma)$, $\gamma \in \Gamma$, we construct a regular function which "extends" $c(\gamma)$ in some sense which coincides very rarely with the usual sense.

Let us proceed to the definition of these two operations. We start with an integer $r \in \mathbb{N}$ and two functions λ and μ which satisfy the following hypotheses:

(11.1) $\lambda \in L^\infty(\mathbb{R}^n)$, λ has compact support and the integral of λ equals 1;

(11.2) $\mu \in H^r(\mathbb{R}^n)$, μ has compact support, the integral of μ equals 1 and $\partial^\alpha \hat{\mu}(2k\pi) = 0$ when $k \in \mathbb{Z}^n$, $k \neq 0$ and $0 \leq |\alpha| \leq r$;

(11.3) λ and μ are related by

$$\iint \lambda(x)\mu(y)(x-y)^\alpha \, dx \, dy = 0 \qquad \text{when} \qquad 1 \leq |\alpha| \leq r.$$

Following [4], we define the "restriction" or sampling operator and the "extension" operator by

(11.4) $(r_h u)(hk) = h^{-n} \int \lambda(h^{-1}x - k)u(x) \, dx,$

where $u \in L^2(\mathbb{R}^n)$ and $k \in \mathbb{Z}^n$, and

(11.5) $(p_h u_h)(x) = \sum_{k \in \mathbb{Z}^n} u_h(hk)\mu(h^{-1}x - k),$

where $u_h \in l^2(\Gamma)$.

J.P. Aubin shows in [4] that conditions (11.1), (11.2) and (11.3) imply the following property: if $0 \leq s \leq r$ and if u belongs to $H^s(\mathbb{R}^n)$, then $p_h r_h(u)$ converges to u in H^s norm as the sampling step h tends to 0.

We observe that the conditions on λ and μ are not the same. For the right-hand side of (11.5) to belong to $H^r(\mathbb{R}^n)$, the same must be true of μ. On the other hand, $\lambda \in L^\infty$ is enough to define (11.4) when $u \in H^s$ and $0 < s < r$. The definition of r_h is not applicable when u is a distribution in $H^{-s}(\mathbb{R}^n)$, where $0 < s < r$. In finite element methods, the approximation of distributions uses a scheme which is the dual of (11.4) and (11.5).

Finally, when μ is given and satisfies the conditions of (11.2), it is very easy to construct λ so that (11.3) holds. Indeed, this last property can be written $\int x^{\alpha} \lambda(x) \, dx = c(\alpha, \mu)$, $|\alpha| \le r$ and, for each $\delta > 0$, an infinitely differentiable function λ can be found whose support is contained in the ball $|x| \le \delta$ and whose moments of order less than or equal to r are pre-determined.

Returning to r-regular multiresolution approximations, suppose the function $g(x)$ which appears in Definition 1 has compact support. We have shown that $(\partial^{\alpha} \hat{\phi})(2k\pi) = 0$, for $k \in \mathbf{Z}^n$, $k \ne 0$ and $|\alpha| \le r$, and, by multiplying $\hat{\phi}$ by a constant of modulus 1, we can suppose that $\hat{\phi}(0) = 1$. But $\hat{\phi}$ and $\hat{g}$ are related by the identity

$$(11.6) \qquad \hat{g}(\xi) = \hat{\phi}(\xi) \left(\sum |\hat{g}(\xi + 2k\pi)|^2 \right)^{1/2}.$$

We have also shown that the series on the right-hand side is an infinitely differentiable function and strictly positive. So we have

$$\partial^{\alpha} \hat{g}(2k\pi) = 0 \qquad \text{when } k \in \mathbf{Z}^n, \ k \ne 0 \text{ and } |\alpha| \le r.$$

Thus, if the function g has compact support, it has exactly the properties required by (11.2) to define an "extension operator".

The sampling step (in the case of multiresolution approximations) is $h = 2^{-j}$ and the corresponding lattice is $\Gamma_j = 2^{-j} \mathbf{Z}^n$. So the "extension" operator, denoted by P_j, transforms a square-summable sequence $\alpha(\gamma)$, $\gamma \in \Gamma_j$ into $f(x) = \sum_{\gamma \in \Gamma_j} \alpha(\gamma) g(2^j (x - \gamma))$, a general function of V_j.

As we have implied, it is then easy to associate with g a second function of compact support (which can be chosen as regular as required) and to construct a "restriction" or sampling operator in the sense of the finite element method. In this chapter, we have not attempted to construct such an operator. But we should indicate that the "restriction" of $f \in V_j$ to Γ_j is not the sequence $\alpha(\gamma)$ used to construct f. That would be the case if the function λ used in the definition of the restriction were of compact support and had the property that $\bar{\lambda}(x - k)$, $k \in \mathbf{Z}^n$ was the basis dual to $g(x - k)$. The dual basis $h(x - k)$ always exists in the case of an r-regular multiresolution approximation and a straightforward adaptation of the calculations of Section 4 gives

$$\hat{h}(\xi) = \hat{g}(\xi) \left(\sum |\hat{g}(\xi + 2k\pi)|^2 \right)^{-1}.$$

If we go back to the example of the multiresolution approximation made up of nested spaces of spline functions, we see that the function h constructed in this way can never have compact support.

In Chapter 3, we shall show that, for each $r \in \mathbf{N}$, there exists an r-regular multiresolution approximation of $L^2(\mathbf{R}^n)$ such that the corresponding function ϕ has compact support. Then the extension and restriction operators P_j and R_j defined, respectively, by

$$P_j\alpha(x) = \sum \alpha(2^{-j}k)\phi(2^j x - k)$$

and

$$(R_j f)(2^{-j}k) = 2^{nj} \int f(u)\bar{\phi}(2^j u - k)\,du$$

have the remarkable property that $R_j P_j = I$ and that $R_j : L^2(\mathbb{R}^n) \to l^2(\Gamma_j)$ is the adjoint of $P_j : l^2(\Gamma_j) \to L^2(\mathbb{R}^n)$.

These exceptional properties imply that the same operators should (and must) be preserved to get discrete versions of distributions

$$f \in H^{-s}(\mathbb{R}^n), 0 < s \le r.$$

The theory of multiresolution approximation thus appears to be a new type of finite element method. But the points of view are different. Just as physicists have done in using "lattice approximation" in quantum field theory calculations, we have imposed the compatibility relation $\Gamma_j \subset \Gamma_{j+1}$ on the lattices we use. This inclusion is destined to play a crucial role in Chapter 3, because it will enable us to define the *innovation space* W_j by $V_j \oplus W_j = V_{j+1}$.

The *dynamics of successive approximations* are essential in the theory of multiresolution approximations, so we have not singled out algorithms leading to basic functions (g or ϕ) whose supports are compact. We have been content with the functions' rapid decrease at infinity.

In finite element methods, the dynamics of succesive approximations are absent unless one uses grid refinements. But such refinements are never repeated infinitely often.

12 Example: the Littlewood-Paley multiresolution approximation

We are going to apply (10.5), the formula for the symbol $\sigma(x,\xi)$ of the operator E_0, to the particular case of the Littlewood-Paley multiresolution approximation of Section 2. Recall that we then have $\hat{\phi}(\xi) = 1$ for $-2\pi/3 \le \xi \le 2\pi/3$, $\hat{\phi}(\xi) = 0$ for $|\xi| \ge 4\pi/3$, $\hat{\phi}(-\xi) = \hat{\phi}(\xi)$, $\hat{\phi} \in \mathcal{D}(\mathbb{R})$, $\hat{\phi}(\xi) \ge 0$ and, finally, $(\hat{\phi}(\xi))^2 + (\hat{\phi}(2\pi - \xi))^2 = 1$ for $0 \le \xi \le 2\pi$.

Formula (10.5) has only three terms. The central term $(\hat{\phi}(\xi))^2$ is the symbol of a convolution operator, which we denote by S. The two other terms correspond to $k = -1$ and $k = 1$. We set $\eta(\xi) = \hat{\phi}(\xi - 2\pi)\hat{\phi}(\xi)$. The function $\eta(\xi)$ has support in $2\pi/3 \le \xi \le 4\pi/3$ and satisfies $\eta(2\pi - \xi) = \eta(\xi)$. We let Δ^+ denote the corresponding convolution operator. Similarly, we set $\theta(\xi) = \hat{\phi}(\xi + 2\pi)\hat{\phi}(\xi)$, which has support in

$-4\pi/3 \le \xi \le -2\pi/3$, and write Δ^- for the corresponding convolution operator. By (10.5) we then get the following identity:–

Proposition 9. *In the case of the Littlewood-Paley multiresolution approximation, the operator E_j is given by*

$$(12.1) \qquad E_j = S_j + M_j\Delta_j^- + M_j^{-1}\Delta_j^+ ,$$

where M_j is the operator of pointwise multiplication by $e^{2\pi i 2^j x}$ and where Δ_j^+ and Δ_j^- are defined, via the Fourier transform, by the multipliers $\eta(2^{-j}\xi)$ and $\theta(2^{-j}\xi)$.

The right-hand side thus consists of three terms: a main term and two error terms. The main term is a classical approximate identity given by a convolution operator. The error terms correspond to two frequency intervals $[(2\pi/3)2^j , (4\pi/3)2^j]$ and $[-(4\pi/3)2^j , -(2\pi/3)2^j]$ and to the subspaces U_j and V_j of $L^2(\mathbb{R}^n)$ consisting of those functions whose Fourier transforms have supports in the first or second interval. It is appropriate to observe that $M_j : V_j \to U_j$ is an isometric isomorphism and thus that the image of $M_j\Delta_j^-$ is U_j. Similarly, the image of $M_j^{-1}\Delta_j^+$ is V_j.

The two error terms tend to 0 in all the classical function spaces (except L^∞ and related spaces) in the sense of strong convergence of operators.

We shall see how this very simple identity comes into the construction of a Schauder basis for the disc algebra $A(D)$.

We proceed to the calculation of $D_j = E_{j+1} - E_j$. Putting $\Delta_j = S_{j+1} - S_j$, we get

$$(12.2) \qquad D_j = \Delta_j + R_j ,$$

where

$$(12.3) \qquad R_j = M_{j+1}\Delta_{j+1}^- + M_{j+1}^{-1}\Delta_{j+1}^+ - M_j\Delta_j^- - M_j^{-1}\Delta_j^+ .$$

Here the main term is, again, Δ_j and R_j is the error term. The main term Δ_j is the term used in the usual Littlewood-Paley decomposition, which is why we used the name Littlewood-Paley multiresolution approximation. It is the multiresolution approximation whose properties are the closest to the usual Littlewood-Paley decomposition.

To conclude this section, we shall calculate the function ϕ in the case of the multiresolution approximation given by splines of order r.

Starting with the characteristic function $\chi(x)$ of $[0,1)$, we put $g(x) = \chi * \cdots * \chi$ ($r+1$ terms). Up to a factor of proportionality and up to translation by $k \in \mathbb{Z}$, this is the function in V_0 whose support lies in an interval $[a,b]$ of shortest possible length. We have

$$\hat{g}(\xi) = \left(\frac{1-e^{-i\xi}}{i\xi}\right)^{r+1} .$$

With the help of this function, we define ϕ by

$$\hat{\phi}(\xi) = \hat{g}(\xi) \left(\sum_{-\infty}^{\infty} |\hat{g}(\xi + 2k\pi)|^2 \right)^{-1/2}.$$

To proceed, we distinguish between two cases. If r is odd, we put $r + 1 = 2s$, $s \in \mathbb{N}$, and get $\hat{g}(\xi) = \left((\sin \xi/2)/(\xi/2) \right)^{2s} e^{-i\xi s}$.
We then replace $g(x)$ by $g(x + s)$, without modifying the fundamental property that $g(x - k)$, $k \in \mathbb{Z}$, is a Riesz basis of V_0. This gives $\hat{g}(\xi) = \left((\sin \xi/2)/(\xi/2) \right)^{2s}$.

Now we compute

$$\left(\sum |\hat{g}(\xi + 2k\pi)|^2 \right)^{1/2} = (\sin \xi/2)^{2s} \left(\sum (\xi/2 + k\pi)^{-4s} \right)^{1/2}.$$

We begin with the identity $\sum_{-\infty}^{\infty} (z + k\pi)^{-1} = \cot z$, which we differentiate $q - 1$ times. This then gives

$$(12.4) \qquad \sum_{-\infty}^{\infty} \frac{1}{(z + k\pi)^q} = \frac{P_q(\cos z)}{(\sin z)^q},$$

where P_q is a polynomial of degree $q - 2$. When q is even, it is an obvious consequence of (12.4) that $P_q(t)$ is strictly positive on $[0, 1]$. Returning to the calculation of $\hat{\phi}(\xi)$, with $r + 1 = 2s$, we get

$$\hat{\phi}(\xi) = \left(\frac{\sin \xi/2}{\xi/2} \right)^{2s} (P_{4s}(\cos \xi/2))^{-1/2}.$$

As Proposition 7 requires, $\hat{\phi}(\xi)$ and all its derivatives up to order r vanish at the points $\xi = 2k\pi$, $k \neq 0$.

Now suppose that r is even. By translating $g(x)$ by an integer, if necessary, we can take $\hat{g}(\xi) = \left((\sin \xi/2)/(\xi/2) \right)^{r+1} e^{-i\xi/2}$. That is, $g(x)$ is centred on $x = 1/2$ instead of $x = 0$. The computation of $\hat{\phi}(\xi)$ is similar to what we have just done, and gives

$$\hat{\phi}(\xi) = \left(\frac{\sin \xi/2}{\xi/2} \right)^{r+1} e^{-i\xi/2} (P_{2r+2}(\cos \xi/2))^{-1/2}.$$

We see that ϕ is a real-valued function centred on $x = 1/2$. Just as before, $\hat{\phi}(\xi)$ and all its derivatives up to order r vanish at $\xi = 2k\pi$, $k \neq 0$.

13 Notes and comments

The notion of multiresolution approximation was put in its final form by S. Mallat and the author (autumn 1986, [178]). Mallat relates this concept to algorithms in image processing. If, in the plane, $f(x, y)$ denotes the ideal image (a limit whose precision is absolute), $f_j(x, y) \in$

V_j represents an approximation of the ideal image, an approximation whose resolution is (of order of magnitude) 2^{-j}.

We do not have a general method for constructing all the multiresolution approximations of $L^2(\mathbb{R}^n)$. We shall, however, describe—in Chapter 3—an algorithm due to Mallat which gives all the multiresolution approximations of $L^2(\mathbb{R}^n)$ that we know. The algorithm will be used to prove the existence, for each $r \in \mathbb{N}$, of an orthonormal basis composed of "r-regular" wavelets of compact support (I. Daubechies [88]).

The concept of multiresolution approximation is a generalization of an algorithm introduced by G. Deslauriers and S. Dubuc which they called dyadic interpolation. Their work preceded ours and was designed to construct fractal curves ([100], [101]). Here is a brief exposition of Deslauriers' and Dubuc's ideas, deliberately written in the language appropriate to Mallat's algorithm in Chapter 3.

We start from the obvious inclusion $\mathbb{Z} \subset \mathbb{Z}/2$ which we try to imitate as closely as possible by a "functional inclusion" $J : l^\infty(\mathbb{Z}) \to l^\infty(\mathbb{Z})$. To do this, we let $\tau_k : l^\infty(\mathbb{Z}) \to l^\infty(\mathbb{Z})$, $k \in \mathbb{Z}$, denote the translation operator (defined by $\tau_k f(x) = f(x-k)$) and, by abuse of language, we also let τ_k denote translation by k acting on $l^\infty(\mathbb{Z}/2)$.

We then require J to satisfy the following three properties:

(a) $J : l^\infty(\mathbb{Z}) \to l^\infty(\mathbb{Z})$ is a continuous linear operator which remains continuous when $l^\infty(\mathbb{Z})$ and $l^\infty(\mathbb{Z}/2)$ are given the weak topologies $\sigma(l^\infty(\mathbb{Z}), l^1(\mathbb{Z}))$ and $\sigma(l^\infty(\mathbb{Z}/2), l^1(\mathbb{Z}/2))$;

(b) $J\tau_k = \tau_k J$ for each $k \in \mathbb{Z}$;

(c) for every sequence f in $l^\infty(\mathbb{Z})$, $J(f)$ is a sequence in $l^\infty(\mathbb{Z}/2)$ whose restriction to $\mathbb{Z}$ coincides with f.

The operator J thus extends to $\mathbb{Z}/2$ sequences which are defined only on $\mathbb{Z}$. Let ε_k denote the sequence in $l^\infty(\mathbb{Z})$ which equals 1 at k and is 0 elsewhere. We define the symbol $\omega \in l^1(\mathbb{Z})$ of the operator J by $J(\varepsilon_0)(l/2) = \omega(l)$, $l \in \mathbb{Z}$; (b) then implies that $J(\varepsilon_k)(l/2) = \omega(l-2k)$. By linearity and continuity, we finally get

$$(13.1) \qquad J(f)(l/2) = \sum_{-\infty}^{\infty} f(k)\omega(l-2k).$$

The symbol ω defines the operator J completely. Property (c) is equivalent to $\omega(0) = 1$ and $\omega(2k) = 0$, $k \in \mathbb{Z}$, $k \neq 0$. Conversely, every sequence $\omega \in l^1(Z)$ with these properties defines an operator J satisfying (a), (b) and (c).

Having established this, we transform J, by a simple change of scale, into an operator $J_m : l^\infty(\mathbb{Z}/2^m) \to l^\infty(\mathbb{Z}/2^{m+1})$ satisfying (a), commuting with the translations $2^{-m}k$, $k \in \mathbb{Z}$ and extending the sequences of

$l^\infty(\mathbf{Z}/2^m)$ to $\mathbf{Z}/2^{m+1}$. Moreover, J_m is defined by the same sequence $\omega(l)$, $l \in \mathbf{Z}$ as is J, in the following sense: $J_m(\varepsilon_0)$ is the sequence whose value at $l/2^{m+1}$ equals $\omega(l)$, for each $l \in \mathbf{Z}$.

The next step consists of "approximating" $\mathbf{R}$ by the increasing sequence of nested subgroups $\mathbf{Z}/2^m$, $m \in \mathbf{N}$. Taking an arbitrary sequence $f \in l^\infty(\mathbf{Z})$, we can extend to sequences $f_1 \in l^\infty(\mathbf{Z}/2), \ldots, f_m \in l^\infty(\mathbf{Z}/2^m)$. We have $J_m(f_m) = f_{m+1}$, where

$$(13.2) \qquad f_{m+1}(l/2^{m+1}) = \sum_{-\infty}^{\infty} f_m(k/2^m)\omega(l - 2k).$$

The convergence problem consists of knowing whether there exists a continuous function f_∞ on $\mathbf{R}$ whose restriction to $\mathbf{Z}/2^m$ coincides with f_m for every $m \in \mathbf{N}$.

To deal with this problem, it is enough to look at what happens in the particular case $f = \varepsilon_0$. We then write g for the corresponding f_∞. If we suppose that $g(x)$ is continuous, we also have $g \in L^1(\mathbf{R})$. This gives, in full generality, $f_\infty(x) = \sum f(k)g(x - k)$.

The convergence problem has not been resolved so far, but we do have sufficient conditions as a result of the work of Deslauriers and Dubuc. Their work also includes the calculation of the Fourier transform of g (always supposing g exists). They observe that the measure $g(x)\, dx$ is the weak limit of the "Riemann sums" $\mu_m = 2^{-m} \sum g(2^{-m}k)\delta_{2^{-m}k}$, where δ_a denotes Dirac measure at a. Put $\sigma_m = \frac{1}{2} \sum \omega(l)\delta_{2^{-m}l}$. Then $\mu_m = \sigma_1 * \cdots * \sigma_m$, and, setting $\gamma(\xi) = \frac{1}{2} \sum \omega(l)e^{il\xi}$, we get

$$(13.3) \qquad \hat{g}(\xi) = \prod_{1}^{\infty} \gamma(2^{-m}\xi).$$

The infinite product can converge only if $\gamma(0) = 1$, that is, if $J(1) = 1$, where the sequence 1 denotes the constant sequence, all of whose terms are 1.

We finally suppose that the function $g(x)$ that we have constructed satisfies (2.6), for a certain integer $r \in \mathbf{N}$. We denote by V_0 the linear space of all limit functions $f_\infty(x)$ obtained from sequences $f(k)$ in $l^2(\mathbf{Z})$ instead of $l^\infty(\mathbf{Z})$. In other words, $f \in V_0$ if and only if

$$(13.4) \qquad f(x) = \sum \alpha_k g(x - k) \qquad \text{where} \qquad \alpha_k = f(k) \in l^2(\mathbf{Z}).$$

Then V_0 is a closed subspace of $L^2(\mathbf{R})$ and $g(x - k)$, $k \in \mathbf{Z}$, is a Riesz basis of V_0. Further, there exists a multiresolution approximation V_j, $j \in \mathbf{Z}$, of $L^2(\mathbf{R})$ based on V_0.

This multiresolution approximation has a particular property described by (13.4), namely that $g(0) = 1$ and $g(k) = 0$ for $k \in \mathbf{Z}$, $k \neq 0$.

Conversely, let V_j, $j \in \mathbf{Z}$ be an r-regular multiresolution approximation of $L^2(\mathbb{R})$, for $r \geq 1$, such that the function $g(x)$ satisfies, besides (2.4) and (2.6), the condition $g(k) = 0$, if $k \in \mathbb{Z}$, $k \neq 0$, and $g(0) = 1$. Then the multiresolution approximation arises from a dyadic interpolation algorithm with $\omega(l) = g(l/2)$, $l \in \mathbb{Z}$.

In particular, the multiresolution approximation given by nested spaces of splines of odd order arises from a dyadic interpolation algorithm where the sequence $\omega(l)$, $l \in \mathbb{Z}$, decreases exponentially.

The Littlewood-Paley multiresolution approximation which constitutes our third example also arises from a dyadic interpolation algorithm where $\omega(l)$, this time, is a sequence which decreases rapidly.

Deslauriers and Dubuc restrict their attention to "short sequences" $\omega(l)$ which take only four values $a = \omega(3)$, $b = \omega(1)$, $c = \omega(-1)$ and $d = \omega(-3)$, apart from $\omega(0) = 1$. They are interested in obtaining functions $g(x)$ with a fractal structure, by careful choice of a, b, c, and d.

It will be interesting—in the next chapter—to compare the operators J, defined by Deslauriers and Dubuc, with the analogues introduced by Mallat. The latter are partial isometries $J : l^2(\mathbb{Z}) \to l^2(\mathbb{Z})$ commuting with translations and satisfying a certain regularity condition, which allows us to "pass from the discrete to the continuous".

3

Orthonormal wavelet bases

1 Introduction

Let $m \in \mathbb{N}$ be an integer. A function $\psi(x)$ of a real variable is called a *(basic) wavelet of class m* if the following four, apparently contradictory, properties hold:

(a) if $m = 0$, $\psi(x)$ belongs to $L^\infty(\mathbb{R})$; if $m \geq 1$, $\psi(x)$ and all its derivatives up to order m belong to $L^\infty(\mathbb{R})$;

(b) $\psi(x)$ and all its derivatives up to order m decrease rapidly as $x \to \pm\infty$;

(c) $\int_{-\infty}^{\infty} x^k \psi(x)\, dx = 0$ for $0 \leq k \leq m$;

(d) the collection of functions $2^{j/2}\psi(2^j x - k)$, $j, k \in \mathbb{Z}$, is an orthonormal basis of $L^2(\mathbb{R})$.

The functions $2^{j/2}\psi(2^j x - k)$, $j, k \in \mathbb{Z}$, are the *wavelets* (generated by the "mother" ψ) and the conditions (a), (b) and (c) express, respectively, the *regularity*, the *localization* and the *oscillatory* character that we want to give the "mother wavelet". By a simple change of scale, these conditions are satisfied by the wavelets themselves.

To be precise, let I denote the dyadic interval $[k2^{-j}, (k+1)2^{-j})$ and set $\psi_I(x) = 2^{j/2}\psi(2^j x - k)$. Then ψ_I is *essentially concentrated on the interval I*. That is, if MI, $M \geq 1$, denotes the interval which has the same centre as I and length M times that of I, we have

$$\left(\int_{(MI)^c} |\psi_I(x)|^2\, dx \right)^{1/2} = \varepsilon(M),$$

where $\varepsilon(M)$ decreases rapidly as M tends to infinity. (E^c denotes the complement in $\mathbb{R}$ of a subset E of $\mathbb{R}$.)

Condition (c) clearly implies that *all the moments of order* $k \leq m$ *of* ψ_I *are zero*—the *oscillation* or *cancellation* condition.

Explicitly, condition (b) means that, for every $N \geq 1$ and for $0 \leq l \leq m$,

$$\left| \left(\frac{d}{dx} \right)^l \psi_I(x) \right| \leq C_N 2^{j/2} 2^{jl} \left(1 + |2^j x - k| \right)^{-N}.$$

This expresses the *regularity* and *localization* of the wavelets ψ_I.

Let $\mathcal{I}$ denote the collection of dyadic intervals I. The functions ψ_I are the wavelets generated by ψ, and the representation of a function f on the real line by a wavelet series is, by definition, the identity

$$(1.1) \qquad f(x) = \sum_{I \in \mathcal{I}} \alpha(I) \psi_I(x) \qquad \text{where} \qquad \alpha(I) = (f, \psi_I).$$

This identity works perfectly if f belongs to the space $L^2(\mathbb{R})$, because the convergence of the right-hand side to f takes place in $L^2(\mathbb{R})$ norm. The identity also works if f is in $L^p(\mathbb{R})$, for $1 < p < \infty$. The right-hand side converges to f and the series on the right-hand side is commutatively convergent—the order of the terms is not relevant to the convergence, which is reasonable, since $\mathcal{I}$ is not an ordered set.

But (1.1) does not hold when f belongs to $L^\infty(\mathbb{R})$ or $L^1(\mathbb{R})$.

For example, if $f(x) = 1$ identically, all the wavelet coefficients are zero and (1.1) gives $1 = 0$, an absurdity. Similarly, if $f(x)$ is any function in $\mathcal{D}(\mathbb{R})$, say, whose integral over $\mathbb{R}$ is 1, then the right-hand side cannot converge in $L^1(\mathbb{R})$ norm. Otherwise we could integrate the right-hand side term by term and once again get $1 = 0$. This difficulty will be the starting point of Chapter 5, whereas convergence of (1.1) in $L^p(\mathbb{R})$, $1 < p < \infty$, is related to the general theory of Calderón-Zygmund operators, which will be covered in Chapter 7.

All the difficulties we have raised disappear if we have at our disposal a second function $\phi(x)$, called the "father wavelet", ($\psi(x)$ was the "mother wavelet") satisfying conditions (a) and (b), just like ψ, and having two other properties. Instead of (c), $\phi(x)$ satisfies $\int_{-\infty}^{\infty} \phi(x) \, dx = 1$ and instead of (d) we require that the functions $\phi(x - k)$, $k \in \mathbb{Z}$, and $\psi_I(x)$, $I \in \mathcal{I}$, $|I| \leq 1$, together form an orthonormal basis of $L^2(\mathbb{R})$. We then have

$$(1.2) \qquad f(x) = \sum_{-\infty}^{\infty} \beta_k \phi(x - k) + \sum_{I \in \mathcal{I}, \, |I| \leq 1} \alpha(I) \psi_I(x),$$

where

$$\beta_k = \int_{-\infty}^{\infty} f(t)\bar{\phi}(t-k)\,dt \qquad \text{and} \qquad \alpha(I) = (f, \psi_I)\,.$$

This time the second series involves only "small" dyadic intervals $I \in \mathcal{I}$. We could well say that $\sum_{-\infty}^{\infty} \beta_k \phi(x-k)$ is a first approximation of $f(x)$, a "blurred image" of $f(x)$, which we "bring into focus" by adding finer and finer details. The first details are of dimension 1 and are given by the terms $\alpha(I)\psi_I(x)$ such that $|I| = 1$; successive details are smaller, of dimension $1/2, 1/4, \ldots$ This touching in of details is regularly spaced, taking account of dimension—the details of dimension 2^{-j} are placed at the points $k2^{-j}$.

Naturally, algorithm (1.2) is much better than (1.1) and enables us to avoid the contradictions that we meet in applying the algorithm to spaces of functions or distributions other than $L^2(\mathbb{R})$.

Algorithm (1.2) logically implies (1.1). We can begin to see this by making a simple change of variable in (1.2), to give

$$(1.3) \qquad f(x) = \sum_{-\infty}^{\infty} \beta_k \phi(2^l x - k) + \sum_{I \in \mathcal{I},\ |I| \le 2^{-l}} \alpha(I)\psi_I(x)\,,$$

where

$$l \in \mathbb{Z} \qquad \text{and} \qquad \beta_k = 2^l \int_{-\infty}^{\infty} f(t)\bar{\phi}(2^l t - k)\,dt\,.$$

For f in $L^2(\mathbb{R})$ we let l tend to $-\infty$. The limit gives (1.1).

On the other hand, we don't know whether (1.1) implies (1.2): the problem is with the existence and the regularity of the function ϕ.

If we do have algorithm (1.2), that means that the wavelets ψ_I arise from an m-regular multiresolution approximation. Indeed, let V_l denote the closed subspace of $L^2(\mathbb{R})$ which has the functions $2^{l/2}\phi(2^l x - k)$, $k \in \mathbb{Z}$, as an orthonormal basis and let W_l be the orthogonal complement of V_l in V_{l+1}. Then (1.2) expresses exactly that the functions $2^{l/2}\psi(2^l x - k)$, $k \in \mathbb{Z}$, form an orthonormal basis of W_l.

The decomposition of (1.2) can be applied with remarkable flexibility. For example, if ψ and ϕ both have compact support, then (1.2) gives a decomposition of any distribution of order less than m: the series on the right-hand side converges to $f(x)$ in the sense of distributions and the scalar products β_k and $\alpha(I)$ certainly make sense, because $\phi(x-k)$ and ψ_I are test functions.

By the series of (1.2), general distributions are decomposed into series of correctly localized fluctuations of a characteristic form. Further, as we shall see in Chapter 6, the order of the distribution f can be calculated exactly and directly from the size of its wavelet coefficients—standard

Fourier analysis gives only an estimate, which is not exact, except in the case of Sobolev spaces.

We now intend to pass to the n-dimensional case and explain the geometric ideas underlying the construction of orthonormal wavelet bases.

We start with the lattices $\Gamma_j = 2^{-j}\mathbb{Z}^n$ in $\mathbb{R}^n$. They form a nested sequence whose union is dense in $\mathbb{R}^n$. Approximation to $L^2(\mathbb{R}^n)$ by the nested sequence of closed subspaces V_j imitates and reflects the geometric approximation to $\mathbb{R}^n$ by the nested sequence of lattices Γ_j.

Let Λ denote the union of the Γ_j, less the point 0, and let Λ_j denote the set of those points of Γ_{j+1} which do not belong to Γ_j. The sets Λ_j, $j \in \mathbb{Z}$ thus form a partition of Λ, moreover, $\Lambda_j = 2^{-j}\Lambda_0$ and $\Lambda_0 = \mathbb{Z}^n + \frac{1}{2}E$, where E is the set of the $2^n - 1$ sequences $\varepsilon = (\varepsilon_1, \ldots, \varepsilon_n)$, with $\varepsilon_j = 0$ or 1, but excluding the sequence $(0, 0, \ldots, 0)$.

Pursuing the analogy between functional analysis in $L^2(\mathbb{R}^n)$ and geometry in $\mathbb{R}^n$, we can predict that the sets Λ_j will serve to sample the orthogonal complement W_j of V_j in V_{j+1}. In other words, the identity $V_{j+1} = V_j \oplus W_j$ will correspond to the fact that Γ_{j+1} is the disjoint union of Γ_j with Λ_j. Returning to formula (1.3), an orthonormal basis of V_j will be obtained by taking the union of the orthonormal basis ϕ_γ, $\gamma \in \Gamma_j$, of V_j, constructed in Chapter 2, with the orthonormal basis ψ_λ, $\lambda \in \Lambda_j$, of W_j, which we shall see how to construct in this chapter.

As in the one-dimensional case, the localization, size and regularity of a wavelet ψ_λ are defined by the geometric position of λ and by the integer j for which λ is in Λ_j. More precisely,

(a) ψ_λ is "centred on λ", which means that, on the scale 2^{-j}, $\psi_\lambda(x)$ decreases rapidly as x moves away from λ;

(b) all the derivatives of ψ_λ of order up to and including r have the same rapid rate of decrease, taking account of the scale 2^{-j}—the integer r describes the regularity of the multiscaled analysis used in constructing the wavelets;

(c) $\int x^\alpha \psi_\lambda(x)\,dx = 0$ for each $\lambda \in \Lambda$ and every $\alpha \in \mathbb{N}^n$ such that $|\alpha| \le r$;

(d) if $\lambda = 2^{-j}k + 2^{-j-1}\varepsilon$, where $j \in \mathbb{Z}$, $k \in \mathbb{Z}^n$ and $\varepsilon = (\varepsilon_1, \ldots, \varepsilon_n) \in E$, then $\psi_\lambda(x) = 2^{nj/2}\psi^\varepsilon(2^j x - k)$, where the ψ^ε are the $2^n - 1$ basic wavelets which will be described explicitly in this chapter;

(e) the wavelets $\psi_\lambda(x)$, $\lambda \in \Lambda$, form an orthonormal basis of $L^2(\mathbb{R}^n)$.

Many authors have used constructions which are similar to orthogonal wavelet bases. For example, A. Chang, R. Fefferman and A. Uchiyama have decomposed the space BMO into series similar to (1.1), but whose terms are "almost orthogonal" rather than orthogonal. The absence of orthogonality is irrelevant for their purposes. M. Frazier and B. Jaw-erth, as well as I. Daubechies and A. Grossmann ([114] and [90]) have introduced algorithms similar to (1.1) in which the coefficients are, in-deed, calculated as scalar products, but they use a redundant system of functions ψ_I. In this case, the decomposition is not unique, although a particular one is preferred. The lack of uniqueness prevents us proving that the wavelet coefficients satisfy any necessary and sufficient condi-tions for the series $\sum \xi(I)\psi_I(x)$ to converge in a given function space. We can find either sufficient conditions, in the case of general coeffi-cients, or necessary and sufficient conditions, in the case of particular coefficients given by $\alpha(I) = (f, \psi_I)$.

In his construction of an unconditional basis of the (real) Hardy space H^1, L. Carleson ([46]) used a set of functions of the form $2^{j/2}\psi(2^j x - k)$, $j, k \in \mathbb{Z}$, where ψ was a smooth version of the function $h(x)$ used to define the Haar system. This family of functions did not constitute an orthonormal basis, which made it necessary for Carleson to devise very subtle estimates for the dual basis.

For his part, P. Wojtaszczyk has shown that the Franklin system ([237]) is an unconditional basis of H^1. The Franklin system does form an orthonormal basis of L^2, but its algebraic structure is not simple. Wojtaszczyk's work depends on delicate asymptotic estimates on the functions of the Franklin system, originally proved by Z. Ciesielski ([55]).

The "inventor of wavelets" was , in fact, J.O. Strömberg. He showed, in [223], that, for each integer $m \geq 1$, there exists a C^m-function $\psi(x)$ which decreases exponentially at infinity and is such that the sequence $2^{j/2}\psi(2^j x - k)$, $j, k \in \mathbb{Z}$, is an orthonormal basis of $L^2(\mathbb{R})$. In the same paper, Strömberg describes the corresponding function ϕ and thus obtains the multidimensional wavelets. We describe Strömberg's con-struction in Section 12, but it has the drawback that it only applies to situations involving spline functions. We have therefore given priority to the method presented in Section 2.

Orthonormal wavelet bases provide a very simple method of obtaining results about unconditional bases in classical function spaces.

For any (C^∞) manifold X, with or without boundary, Z. Ciesielski and T. Figiel ([58]) have constructed an orthonormal basis of $L^2(X)$ which is also an unconditional basis for all the Sobolev and Besov spaces of order less than a certain integer m ($m \geq 1$ is arbitrary, but the basis depends on m).

S.V. Bochkariev ([20]) has proved that the Franklin system can be used to construct a Schauder basis of the disc algebra $A(D)$ or of the corresponding polydisc algebra. Once again, we shall see that the use of wavelets, while preserving Bochkariev's ideas, avoids the technical difficulties of his proof.

Lastly, orthonormal wavelet bases arise in mathematical physics. Their discovery had, moreover, been anticipated and their existence had been announced by several research groups working in constructive field theory ([8], [9] and [10]). The use of wavelets leads to the simplification and clarification of calculations of renormalization.

2 The construction of wavelets in dimension 1

Decomposition into wavelet series is perhaps destined to provide stiff competition for both Fourier series and integrals as well as for traditional Fourier analysis.

But at the moment, it is worth observing, with due modesty, that the Fourier transform remains indispensable in the construction of the basic wavelet ψ (or "mother wavelet") arising from an r-regular multiresolution approximation.

So let V_j, $j \in \mathbb{Z}$, be an r-regular multiresolution approximation of $L^2(\mathbb{R})$. In Chapter 2, we learned how to construct a function $\phi \in V_0$ with the following two properties:

for every $N \geq 1$ and $0 \leq k \leq r$,

$$(2.1) \qquad \left| \left(\frac{d}{dx} \right)^k \phi(x) \right| \leq C_N (1 + |x|)^{-N} ;$$

$(2.2) \qquad \phi(x - k)$, $k \in \mathbb{Z}$, is an orthonormal basis of V_0.

The existence of the function ϕ is the starting point of the construction of ψ (the "mother wavelet" is a daughter of the "father wavelet"!)

Let us start by describing the subspace $\mathcal{F}V_0$ consisting of the Fourier transforms of functions $f \in V_0$.

Lemma 1. *The functions* $g \in \mathcal{F}V_0$ *can be written in the form* $g(\xi) = m(\xi)\hat{\phi}(\xi)$, *where* $m(\xi)$ *is* 2π-periodic, *belongs to* $L^2[0, 2\pi]$ *and where* $\|g\|_2 = \left(\int_0^{2\pi} |m(\xi)|^2 \, d\xi \right)^{1/2}$.

The proof is immediate, since we can apply the Fourier transform to the identity $f(x) = \sum \alpha_k \phi(x - k)$ to give

$$\hat{f}(\xi) = \left(\sum \alpha_k e^{-ik\xi} \right) \hat{\phi}(\xi) = m(\xi)\hat{\phi}(\xi).$$

It is then sufficient to note that

$$\|f\|_2 = \left(\sum |\alpha_k|^2 \right)^{1/2} = \frac{1}{\sqrt{2\pi}} \left(\int_0^{2\pi} |m(\xi)|^2 \right)^{1/2}.$$

We shall use Lemma 1 as a dictionary which helps us translate calculations on functions in V_0 into calculations on functions $m(\xi)$ which are 2π-periodic—a more familiar context.

It is time we indicated what we intend to do. So let V_j, $j \in \mathbb{Z}$, be an r-regular multiresolution approximation of $L^2(\mathbb{R})$. The orthogonal complement of V_j in V_{j+1} is denoted by W_j. We observe that, if $f(x) \in W_j$, then $f(2x) \in W_{j+1}$ and vice versa.

We shall prove the following result:

Theorem 1. *There exists a function $\psi \in W_0$ with the following properties:*

for every $m \geq 1$ and $0 \leq q \leq r$,

(2.3)
$$\left| \left(\frac{d}{dx} \right)^q \psi(x) \right| \leq C_m (1 + |x|)^{-m} \, ;$$

(2.4) *the sequence $\psi(x - k)$, $k \in \mathbb{Z}$, is an orthonormal basis of W_0.*

As a consequence, by a simple change of scale, we find that the sequence $2^{j/2} \psi(2^j x - k)$, $k \in \mathbb{Z}$, for fixed $j \in \mathbb{Z}$, is an orthonormal basis of W_j. Since $L^2(\mathbb{R})$ is the Hilbert sum of the subspaces W_j, $j \in \mathbb{Z}$, we obtain an orthonormal basis of $L^2(\mathbb{R})$ by taking the union over $j \in \mathbb{Z}$ of the sequences $2^{j/2} \psi(2^j x - k)$, $k \in \mathbb{Z}$.

Further, the method of constructing ψ will automatically give

$$\int x^q \psi(x) \, dx = 0 \qquad \text{for } 0 \leq q \leq r.$$

We shall prove the theorem by constructing $2^{-1/2} \psi(x/2) \in W_{-1}$ rather than $\psi(x) \in W_0$. To this end, we start by characterizing the subspaces $\mathcal{F}V_{-1}$ and $\mathcal{F}W_{-1}$ of $\mathcal{F}V_0$. Note that $\phi(x/2)$ belongs to V_{-1}. Since $V_{-1} \subset V_0$, we get

(2.5)
$$\frac{1}{2} \phi \left(\frac{x}{2} \right) = \sum_{-\infty}^{\infty} \alpha_k \phi(x + k)$$

$$\text{where}$$

$$2\alpha_k = \int_{-\infty}^{\infty} \phi \left(\frac{x}{2} \right) \bar{\phi}(x + k) \, dx.$$

The α_k thus decrease rapidly and the Fourier transform gives

$$\hat{\phi}(2\xi) = m_0(\xi) \hat{\phi}(\xi) \qquad \text{where} \qquad m_0(\xi) = \sum_{-\infty}^{\infty} \alpha_k e^{ik\xi}.$$

Lemma 2. *With the above notation, $|m_0(\xi)|^2 + |m_0(\xi + \pi)|^2 = 1$.*

Indeed, $\sum_{-\infty}^{\infty} |\hat{\phi}(\xi + 2k\pi)|^2 = 1$, so that $\sum_{-\infty}^{\infty} |\hat{\phi}(2\xi + 2k\pi)|^2 = 1$. But

$\hat{\phi}(2\xi) = m_0(\xi)\hat{\phi}(\xi)$, where $m_0(\xi)$ is 2π-periodic. Hence $\sum_{-\infty}^{\infty} |m_0(\xi + k\pi)|^2 |\hat{\phi}(\xi + k\pi)|^2 = 1$. Separating the even values of k from the odd values gives Lemma 2.

The definition of V_{-1} gives $\mathcal{F}V_{-1} = \{m(2\xi)\hat{\phi}(2\xi) : m \in L^2[0, 2\pi]\}$. But $\hat{\phi}(2\xi) = m_0(\xi)\hat{\phi}(\xi)$, so $\mathcal{F}V_{-1} = \{m(2\xi)m_0(\xi)\hat{\phi}(\xi)\}$. Let U denote the unitary operator taking $\mathcal{F}V_0$ onto $L^2[0, 2\pi]$, defined by

$$U(m(\xi)\hat{\phi}(\xi)) = m(\xi).$$

Rather than calculate the orthogonal complement of $\mathcal{F}V_{-1}$ in $\mathcal{F}V_0$, we shall determine the orthogonal complement of $U\mathcal{F}V_{-1}$ in $U\mathcal{F}V_0$, that is, we shall look for the orthogonal complement in $L^2[0, 2\pi]$ of the space of functions $m(2\xi)m_0(\xi)$, where $m(\xi)$ is 2π-periodic and belongs to $L^2[0, 2\pi]$. If $l(\xi)$ belongs to the orthogonal complement, we have

$$\int_0^{2\pi} m(2\xi)m_0(\xi)\bar{l}(\xi)\, d\xi = 0\,,$$

for every $m(\xi)$. This amounts to writing

(2.6) $\qquad m_0(\xi)\bar{l}(\xi) + m_0(\xi + \pi)\bar{l}(\xi + \pi) = 0\,.$

This identity means that the vector $(l(\xi), l(\xi+\pi))$ is orthogonal, in the Hermitian space $\mathbb{C}^2$, to the unit vector $(m_0(\xi), m_0(\xi+\pi))$ and is thus proportional to the vector $e^{-i\xi}(\bar{m}_0(\xi + \pi), -\bar{m}_0(\xi))$. In other words, $l(\xi) = e^{-i\xi}\bar{m}_0(\xi + \pi)\lambda(\xi)$, where $\lambda(\xi) = (l(\xi)m_0(\xi + \pi) - l(x + \pi)m_0(\xi))e^{i\xi}$. We note that $\lambda(\xi)$ is π-periodic. Further, the mapping which takes the function $l \in L^2[0, 2\pi]$ to $\lambda \in L^2[0, \pi]$ is an isometry. An orthonormal basis of $U\mathcal{F}W_{-1}$ is thus given by the functions $e^{-i\xi}\bar{m}_0(\xi + \pi)e^{2ki\xi}$. This leads us to define $\psi \in W_0$ by

(2.7) $\qquad \hat{\psi}(2\xi) = e^{-i\xi}\bar{m}_0(\xi + \pi)\hat{\phi}(\xi)\,.$

Then the functions $\psi(x - k)$, $k \in \mathbb{Z}$, form the required basis of W_0.

Working back, we may remark that

(2.8) $\qquad \dfrac{1}{2}\psi\left(\dfrac{x + 1}{2}\right) = \sum_{-\infty}^{\infty} \bar{\alpha}_k(-1)^k \phi(x - k)\,,$

so ψ has the same order of regularity and the same rate of decrease as ϕ.

Take the example of the Haar system. We start with the most rudimentary multiresolution approximation, where V_0 is the set of $L^2(\mathbb{R})$ functions constant on each interval $[k, k + 1)$. In this case, ϕ is the characteristic function of $[0, 1)$ and (2.5) gives $\alpha_0 = 1/2$, $\alpha_{-1} = 1/2$ and all the other α_k are zero. Then (2.8) defines $\psi(x)$ as the function with support $[0, 1)$, equal to -1 on $[0, 1/2)$ and 1 on $[1/2, 1)$. The family

of corresponding wavelets forms the *Haar system*, and its regularity is
$r = 0$.

We continue our examples by taking as starting point the multiresolution approximation formed by the continuous, piecewise affine, spline
functions (the restriction of $f \in V_0$ to each interval $[k, k+1]$ is an affine
function). In this case,

$$\hat{\phi}(\xi) = \left(\frac{\sin \xi/2}{\xi/2}\right)^2 \left(1 - \frac{2}{3}\sin^2 \frac{\xi}{2}\right)^{-1/2}$$

and $\phi(x)$ is a continuous function on $\mathbb{R}$ satisfying $\phi(x) = O(e^{-\beta|x|})$ as
$|x| \to \infty$, with $\beta = \log(2 + \sqrt{3})$. Moreover, the restrictions of $\phi(x)$ to
the intervals $[k, k+1]$, $k \in \mathbb{Z}$, are linear. We have

$$m_0(\xi) = \cos^2 \frac{\xi}{2} \left(1 - \frac{2}{3}\sin^2 \xi\right)^{-1/2} \left(1 - \frac{2}{3}\sin^2 \frac{\xi}{2}\right)^{1/2},$$

which gives

$$\hat{\psi}(\xi) = e^{-i\xi/2} \sin^2 \frac{\xi}{4} \left(\frac{\sin \frac{\xi}{4}}{\frac{\xi}{4}}\right)^2 \left(\frac{1 - \frac{2}{3}\cos^2 \frac{\xi}{4}}{1 - \frac{2}{3}\sin^2 \frac{\xi}{4}}\right)^{1/2} \left(1 - \frac{2}{3}\sin^2 \frac{\xi}{2}\right)^{-1/2}$$

([186], p. 16).

Figure 1, on the next page, is a picture of the function $\psi(x)$, which
indicates that the exponential decrease of $\psi(x)$ is numerically effective.

We pass to the more general case of splines of order r. If r is odd,
then $\hat{\phi}(\xi)$ is everywhere *positive or zero*. So a straightforward calculation
gives

$$\hat{\psi}(\xi) = ((\hat{\phi}(\xi/2))^2 - (\hat{\phi}(\xi))^2)^{1/2} e^{-i\xi/2}.$$

That is, $\psi(x) = u(x - 1/2)$, where $u(x)$ is a real, even function of
exponential decrease.

If, on the other hand, r is even, we have

$$\hat{\phi}(\xi) = e^{-i\xi/2} \frac{\sin \xi/2}{\xi/2} A(\xi), \qquad \text{where} \qquad A(\xi) \geq 0,$$

which leads to

$$m_0(\xi) = \cos(\xi/2) e^{-i\xi/2} M(\xi), \qquad \text{where} \qquad M(\xi) \geq 0$$

and so to

$$\hat{\psi}(\xi) = -i \frac{\sin^2 \xi/4}{\xi/4} B(\xi) e^{-i\xi/2}, \qquad \text{where} \qquad B(\xi) \geq 0.$$

We remark that $\psi(x) = v(x - 1/2)$ where, this time, $v(x)$ is a real and
odd function—as in the example of the Haar system.

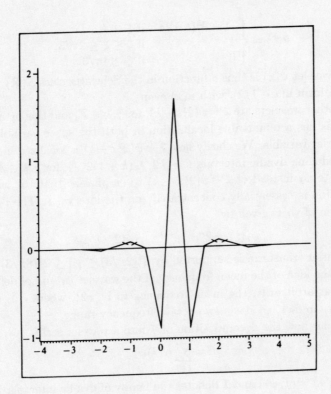

Figure 1

We shall consider a third example of one-dimensional wavelets, based on the Littlewood-Paley multiresolution approximation.

This time, $\hat{\phi}(\xi) = \theta(\xi) \geq 0$, so

$$\hat{\psi}(\xi) = ((\hat{\phi}(\xi/2))^2 - (\hat{\phi}(\xi))^2)^{1/2} e^{-i\xi/2} = \theta_1(\xi) e^{-i\xi/2}.$$

The function $\theta_1(\xi)$ has the following properties: it belongs to $\mathcal{D}(\mathbb{R})$ and is even, so it is completely determined by its restriction to $[0, +\infty)$. This restriction actually has support in $[2\pi/3, 8\pi/3]$, which can be regarded as the union $J \cup K$ of two intervals, $J = [2\pi/3, 4\pi/3]$ and $K = [4\pi/3, 8\pi/3]$. So $\theta_1(\xi) \geq 0$, for all ξ, $\theta_1^2(\xi) + \theta_1^2(2\xi) = 1$, when $\xi \in J$ and, lastly, $\theta_1^2(\xi) + \theta_1^2(4\pi - \xi) = 1$, when $\xi \in K$. It is easy to show that these properties of θ_1 are enough to construct the function θ satisfying the conditions of the Littlewood-Paley multiresolution approximation: we simply require $\theta(\xi)$ to be the even function defined by

$$\theta(\xi) = \begin{cases} (1 - \theta_1^2(\xi))^{1/2} & \text{if } \xi \in J; \\ 1 & \text{if } 0 \leq \xi \leq 2\pi/3; \\ 0 & \text{if } \xi \geq 4\pi/3. \end{cases}$$

The wavelet $\psi(x)$ is thus a function in the Schwartz class $\mathcal{S}(\mathbb{R})$ which is of the form $u(x - 1/2)$, with $u(x)$ even.

The other wavelets are $2^{j/2}\psi(2^j x - k)$, for $j, k \in \mathbb{Z}$, and this orthonormal basis has a remarkable localization in both the space variable and the Fourier variable. We clarify how $2^{j/2}\psi(2^j x - k)$ is localized in space by introducing dyadic intervals $I = [k2^{-j}, (k+1)2^{-j})$, for $j, k \in \mathbb{Z}$. We now write ψ_I instead of $2^{j/2}\psi(2^j x - k)$ to emphasize that the wavelet in question is "essentially concentrated" on the interval I. The Fourier transform of ψ_I is given by

$$(2.9) \qquad \hat{\psi}_I(\xi) = 2^{-j/2}e^{-ik2^{-j}\xi}\hat{\psi}(2^{-j}\xi).$$

The Fourier transform is supported by $2^j(2\pi/3) \leq |\xi| \leq 2^j(8\pi/3)$, suggesting the idea of the mean frequency of the wavelet ψ_I and of defining, somewhat arbitrarily, the mean frequency to be $c2^j$, where $c > 0$ is a constant. In fact, ψ_I has a two octave frequency range.

Consider now the decomposition of f into a wavelet series

$$f = \sum_{I \in \mathcal{I}} \alpha(I)\psi_I,$$

where $\alpha(I) = (f, \psi_I)$ and $\mathcal{I}$ denotes the family of dyadic intervals. Then

$$\alpha(I) = \int f(x)\bar{\psi}_I(x)\,dx = \frac{1}{2\pi}\int \hat{f}(\xi)\bar{\hat{\psi}}_I(\xi)\,d\xi$$
$$= \frac{1}{2^{j/2}}\frac{1}{2\pi}\int \hat{f}(\xi)e^{ik2^{-j}\xi}\bar{\hat{\psi}}(2^{-j}\xi)\,d\xi$$
$$= \frac{1}{2^{j/2}}\frac{1}{2\pi}\int \hat{f}(\xi)e^{i(k+\frac{1}{2})2^{-j}\xi}\theta_1(2^{-j}\xi)\,d\xi.$$

Following Morlet, we let $\Delta_j : L^2(\mathbb{R}) \to L^2(\mathbb{R})$, $j \in \mathbb{Z}$, be the filtering operators defined by

$$\mathcal{F}(\Delta_j f)(\xi) = \hat{f}(\xi)\theta_1(2^{-j}\xi).$$

The operators Δ_j are self-adjoint and satisfy $Id = \sum_{-\infty}^{\infty} \Delta_j^2$, where Id is the identity operator. All this gives the wavelet coefficient $\alpha(I)$ in the form

$$(2.10) \qquad \alpha(I) = 2^{-j/2}(\Delta_j f)\left(2^{-j}\left(k + \frac{1}{2}\right)\right).$$

The Littlewood-Paley decomposition of f consists of writing $f = \sum_{-\infty}^{\infty} \Delta_j^2(f)$, whereas the wavelet decomposition of f leads to sampling the dyadic blocks $\Delta_j(f)$ (not $\Delta_j^2(f)$) while adhering to Shannon's rule, described in Chapter 1. The sampling points are $(k + 1/2)2^{-j}$, $k \in \mathbb{Z}$, and the sampling step is of the order of magnitude specified by Shannon's rule. In fact, we subsample the function $f_j = \Delta_j(f)$ for the following reason. The support of the Fourier transform $\hat{f}_j$ of f_j is contained in the union of the interval $[2^j(2\pi/3), 2^j(8\pi/3)]$ and its symmetric image about 0. If we restrict f_j to the arithmetic progression $(k + \frac{1}{2})2^{-j}$, $k \in \mathbb{Z}$, we conform to Shannon's rule only approximately. Indeed, the two intervals making up the support of $\hat{f}_j$ are congruent to each other modulo $2\pi 2^j$. More precisely, we have

$$[-8\pi/3, -4\pi/3] + 4\pi = [4\pi/3, 8\pi/3],$$

whereas

$$[-4\pi/3, -2\pi/3] + 2\pi = [2\pi/3, 4\pi/3].$$

Thus the information given by the sampling $f_j(2^{-j}(k + \frac{1}{2}))$ is not sufficient to recover the function f_j. In fact, something even subtler is happening, because the intervals $[2^j(2\pi/3), 2^j(8\pi/3)]$ overlap. Hence the samplings of f_{j-1} and of f_{j+1} indirectly give information about f_j. So there is no strict logical connection between the decomposition into wavelet series and Shannon's sampling rules. We have to be content with a heuristic relationship.

Figure 2 is the graph of the function ψ constructed above.

It is worth making a final point before leaving the case of dimension 1.

According to the introduction to this chapter, the functions $f \in V_j$ are "naturally sampled" at the points $\lambda \in \Gamma_j = 2^{-j}\mathbb{Z}$. The functions $f \in W_j$ (the orthogonal complement of V_j in V_{j+1}) are then "naturally sampled" at the points $\lambda \in \Lambda_j = \Gamma_{j+1} \backslash \Gamma_j$. These points are exactly $\lambda = 2^{-j}(k + \frac{1}{2})$. So, if we trust this heuristic approach, the wavelets are centred on the points $\lambda = 2^{-j}(k + \frac{1}{2}) \in \Lambda_j$, $k \in \mathbb{Z}$. But this is exactly what we have proved in the three cases where we have calculated the wavelets $2^{j/2}\psi(2^j x - k)$. In the case of splines of odd order and in the Littlewood-Paley case, the wavelet $2^{j/2}\psi(2^j x - k)$ is effectively centred on $\lambda = 2^{-j}(k + \frac{1}{2})$ and its graph is symmetric about the vertical line through λ. Lastly, the wavelet attains its maximum precisely at λ.

In the case of splines of even order, the wavelet $2^{j/2}\psi(2^j x - k)$ is again centred on λ, but this time the symmetry works differently: the graph of our wavelet has the point $(\lambda, 0)$ as centre of symmetry.

We shall finish this section by *calculating wavelet coefficients of particular functions* and by expressing those functions as wavelet series.

We start with the function $f(x) = \log|x|$ which belongs to the John and Nirenberg space BMO, as we shall see in Chapter 5.

Let us calculate the wavelet coefficients of $\log|x|$ with respect to the

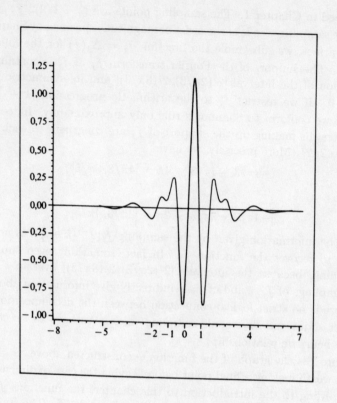

Figure 2

Littlewood-Paley wavelet basis. The best way to do this is with the help of the Fourier transform. It is known that the Fourier transform—in the sense of distributions—of $\log|x|$ is $-\pi\,\mathrm{fp}\,|\xi|^{-1}$, where fp denotes the Hadamard finite part. Consulting (2.10), we see that we must calculate $\Delta_j f$ when $f(x) = \log|x|$. The Fourier transform of $\Delta_j f$ is $-\pi\theta_1(2^{-j}\xi)/|\xi| = -\pi 2^{-j}\theta_2(2^{-j}\xi)$, where $\theta_2(u) = \theta_1(u)/|u|$ has all the qualitative properties of θ_1, namely, θ_2 is an infinitely differentiable, even function which is positive or zero and has support contained in the union of the two intervals $[-8\pi/3, -2\pi/3]$ and $[2\pi/3, 8\pi/3]$.

Let $\omega \in \mathcal{S}(\mathbb{R})$ denote the function whose Fourier transform is θ_2. With this notation, $\Delta_j f(x) = -\pi\omega(2^j x)$. The wavelet coefficients of $\log|x|$ are thus $-\pi 2^{-j/2}\omega(k + \frac{1}{2})$. The point to notice is that the variables k and j are separated.

The wavelet series of $\log|x|$ is thus

(2.11) $$\log|x| = -\pi \sum_{-\infty}^{\infty} \sum_{-\infty}^{\infty} \omega(k + \frac{1}{2})\psi(2^j x - k).$$

It is an easy matter to calculate the Fourier transform of

$$\sigma(x) = \sum_{-\infty}^{\infty} \omega(k + \frac{1}{2})\psi(x - k).$$

Then it is not hard to verify that $\sigma(x)$ belongs to $\mathcal{S}(\mathbb{R})$ and that the Fourier transform of σ has the same support as that of ψ. So we get

$$\log|x| = -\pi \sum_{-\infty}^{\infty} \sigma(2^j x)$$

which is, after all, a pretty unremarkable decomposition of $\log|x|$.

What is less unremarkable is the nature of this identity. It does not express a numerical equality. Here's why. If we replace x by $2x$, the right-hand side does not change at all, but $\log|2x| = \log|x| + \log|2|$. A numerical interpretation would thus lead to the contradiction $\log 2 = 0$.

Now for the explanantion of this paradox. When f belongs to $L^1(\mathbb{R})$, then $\sum_0^\infty f(2^j x)$ defines a function in $L^1(\mathbb{R})$, since $\|f(2^j x)\|_1 = 2^{-j}\|f\|_1$. On the other hand, $\sum_{-\infty}^{-1}$ poses a problem. The problem disappears if f satisfies a Hölder condition of exponent $\gamma > 0$—that is, if $|f(x') - f(x)| \le C|x' - x|^\gamma$, for some constant $C > 0$ and any $x, x' \in \mathbb{R}$—and if we are allowed to add (or subtract) floating constants at will. For example, even if the series $\sum_{-\infty}^{-1} f(2^j x)$ diverges for each x, the series $\sum_{-\infty}^{-1}(f(2^j x) - f(2^j x_0))$ converges uniformly on each compact subset of $\mathbb{R}$.

The identity $\log|x| = -\pi \sum_{-\infty}^{\infty} \sigma(2^j x)$ thus means that, for a certain normalizing constant a,

$$\log|x| = a - \pi \sum_{-\infty}^{\infty}(\sigma(2^j x) - \sigma(2^j)).$$

We shall make more systematic use of this kind of "additive renormalization" in our treatment of wavelet series of BMO functions.

Now we consider the function $\operatorname{sgn} x = x/|x|$. Its wavelet coefficients are easily calculated if we put

$$\varepsilon(k) = \int_{-k}^{\infty} \psi(x)\,dx = -\int_{-\infty}^{-k} \psi(x)\,dx.$$

We then get $\alpha(I) = 2^{-j/2}\varepsilon(k)$ and the sequence $\varepsilon(k)$ decreases rapidly as $|k| \to \infty$. Thus there is no great difference between the function $\operatorname{sgn} x$ and the function $\log|x|$, as far as the order of magnitude of their wavelet coefficients is concerned.

What explanantion does this mystery have?

Let H denote the Hilbert transform, that is, the operator of convolution by the distribution $\mathrm{PV}(1/\pi x)$. With $\mathcal{F}$ denoting the Fourier transform, we get $\mathcal{F}(Hf)(\xi) = -i\,\mathrm{sgn}\,\xi \hat{f}(\xi)$, so H is an isometry of $L^2(\mathbb{R})$. Further, H commutes with translations and dilations. Let $\mathcal{S}_0(\mathbb{R}) \subset \mathcal{S}(\mathbb{R})$ be the subspace of $\mathcal{S}(\mathbb{R})$ consisting of the functions all of whose moments are zero. Then $\mathcal{F} : \mathcal{S}_0(\mathbb{R}) \to \mathcal{S}_0(\mathbb{R})$ is a topological isomorphism. With the help of these remarks, or by direct calculation, we can check that the functions $\tilde{\psi}_I = H(\psi_I)$ are also wavelets (arising out of a different multiresolution approximation).

Now $H(\mathrm{sgn}\,x) = 2\log|x|$ in a sense which will be explained in Chapter 7. So, using the notation $\langle \cdot, \cdot \rangle$ to denote the duality between test functions and distributions,

$$\langle 2\log|x|, \psi_I \rangle = \langle H(\mathrm{sgn}\,x), \psi_I \rangle = -\langle \mathrm{sgn}\,x, \tilde{\psi}_I \rangle.$$

Thus the wavelet coefficients of the function $2\log|x|$ (with respect to the wavelet basis ψ_I) are the same as the wavelet coefficients of $\mathrm{sgn}\,x$ (with respect to the basis $\tilde{\psi}_I$). This explains why there is no qualitative difference between the wavelet coefficients of the two functions.

Our final calculation is of the wavelet coefficients of the function $|x|^{-\alpha}$, where $0 < \alpha < 1$. We continue to use the Littlewood-Paley wavelet basis.

This time we do not need to pass to the Fourier transform. The coefficients are given directly by

$$\alpha(I) = 2^{j/2} \int |x|^{-\alpha} \psi(2^j x - k)\,dx\,.$$

The change of variable $2^j x = u$ gives $\alpha(I) = 2^{j\alpha}2^{-j/2}\omega_\alpha(k)$, where the sequence $\omega_\alpha(k)$ decreases rapidly as $|k| \to \infty$.

The wavelet series is

$$|x|^{-\alpha} = \sum_{-\infty}^{\infty}\sum_{-\infty}^{\infty} \omega_\alpha(k)2^{j\alpha}\psi(2^j x - k) = \sum_{-\infty}^{\infty} 2^{j\alpha}\psi_\alpha(2^j x)\,.$$

The series decomposition of $|x|^{-\alpha}$ is unremarkable, and the function ψ_α belongs to $\mathcal{S}_0(\mathbb{R})$. The series converges in $L^1[-T, T]$, for every $T > 0$.

For the moment, we shall content ourselves with the following observation. An important difference between wavelet series and Fourier analysis, whether by series or integral, is the effect of the singularities of the function on the coefficients. Let $\varepsilon > 0$ be fixed. In the case of the function $|x|^{-\alpha}$ and its decomposition into a wavelet series, the coefficients $\alpha(I)$, relative to intervals I not intersecting $[-\varepsilon, \varepsilon]$, decrease as $O(|I|^N)$ for each $N \geq 1$ as $|I|$ tends to 0. Here's why: $|k2^{-j}| \geq \varepsilon$ and $2^{j\alpha}2^{-j/2}\omega_\alpha(k) = O(2^{-Nj})$ for each $N \geq 1$ as $j \to \infty$.

The intervals are far away from the singularity, and the wavelet coefficients are as small as for an infinitely differentiable function.

Decomposition into wavelets thus allows singularities to be located by observing the places where there are abnormally large wavelet coefficients. Obviously, nothing of the kind happens for the Fourier transform.

We shall return to this point when we study periodic wavelets. We shall then compare wavelet series with Fourier series and rediscover, in a more precise fashion, what we have just found.

3 Construction of wavelets in dimension 2 by the tensor product method

We have the choice of two routes.

We may construct certain wavelet bases from those obtained in dimension 1 by choosing multiresolution approximations given by tensor products.

But we may just as well start with an arbitrary multiresolution approximation in dimension 2 and apply the analogue of the construction which we used in the one-dimensional case.

In both approaches, the two-dimensional wavelets are given by the following expression:–

$$(3.1) \qquad 2^j \psi(2^j x - k, 2^j y - l), \qquad j, k, l \in \mathbf{Z},$$

but ψ will no longer be a single function: on the contrary, ψ will belong to a finite set F consisting of three elementary wavelets.

This is because the two dimensional wavelets $2^j \psi(2^j x - k, 2^j y - l)$ have to form a Hilbert basis of W_j as (k, l) runs through $\mathbf{Z}^2$. In other words, the complete set

$$(3.2) \qquad 2^j \phi(2^j x - k, 2^j y - l), \qquad 2^j \psi(2^j x - k, 2^j y - l)$$

has to give a basis of V_{j+1} as (k, l) describes $\mathbf{Z}^2$. A geometric picture is given by letting the functions $2^j \phi(2^j x - k, 2^j y - l)$ correspond to the points $\gamma = (2^{-j}k, 2^{-j}l) \in \Gamma_j = 2^{-j}\mathbf{Z}^2$. The functions $2^{j+1}\phi(2^{j+1}x - k, 2^{j+1}y - l)$, which correspond to the points $\gamma = (2^{-j-1}k, 2^{-j-1}l) \in \Gamma_{j+1}$, form a Hilbert basis for the whole space V_{j+1}. If we continue this analogy, the wavelets $2^j \psi(2^j x - k, 2^j y - l)$ must be associated with the points $\lambda \in \Gamma_{j+1} \backslash \Gamma_j = \Lambda_j$. These points are of the form $\lambda = 2^{-j}(k + \alpha_1/2, l + \alpha_2/2)$, $k, l \in \mathbf{Z}$, $\alpha_1, \alpha_2 = 0$ or 1, but not both 0 simultaneously. This means that there are three possibilities for $\alpha = (\alpha_1, \alpha_2)$ and thus three elementary wavelets ψ.

After this heuristic introduction, we can be more precise. We shall construct two-dimensional wavelet bases.

Let $\mathcal{V}_j$, $j \in \mathbb{Z}$ be an r-regular multiresolution approximation of $L^2(\mathbb{R})$. We apply the procedure of Section 2 to get corresponding functions ϕ and ψ.

We now define $V_j \subset L^2(\mathbb{R}^2)$ as the closure in $L^2(\mathbb{R}^2)$ norm of the algebraic tensor product $\mathcal{V}_j \otimes \mathcal{V}_j$. An orthonormal basis of V_0 is then made up of the products $\phi(x-k)\phi(y-l)$, $(k,l) \in \mathbb{Z}^2$. In other words, putting $\phi(x,y) = \phi(x)\phi(y)$, we get the orthonormal basis of V_0 as the orbit of the function ϕ under the action of $\mathbb{Z}^2$.

Let $\mathcal{W}_0 \subset \mathcal{V}_1$ denote the orthogonal complement of $\mathcal{V}_0$ in $\mathcal{V}_1$. It is then clear that

$$(3.3) \qquad V_1 = V_0 \oplus \overline{(\mathcal{V}_0 \otimes \mathcal{W}_0)} \oplus \overline{(\mathcal{W}_0 \otimes \mathcal{V}_0)} \oplus \overline{(\mathcal{W}_0 \otimes \mathcal{W}_0)}.$$

Indeed, $V_1 = \overline{(\mathcal{V}_0 \oplus \mathcal{W}_0) \otimes (\mathcal{V}_0 \oplus \mathcal{W}_0)}$, and it is enough to expand the tensor product, which has the algebraic properties of multiplication, and is thus distributive over addition.

Let W_0 denote the orthogonal complement of V_0 in V_1. We then have $W_0 = W_{0,1} \oplus W_{1,0} \oplus W_{1,1}$, where

$$W_{0,1} = \overline{(\mathcal{V}_0 \otimes \mathcal{W}_0)}, \quad W_{1,0} = \overline{(\mathcal{W}_0 \otimes \mathcal{V}_0)} \quad \text{and} \quad W_{1,1} = \overline{(\mathcal{W}_0 \otimes \mathcal{W}_0)}.$$

Finally, we get an orthonormal basis of W_0 by taking the union of the three families $\phi(x-k)\psi(y-l)$, $\psi(x-k)\phi(y-l)$ and $\psi(x-k)\psi(y-l)$, $(k,l) \in \mathbb{Z}^2$, which are, respectively, the orthonormal bases of $W_{0,1}$, $W_{1,0}$ and $W_{1,1}$.

A different way of explaining this procedure is to use the following picture. If V_j, $j \in \mathbb{Z}$ is a multiresolution approximation of $L^2(\mathbb{R}^2)$, we call W_j the *innovation space*, because it contains the extra information needed to pass from an approximation of scale 2^{-j} to one of scale 2^{-j-1}. Identity (3.3) shows that innovating in two variables means innovating in y without innovating in x, or vice versa, or innovating in both x and y.

We shall apply this construction of two-dimensional wavelets to an example in which the nested subspaces $V_j \subset L^2(\mathbb{R}^2)$ have a simple geometric meaning. As usual, it is enough to define V_0. For $r = 0$, the functions $f \in V_0$ are, quite simply, those which are constant on every square $k \le x < k+1$, $l \le y < l+1$. The resulting wavelet basis is the two-dimensional Haar system. For $r \ge 1$, the functions $f(x,y)$ belonging to V_0 are those which satisfy the global regularity condition $(\partial/\partial x)^p (\partial/\partial y)^q f(x,y) \in L^\infty(\mathbb{R}^2) \cap L^2(\mathbb{R}^2)$ and whose restriction to each square $k \le x < k+1$, $l \le y < l+1$ coincides with a polynomial $P_{k,l}(x,y)$ of degree r or less. Then the other V_j are defined by the usual rule.

In this example, $V_0 = \mathcal{V}_0 \widehat{\otimes} \mathcal{V}_0$, where $\mathcal{V}_0 \subset L^2(\mathbb{R})$ is the base space for the multiresolution approximation by splines of order r. The symbol $\hat{\otimes}$ means that the algebraic tensor product is completed in the $L^2(\mathbb{R}^2)$ norm.

The construction of two-dimensional wavelets applies in this situation. We begin with the basic wavelet $\psi(x)$ of the splines of order r. The function $\psi(x)$ is centred on $x = 1/2$, and the function $\psi(x+1/2)$, centred on $x = 0$, is even when r is odd and vice versa.

We let ϕ denote the function in $\mathcal{V}_0$ such that $\phi(x - k)$, $k \in \mathbb{Z}$, is an orthonormal basis of $\mathcal{V}_0$ and is constructed by the canonical procedure of Chapter 2. Then the two-dimensional wavelets are $2^j\phi(2^jx-k)\psi(2^jy-l)$, $2^j\psi(2^jx - k)\phi(2^jy - l)$ and $2^j\psi(2^jx - k)\psi(2^jy - l)$, where $(k,l) \in \mathbb{Z}^2$.

When r is even, the centres of symmetry of the three wavelets are the points $\lambda = 2^{-j}(k + 1/2, l + 1/2)$ and not the points $2^{-j}(k + \varepsilon_1/2, l + \varepsilon_2/2)$, $(\varepsilon_1, \varepsilon_2) \in \{0,1\}^2 \setminus \{(0,0)\}$, that one mught expect on rereading the introduction to this chapter. When r is odd, however, the centres of symmetry of our three wavelets are indeed the points indicated by the heuristic considerations of the introduction.

4 The algorithm for constructing multi-dimensional wavelets

Let us remind ourselves what the problem is. Suppose that $V_j \subset L^2(\mathbb{R}^n)$ denotes an r-regular multiresolution approximation. We let $W_0 \subset V_1$ denote the orthogonal complement of V_0 in V_1 and we look for functions $\psi_\varepsilon \in W_0$, $\varepsilon \in E$, such that the functions $\psi_\varepsilon(x - k)$, $\varepsilon \in E$, $k \in \mathbb{Z}^n$, form an orthonormal basis of W_0. Further, the functions ψ_ε must satisfy the property

(4.1) $$|\partial^\alpha \psi_\varepsilon(x)| \leq C_m(1 + |x|)^{-m},$$

for every $\alpha \in \mathbb{N}^n$ such that $|\alpha| \leq r$, every $\varepsilon \in E$ and every $m \in \mathbb{N}$.

We shall describe an abstract version of this problem. Then we shall give an algorithm, due to K. Gröchenig, which solves the abstract version. Only then do we return to the explicit construction of wavelets.

We start with the function $2^{n/2}\phi(2x) \in V_1$ whose orbit under the action of the discrete abelian group $\Gamma_1 = \frac{1}{2}\mathbb{Z}^n$ is an orthonormal basis of V_1. At the abstract level, all we retain is the unitary action U_γ : $l^2(\Gamma_1) \to l^2(\Gamma_1)$ of a discrete abelian group Γ_1, operating by translation on the canonical Hilbert space $H = l^2(\Gamma_1)$.

Next, we consider the inclusion $V_0 \subset V_1$ as well as the action of $\mathbb{Z}^n$ on V_0. The function ϕ becomes a vector v_0 in $H = l^2(\Gamma_1)$ whose orbit under the action of a subgroup $\Gamma_0 \subset \Gamma_1$ ($\Gamma_0 = \mathbb{Z}^n$ in the concrete case) is an orthonormal sequence $U_\gamma(v_0)$, $\gamma \in \Gamma_0$. The index of the subgroup Γ_0 in Γ_1 is finite and will be denoted by q in the abstract setting ($q = 2^n$ in the concrete case).

In fact, the problem we are trying to solve is a version of the "incomplete basis" theorem.

We want to construct $q - 1$ vectors $v_1, \ldots, v_{q-1}$ in H such that the set $\{U_\gamma(v_j) : 0 \le j < q, \quad \gamma \in \Gamma_0\}$ is a Hilbert basis of H. That is, we are trying to add $q - 1$ similar orbits to the orbit of v_0 under the action of Γ_0, so that the q orbits form an orthonormal *basis* of H, the first orbit being only an orthonormal *sequence*.

To do this, we decompose the action of Γ_0 on H into irreducible components. Let $\gamma_0, \gamma_1, \ldots, \gamma_{q-1}$ be the q residues modulo Γ_0 of the elements of Γ_1, with $\gamma_0 = 0$. Then the cosets $\Gamma_0 + \gamma_j$, $0 \le j < q$, form a partition of Γ_1. We decompose H, according to the partition, into $H_0 \oplus H_1 \oplus \cdots \oplus H_{q-1}$, where the elements of H_j are the sequences in $l^2(\Gamma_1)$ concentrated on $\Gamma_0 + \gamma_j$.

We finally reduce to a situation where all the H_j are identified with H_0 and, by the same token, the unitary action of Γ_0 on H_j becomes the action by translation of Γ_0 on $l^2(\Gamma_0)$. The elements of H become column vectors with q components, and the problem has taken a new form.

We consider a discrete abelian group Γ and an integer $q \ge 2$. H denotes the product space $(l^2(\Gamma))^q$ and $U_\gamma : H \to H$ the unitary action by translation of Γ on H.

Let $a_1 \in H$ be a (column) vector $(a_{1,1}(\gamma), \ldots, a_{1,q}(\gamma))$, the orbit of which (under the action of Γ) is an orthonormal sequence. We need to find $q - 1$ vectors $a_2, \ldots, a_q$ in H such that the sequence of vectors $U_\gamma(a_j)$, $1 \le j \le q$, $\gamma \in \Gamma$, is an orthonormal basis of H.

We let G denote the dual group of Γ, that is, G is the compact abelian group of homomorphisms $\chi : \Gamma \to \mathbb{T}$ of Γ into the multiplicative group of complex numbers of modulus 1. Such a homomorphism is denoted by $\gamma(x)$, $\gamma \in \Gamma$, $x \in G$.

With the components $a_{j,k}(\gamma)$, $1 \le j, k \le q$, of the vectors $a_1, \ldots, a_q$, we associate the functions

$$A_{j,k}(x) = \sum_{\gamma \in \Gamma} a_{j,k}(\gamma)\gamma(x).$$

With this notation, we have

Proposition 1. *The sequence of vectors $U_\gamma(a_j)$, $1 \le j \le q$, $\gamma \in \Gamma$, is an orthonormal basis of H if and only if, for each $x \in G$, $(A_{j,k}(x))_{1 \le j,k \le q}$ is a unitary matrix.*

In fact we shall prove a more precise result: it is enough for the vectors

$U_\gamma(a_j)$, $\gamma \in \Gamma$, $1 \le j \le q$, to be orthonormal in order for them to form an orthonormal basis of H.

Let us write down the orthonormality conditions. Putting $z \cdot w = z_1 w_1 + \cdots + z_q w_q$, for $z, w \in \mathbb{C}^q$, we get

$$(4.2) \qquad \sum_{y \in \Gamma} a_{j_1}(y - \gamma_1) \cdot \bar{a}_{j_2}(y - \gamma_2) = 0,$$

unless $j_1 = j_2$ and $\gamma_1 = \gamma_2$, in which case we get 1. Writing $A_j(x) = \sum_{\gamma \in \Gamma} a_j(\gamma)\gamma(x)$, we find that the Fourier series of $A_{j_1}(x) \cdot \bar{A}_{j_2}(x)$ is $\sum_{\gamma \in \Gamma} b_{(j_1,j_2)}(\gamma)\gamma(x)$, where $b_{(j_1,j_2)}(\gamma) = \sum_{y \in \Gamma} a_{j_1}(\gamma + y) \cdot \bar{a}_{j_2}(y)$. Thus (4.2) exactly expresses the orthogonality of the column vectors of the matrix $((A_{j,k}(x)))_{1 \le j,k \le q}$, for every $x \in G$.

The same calculation shows that the length of each column vector is 1, since the $U_\gamma(a_j)$, $\gamma \in \Gamma$, form an orthonormal sequence.

We now prove the result in the other direction. We already know that the vectors $U_\gamma(a_j)$, $\gamma \in \Gamma$, $1 \le j \le q$, form an orthonormal sequence. Suppose that this sequence is not complete, and take $b \in H$ to be a vector orthogonal to all the $U_\gamma(a_j)$. We thus have $\sum_{\gamma \in \Gamma} a_j(\gamma - \gamma_0) \cdot \bar{b}(\gamma) = 0$ for each $\gamma_0 \in \Gamma$. Passing to the corresponding Fourier series, we find that, for each $x \in G$, $B(x) = \sum_{\gamma \in \Gamma} b(\gamma)\gamma(x)$ is orthogonal to each column vector of the matrix $(A_{j,k}(x))_{1 \le j,k \le q}$. But these vectors form an orthonormal basis of $\mathbb{C}^q$. Hence $B(x) = 0$ almost everywhere and all the $b(\gamma)$ are zero.

We are thus confronted by the problem of finding the remaining $q - 1$ columns of a unitary matrix, the first column of which is given. It is known that this problem encounters topological obstructions, except when $q = 2, 4$ or 8.

Following Gröchenig's idea, we shall see that the topological obstructions can, nonetheless, be avoided since the group G is the n-dimensional torus, whereas $q = 2^n$. The regularity of the coefficients $A_{j,k}(x)$ implies that, as x describes G, the image of the first vector is a compact set of measure zero in the unit sphere of $\mathbb{R}^q$. As we shall see, this fact allows us to complete the unitary matrix of Proposition 1.

But we shall first make Proposition 1 more explicit by returning to the language of the wavelet construction. Then we shall give several examples where Gröchenig's algorithm is unnecessary. Only then shall we describe that algorithm.

We start with the function $\phi \in V_0$ which was constructed in Chapter 2. $m_0(\xi)$ is defined as the $2\pi\mathbb{Z}^n$-periodic function such that $\hat{\phi}(2\xi) = m_0(\xi)\hat{\phi}(\xi)$. This identity is equivalent to

$$2^{-n}\phi(x/2) = \sum_{k \in \mathbb{Z}^n} \alpha_k \phi(x + k).$$

We thus have

$$\alpha_k = 2^{-n} \int \phi(x/2)\bar{\phi}(x+k)\, dx$$

and it follows that $\alpha_k = O(|k|)^{-m}$ for each m. Hence the function $m_0(\xi)$ is infinitely differentiable.

Using $m_0(\xi)$ and its Fourier series $\sum \alpha_k e^{ik\cdot\xi}$, we form the matrix of Proposition 1. The decomposition of the series is obtained from the partition of $\mathbb{Z}^n$ into the classes $2\mathbb{Z}^n + \varepsilon$, $\varepsilon \in \{0,1\}^n = R$. This gives

$$m_0(\xi) = \sum_{\eta \in R} e^{i\eta\cdot\xi} m_{(0,\eta)}(2\xi),$$

where the functions $m_{(0,\eta)}$ are also infinitely differentiable and $2\pi\mathbb{Z}^n$-periodic.

The wavelets ψ_ε, $\varepsilon \in \{0,1\}^n \backslash \{(0,\dots,0)\}$, are defined by

$$\hat{\psi}_\varepsilon(2\xi) = m_\varepsilon(\xi)\hat{\phi}(\xi).$$

The functions $m_\varepsilon(\xi)$ are infinitely differentiable $2\pi\mathbb{Z}^n$-periodic functions, at least if the ψ_ε decrease rapidly at infinity. Then we decompose each of the functions $m_\varepsilon(\xi)$ as $\sum_{\eta \in R} e^{i\eta\cdot\xi} m_{(\varepsilon,\eta)}(2\xi)$.

Setting $E = R \backslash \{(0,\dots,0)\}$, we can rephrase the statement of Proposition 1 to give

Corollary 1. *The functions $\phi(x-k)$ and $\psi_\varepsilon(x-k)$, $\varepsilon \in E$, $k \in \mathbb{Z}^n$, form an orthonormal basis of V_1 if and only if the matrix*

(4.3) $$U(\xi) = 2^{n/2}((m_{(\varepsilon,\eta)}(\xi)))_{(\varepsilon,\eta) \in R \times R}$$

is unitary.

Corollary 1 can be stated in a slightly different form by using harmonic analysis on the finite group $(\mathbb{Z}/2\mathbb{Z})^n$.

We take as our starting point the observation that

(4.4) $$m_{(\varepsilon,\eta)}(\xi) = 2^{-n} \sum_{r \in R} e^{-i\pi r\cdot\eta} m_\varepsilon(\xi + r\pi).$$

We then interpret the right-hand side of (4.4) as a Fourier series, that is, we freeze the variable ξ and consider the sequence $m_\varepsilon(\xi + r\pi)$ as the sequence of Fourier coefficients of the function $m_{(\varepsilon,\eta)}(\xi)$ regarded as a function of η. It comes to the same thing, moreover, to remark that the matrix $2^{-n/2}((e^{-i\pi r\cdot\eta}))_{(r,\eta) \in R \times R}$ is unitary.

The product of two unitary matrices is unitary (otherwise apply Plancherel's identity for the group $(\mathbb{Z}/2\mathbb{Z})^n$). We thus get

Corollary 2. *The functions $\phi(x-k)$ and $\psi_\varepsilon(x-k)$, $\varepsilon \in E$, $k \in \mathbb{Z}^n$, form an orthonormal basis of V_1 if and only if the matrix*

(4.5) $$\tilde{U}(\xi) = ((m_\varepsilon(\xi + \eta\pi)))_{(\varepsilon,\eta) \in R \times R}$$

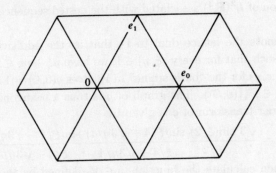

Figure 3

is unitary.

In dimension 1, we are led to take $m_1(\xi) = e^{-i\xi}\bar{m}_0(\xi + \pi)$ as we have, indeed, done. In dimension 2, if the function $m_0(\xi) = m_0(\xi_1, \xi_2)$ is real, a possible choice for the other three columns of $\tilde{U}(\xi)$ is

$$m_1(\xi) = e^{-i\xi_2}m_0(\xi_1 + \pi, \xi_2),$$
$$m_2(\xi) = e^{-i(\xi_1+\xi_2)}m_0(\xi_1, \xi_2 + \pi),$$

and finally

$$m_3(\xi) = e^{-i\xi_2}m_0(\xi_1 + \pi, \xi_2 + \pi).$$

5 Computing two-dimensional wavelets

We intend to use Corollary 2 of the preceding section to construct two wavelet bases in two dimensions which are not reducible to tensor products of one-dimensional wavelets.

In both cases, we start with a multiresolution approximation leading to a real-valued function $m_0(\xi) = m_0(\xi_1, \xi_2)$. We can then use the remark made just after Corollary 2 to construct our wavelets.

The first example is associated with the hexagonal tiling of the plane. We subdivide the hexagons into equilateral triangles. The group acting on V_0 by translation is not $\mathbb{Z}^2$ any longer, but the lattice generated by the vectors $e_0 = (1, 0)$ and $e_1 = (1/2, \sqrt{3}/2)$.

The lattice defines a division of the plane into equilateral triangles. So we let V_0 denote the closed subspace of $L^2(\mathbb{R}^2)$ consisting of the functions which are continuous on the plane and whose restriction to each equilateral triangle of the tiling is an affine function. We then define V_j by the condition that $f(x, y) \in V_0$ if and only if $f(2^j x, 2^j y) \in V_j$.

It is easy to check that the sequence V_j, $j \in \mathbb{Z}$, is a multiresolution approximation of $L^2(\mathbb{R}^2)$ associated with the nested sequence of lattices $\Gamma_j = 2^{-j}\Gamma$.

Let Γ^* denote the lattice dual to Γ, that is, the additive group of $(u,v) \in \mathbb{R}^2$ such that for every $(\xi,\eta) \in \Gamma$ we have $u\xi + v\eta \in 2\pi\mathbb{Z}$.

Writing $g(x,y)$ for the "basic spline" in V_0 gives $g(0,0) = 1$ and $g(\xi,\eta) = 0$, if $(\xi,\eta) \in \Gamma \setminus \{(0,0)\}$. The graph of g is thus a hexagonal pyramid and the Fourier transform of g is given by

$$\hat{g}(\xi,\eta) = \frac{\sqrt{3}}{2} \frac{\sin(\xi/2)}{\xi/2} \frac{\sin(\xi/4 + \sqrt{3}\eta/4)}{\xi/4 + \sqrt{3}\eta/4} \frac{\sin(\xi/4 - \sqrt{3}\eta/4)}{\xi/4 - \sqrt{3}\eta/4}.$$

We can then calculate the function $\phi \in V_0$ defined by the condition that the family $\phi(x-\xi, y-\eta)$, as (ξ,η) describes Γ, forms an orthonormal basis of V_0 and by the further conditions that ϕ is real-valued, even and has integral equal to 1.

We get

$$\hat{\phi}(\xi,\eta) = \hat{g}(\xi,\eta)(\omega(\xi,\eta))^{-1},$$

where

$$\omega(\xi,\eta) = \left(\sum_{(u,v)\in\Gamma^*} |\hat{g}(\xi + u, \eta + v)|^2 \right)^{1/2}.$$

The wavelets can then be computed using Corollary 2 of Proposition 1 (in Section 4). These wavelets decrease exponentially at infinity.

In this *first example*, the basic wavelets ψ_1, ψ_2 and ψ_3 have singularities which are supported by a grid of lines (which form the sides of the triangles whose vertices are the points of $\frac{1}{2}\Gamma$). The same is true of every construction of two-dimensional wavelets by the tensor product method, if we start with one-dimensional splines of order r.

Our *second example* is due to P.G. Lemarié ([165]) and has a different geometric character. The singularities of the functions in V_0 are concentrated on the vertices of the lattice $\mathbb{Z}^2$. Further, this example is related to a decomposition arising from the Laplacian. More precisely, we define V_0 as the closed subspace of $L^2(\mathbb{R}^2)$ consisting of those functions f in the Sobolev space $H^2(\mathbb{R}^2)$ such that $\Delta^2 f$, in the sense of distributions, is a sum of Dirac measures at the points $(k,l) \in \mathbb{Z}^2$.

Then the hypoellipticity of Δ^2 guarantees that f is infinitely differentiable (and even real-analytic) except, possibly, on $\mathbb{Z}^2$.

We approach the definition of V_0 by using a variational method which has the merit of showing that the sampling operator $T_0 : V_0 \to l^2(\mathbb{Z}^2)$ defined by $T_0 f(k,l) = f(k,l)$ is an isomorphism between V_0 and $l^2(\mathbb{Z}^2)$.

Recall that the Sobolev space $H^2(\mathbb{R}^2)$ consists of functions which are continuous (and even in the Hölder class of exponent γ for $0 < \gamma < 1$) so that the restriction of a function of $H^2(\mathbb{R}^2)$ to a point makes sense.

We now state the variational problem.

Starting from a sequence $\alpha(k, l)$ belonging to $l^2(\mathbb{Z}^2)$, we want to minimize $\int_{\mathbb{R}^2} |\Delta f|^2 \, dx$ (where $dx = dx_1 \, dx_2$) over the set V of functions $f \in H^2(\mathbb{R}^2)$ such that $f(k, l) = \alpha(k, l)$. The set V is a closed affine variety in $H^2(\mathbb{R}^2)$.

The difficulty of this variational problem is that the functional we are trying to minimize, namely $J(f) = \int_{\mathbb{R}^2} |\Delta f|^2 \, dx$ is not coercive on the Hilbert space $H^2(\mathbb{R}^2)$. That is, if C is a finite constant, the condition $J(f) \leq C$ does not define a bounded subset of $H^2(\mathbb{R}^2)$: there is no control on $\|f\|_2^2$. But we finesse this difficulty by replacing $J(f)$ by

$$K(f) = \int_{\mathbb{R}^2} |\Delta f|^2 \, dx + \sum_k \sum_l |f(k, l)|^2 \,.$$

It is clear that it comes to the same thing to minimize $J(f)$ or $K(f)$ on the affine variety V defined by $f(k, l) = \alpha(k, l)$. But the functional $K(f)$ is coercive on $H^2(\mathbb{R}^2)$, because of the following lemma.

Lemma 3. *The norms* $(K(f))^{1/2}$ *and* $(\|\Delta f\|_2^2 + \|f\|_2^2)^{1/2}$, *on the Sobolev space* $H^2(\mathbb{R}^2)$, *are equivalent.*

We first show that $(K(f))^{1/2} \leq C(\|\Delta f\|_2 + \|f\|_2)$. To this end, we shall use the following observation, which we evoked in Chapter 2.

Lemma 4. *Let* $\chi \in \mathcal{D}(\mathbb{R}^2)$. *For each real number* s, *there exists a constant* $C(s, \chi)$ *such that, for each function* $f \in H^s(\mathbb{R}^2)$, *we have*

$$\sum_{k \in \mathbb{Z}^2} \|f(x)\chi(x - k)\|_{H^s}^2 \leq C(s, \chi)\|f\|_{H^s}^2 \,.$$

If, in addition, $s > 1$ and if $\chi(0) = 1$, Sobolev's embedding theorem gives

$$|f(k)| \leq \|f(x)\chi(x - k)\|_\infty \leq C_s \|f(x)\chi(x - k)\|_{H^s}$$

and we thus get

$$\left(\sum_{k \in \mathbb{Z}^2} |f(k)|^2 \right)^{1/2} \leq C'_s \|f\|_{H^s} \,,$$

as stated.

Conversely, we want to show that

$$(5.1) \qquad \|f\|_2 \leq C \left(\|\Delta f\|_2 + \left(\sum_{k \in \mathbb{Z}^2} |f(k)|^2 \right)^{1/2} \right) \,.$$

To this end, we split $f \in H^2(\mathbb{R}^2)$ into $g + h$, where $\hat{g}(\xi, \eta) = \hat{f}(\xi, \eta)$, for $-\pi \leq \xi, \eta \leq \pi$, and $\hat{g}(\xi, \eta) = 0$ otherwise.

Plancherel's formula then gives $\|\Delta h\|_2 \leq \|\Delta f\|_2$. Since $\xi^2 + \eta^2 \geq \pi^2$ on the support of $\hat{h}$, we get

$$\|\Delta h\|_2^2 = \frac{1}{4\pi^2} \iint (\xi^2 + \eta^2)^2 |\hat{h}(\xi,\eta)|^2 \, d\xi \, d\eta$$

$$\geq \frac{\pi^2}{4} \iint |\hat{h}(\xi,\eta)|^2 \, d\xi \, d\eta = \pi^4 \|h\|_2^2 \,.$$

Putting this together gives $\pi^2 \|h\|_2 \leq \|\Delta h\|_2 \leq \|\Delta f\|_2$.

We are left with the task of controlling the L^2 norm of g. Now, on applying the first part of the proof of Lemma 3 to h, we get

$$(5.2) \qquad \left(\sum_k \sum_l |h(k,l)|^2 \right)^{1/2} \leq C \|\Delta f\|_2 \leq C(K(f))^{1/2}.$$

But, by the definition of K, $\left(\sum_k \sum_l |f(k,l)|^2 \right)^{1/2} \leq (K(f))^{1/2}$. As a result,

$$(5.3) \qquad \left(\sum_k \sum_l |g(k,l)|^2 \right)^{1/2} \leq (C+1)(K(f))^{1/2}.$$

The $g(k,l)$ are the Fourier coefficients of the $2\pi\mathbf{Z}^2$-periodic function whose restriction to $[-\pi,\pi)^2$ is exactly $\hat{g}$. Thus, the left-hand side of (5.3) is equal to $(1/2\pi)(\iint |\hat{g}(\xi,\eta)|^2 \, d\xi \, d\eta)^{1/2} = \|g\|_2$, which completes the proof of Lemma 3.

Lemma 3 shows that our variational problem has a unique solution: we are now going to study its properties. To do this, we let $N \subset H^2(\mathbb{R}^2)$ denote the subspace consisting of functions which are zero on the lattice $\mathbf{Z}^2$ and we assume $J(f + \lambda u) \geq J(f)$, for each $u \in N$ and every scalar λ. We get $\int \Delta f \, \Delta u \, dx = 0$. The integral is absolutely convergent, since Δf and Δu belong to $L^2(\mathbb{R}^2)$. Now we integrate by parts, noting that the Sobolev space $H^{-2}(\mathbb{R}^2)$ is the dual of $H^2(\mathbb{R}^2)$, where the duality is the usual one between distributions and test functions. Letting $\langle \cdot, \cdot \rangle$ denote the bilinear form which implements the duality, we thus get $\langle \Delta^2 f, u \rangle = 0$ for every function $u \in N$ and, in particular, for every function $u \in \mathcal{D}(\mathbb{R}^2)$ which is zero on $\mathbf{Z}^2$. It follows that $\Delta^2 f$ is a sum of Dirac measures placed at points of the lattice $\mathbf{Z}^2$.

These considerations lead us to define V_0 (the starting point for the multiresolution approximation) by

$$(5.4) \quad f \in V_0 \iff f \in H^2(\mathbb{R}^2) \text{ and } \Delta^2 f = \sum_k \sum_l \beta(k,l)\delta_{(k,l)} \,,$$

where $\delta_{(k,l)}$ is Dirac measure at the point $(k,l) \in \mathbf{Z}^2$.

Summarizing the results of the variational method gives:

Proposition 2. *For each square-summable sequence $\alpha(k, l)$ there exists a unique function $f \in V_0$ such that $f(k, l) = \alpha(k, l)$, $(k, l) \in \mathbf{Z}^2$. The function minimizes $\iint |\Delta f|^2 \, dx$ on the set M of functions in $H^2(\mathbb{R}^2)$ whose restriction to $\mathbf{Z}^2$ is the given sequence.*

We then define the spaces V_j, $j \in \mathbf{Z}$ by the standard procedure of a dyadic change of scale and find that f belongs to V_j if and only if f is in $H^2(\mathbb{R}^2)$ and $\Delta^2 f$ is a sum of Dirac measures based on points of the lattice $2^{-j}\mathbf{Z}^2$. As the lattices are nested, so are the spaces V_j.

To show that $\bigcap V_j = \{0\}$, note that f cannot belong to the intersection unless $\Delta^2 f = c\delta_{(0,0)}$. Taking Fourier transforms gives $\hat{f}(\xi, \eta) = c(\xi^2 + \eta^2)^{-2}$, which is not in $L^2(\mathbb{R}^2)$ unless $c = 0$.

We shall prove that the union of the spaces V_j, $j \in \mathbf{Z}$, is dense in $L^2(\mathbb{R}^2)$, by computing the orthogonal projection $E_j : L^2(\mathbb{R}^2) \to V_j$ and verifying that E_j tends to the identity in the strong operator topology.

To compute E_j, it is enough to find E_0 and then change scale. So we need to construct the canonical orthonormal basis $\phi(x - k, y - l)$, $(k, l) \in \mathbf{Z}^2$, of V_0. Let us start by determining $\mathcal{F}V_0$, the subspace of $L^2(\mathbb{R}^2)$ consisting of the Fourier transforms of functions of V_0. If $f \in V_0$, we have $\Delta^2 f \in H^{-2}(\mathbb{R}^2)$ and $\Delta^2 f = \sum_k \sum_l \beta(k, l)\delta_{(k,l)}$. Lemma 4 gives $\sum\sum |\beta(k, l)|^2 < \infty$.

We shall show that this condition is not enough to characterize $\Delta^2 V_0$. Indeed, let

$$q(\xi, \eta) = \sum \sum \beta(k, l) \exp(-ik\xi - il\eta).$$

Now f belongs to V_0 if and only if f belongs to $H^2(\mathbb{R}^2)$ and $\Delta^2 f = \sum\sum \beta(k, l)\delta_{(k,l)}$. Applying the Fourier transform gives

$$(1 + \xi^2 + \eta^2)\hat{f}(\xi, \eta) \in L^2(\mathbb{R}^2)$$

and

$$(\xi^2 + \eta^2)\hat{f}(\xi, \eta) = q(\xi, \eta).$$

From these two conditions it follows that, in a certain sense, $q(\xi, \eta)$ vanishes at $(0, 0)$. This last condition should be expressed with care, because $q(\xi, \eta)$ is a square-summable $2\pi\mathbf{Z}^2$-periodic function whose value at a point is not determined. To avoid this difficulty, we define

$$(5.5) \qquad \omega(\xi, \eta) = \left(\sum_k \sum_l ((\xi - 2k\pi)^2 + (\eta - 2l\pi)^2)^{-2} \right)^{-1}.$$

Since $q(\xi, \eta)$ is $2\pi\mathbf{Z}^2$-periodic and $q(\xi, \eta)(\xi^2 + \eta^2)^{-2}$ is still square-summable in a neighbourhood of $(0, 0)$, the same remains true for

$$q(\xi, \eta)((\xi - 2k\pi)^2 + (\eta - 2l\pi)^2)^{-2},$$

in a neighbourhood of $(2k\pi, 2l\pi)$, and for

$$q(\xi, \eta) \sum_k \sum_l ((\xi - 2k\pi)^2 + (\eta - 2l\pi)^2)^{-2} = m(\xi, \eta).$$

The conclusion is that $q(\xi, \eta) = \omega(\xi, \eta) m(\xi, \eta)$, where $m(\xi, \eta)$ is $2\pi \mathbf{Z}^2$-periodic and square-summable.

Observe that the function $\omega(\xi, \eta)$ is real-analytic and $2\pi \mathbf{Z}^2$-periodic and, further, that $\omega(\xi, \eta) \sim (\xi^2 + \eta^2)^2$ in a neighbourhood of $(0, 0)$. The product $\omega(\xi, \eta)(\xi^2 + \eta^2)^{-2}$ is also a real-analytic function, and we are in a position to state the characterization of $\mathcal{F} V_0$.

Proposition 3. *A function belongs to $\mathcal{F} V_0$ if and only it is of the form $(\xi^2 + \eta^2)^{-2} \omega(\xi, \eta) m(\xi, \eta)$, where $m(\xi, \eta)$ is $2\pi \mathbf{Z}^2$-periodic and square-summable.*

This statement amounts to saying that $g(x - k, y - l)$, $(k, l) \in \mathbf{Z}^2$, is a Riesz basis of V_0, where g is defined by $\hat{g}(\xi, \eta) = (\xi^2 + \eta^2)^{-2} \omega(\xi, \eta)$. It is then easy to construct the orthonormal basis $\phi(x - k, y - l)$ by following the recipes of Chapter 2. We compute

$$\omega_2(\xi, \eta) = \left(\sum_k \sum_l |\hat{g}(\xi + 2k\pi, \eta + 2l\pi)|^2 \right)^{1/2}$$

$$= \omega(\xi, \eta) \left(\sum_k \sum_l ((\xi + 2k\pi)^2 + (\eta + 2l\pi)^2)^{-4} \right)^{1/2}.$$

The singularities are only apparent, because $\omega(\xi, \eta)$ vanishes at the points $(2k\pi, 2l\pi)$, compensating for the singularities of the double series. In fact, $c \leq \omega_2(\xi, \eta) \leq c'$ for two constants $c' > c > 0$.

We then define ϕ by

$$\hat{\phi}(\xi, \eta) = \frac{\hat{g}(\xi, \eta)}{\omega_2(\xi, \eta)} = \frac{1}{(\xi^2 + \eta^2)^2} \frac{\omega(\xi, \eta)}{\omega_2(\xi, \eta)}.$$

The function ϕ is even, real and decreases exponentially at infinity. We have $\sum_k \sum_l \phi(x - k, y - l) = 1$ because $\hat{\phi}(0, 0) = 1$ and $\hat{\phi}(2k\pi, 2l\pi) = 0$ for $(k, l) \neq (0, 0)$. We then let $E((x, y), (x', y')) = \sum_k \sum_l \phi(x - k, y - l)\phi(x', y')$ denote the kernel of the orthogonal projection E_0 of $L^2(\mathbf{R}^2)$ onto V_0 (the variables x' and y' are those with respect to which the kernel is integrated when the operator E_0 acts on a function).

We get

$$|E((x, y), (x', y'))| \leq C \exp(-\gamma(|x - x'| + |y - y'|)),$$

for a certain constant $\gamma > 0$. Finally, $\iint E((x, y), (x', y')) \, dx' \, dy' = 1$. Lemma 13 of Chapter 2 applies, and the operators E_j converge strongly to the identity. Hence the union of the V_j, $j \in \mathbf{Z}$, is dense in $L^2(\mathbf{R}^2)$.

The wavelets are easily constructed by applying the remark which follows Corollary 2 of Proposition 1 in Section 4. In this situation, the function m_0 is not only real but also positive or zero.

6 The general existence theorem for wavelet bases

The object of this section is to generalize to dimension $n \geq 2$ what we have already done—in Section 2—for dimension 1.

Let, therefore, V_j, $j \in \mathbb{Z}$, be an r-regular multiresolution approximation of $L^2(\mathbb{R}^n)$. As usual, we let W_j denote the orthogonal complement of V_j in V_{j+1}. Then we have the following existence theorem.

Theorem 2. *There exist $q = 2^n - 1$ functions $\psi_1, \ldots, \psi_q$ in V_1 having the following two properties:*

$$(6.1) \qquad |\partial^\alpha \psi_l(x)| \leq C_N (1 + |x|)^{-N}$$

for every multi-index $\alpha \in \mathbb{N}^n$ such that $|\alpha| \leq r$, each $x \in \mathbb{R}^n$ and every $N \geq 1$;

$(6.2) \quad \{\psi_l(x - k), 1 \leq l \leq q, k \in \mathbb{Z}^n\}$ *is an orthonormal basis of W_0.*

Corollary 1. *The functions $2^{nj/2}\psi_l(2^j x - k)$, $1 \leq l \leq q$, $k \in \mathbb{Z}^n$, $j \in \mathbb{Z}$, form an orthonormal basis of $L^2(\mathbb{R}^n)$.*

Just as in one dimension, we start with the r-regular function ϕ, which decreases rapidly (in the sense of (6.1)) and is such that $\phi(x - k)$, $k \in \mathbb{Z}^n$ is an orthonormal basis of V_0. The wavelets will be obtained by manipulating ϕ in a manner that we shall now describe.

We begin by forming the function $m_0(\xi)$ which is $2\pi\mathbb{Z}^n$-periodic, infinitely differentiable, and which satisfies $\hat{\phi}(2\xi) = m_0(\xi)\hat{\phi}(\xi)$. This identity is a consequence of the equation $2^{-n}\phi(x/2) = \sum \alpha_k \phi(x - k)$, where $\alpha_k = 2^{-n} \int \phi(x/2)\bar{\phi}(x - k)\,dx$ is a rapidly decreasing sequence. This gives $m_0(\xi) = \sum \alpha_k e^{-ik\cdot\xi}$ and thus the properties of $m_0(\xi)$ stated above follow.

We are trying to construct wavelets ψ_l, $1 \leq l \leq q$, by attempting to find q $2\pi\mathbb{Z}^n$-periodic functions $m_l(\xi)$ which will give the ψ_l by the rule $\hat{\psi}_l(2\xi) = m_l(\xi)\hat{\phi}(\xi)$. This identity is a result of ψ_l's belonging to V_1.

Instead of using the subindices l we shall use sequences $\varepsilon \in E$, where $E = \{0,1\}^n \setminus \{(0,\ldots,0)\}$. So from now on we shall write ψ_ε, m_ε, etc., rather than ψ_l, m_l, etc.

Corollary 1 of Proposition 1 now leads us to write

$$(6.3) \qquad m_\varepsilon(\xi) = \sum_{\eta \in \{0,1\}^n} e^{i\eta\cdot\xi} m_{\varepsilon,\eta}(2\xi),$$

so that, instead of looking for the functions m_ε, $\varepsilon \in E$, we shall try to find the $m_{\varepsilon,\eta}$.

The final change of notation is to agree that the function $m_0(\xi)$, associated with the identity $\hat{\phi}(2\xi) = m_0(\xi)\hat{\phi}(\xi)$, should now be written as $m_\varepsilon(\xi)$, where $\varepsilon = (0, \ldots, 0)$.

Now we can form the matrix $U(\xi)$ whose columns are the vectors $m_{\varepsilon,\eta}(\xi)$, $\eta \in R = \{0,1\}^n$. The necessary and sufficient condition for the functions $\phi(x - k)$, $k \in \mathbb{Z}^n$, $\psi_\varepsilon(x - k)$, $\varepsilon \in E$, $k \in \mathbb{Z}^n$, to form an orthonormal basis of V_1 is that the matrix $2^{n/2}U(\xi)$ should be unitary.

We are left with the task of constructing that unitary matrix, given its first column vector. The only constraint on the remaining $2^n - 1$ column vectors is that they should be infinitely differentiable, $2\pi\mathbb{Z}^n$-periodic functions in ξ. In particular, the matrix $U(\xi)$ must be a continuous function of ξ. We thus come up against the following well-known topological obstruction: except in dimensions 2, 4 and 8, the unit sphere has no non-trivial, everywhere continuous, tangent vector-field.

So if we are going to be able to construct our wavelets, it will be due to special circumstances.

In fact, the vector $m_{0,\eta}(\xi)$ does not describe all of the unit sphere of $\mathbb{R}^{2^n}$ (or of $\mathbb{C}^{2^n}$ for $m_0(\xi)$ complex-valued). This is because $m_0(\xi)$ and hence the $m_{0,\eta}(\xi)$, $\eta \in R$ are infinitely differentiable $2\pi\mathbb{Z}^n$-periodic functions of $\xi \in \mathbb{R}^n$.

Now if $F : \mathbb{R}^p \to \mathbb{R}^q$ is Lipschitz and if $q > p$, then $F(\mathbb{R}^p)$ has measure zero. It follows that the image K of $\mathbb{R}^n$ by the mapping $\xi \mapsto (m_{0,\eta}(\xi))_{\eta \in R}$ is a compact subset of measure zero of the unit sphere $S \subset \mathbb{R}^{2^n}$ (or $\mathbb{C}^{2^n}$ in the complex case). Indeed, if $n \geq 2$, the dimension q of our sphere is $2^n - 1$, whereas $p = n$, so we certainly have $q > p$.

Now, in a neighbourhood of K in S, there exists a real-analytic determination of an orthonormal frame. We shall recall the proof of this last remark, which concludes the proof of Theorem 2.

Lemma 5. *Let $q \geq 2$ and let S be the unit sphere of $\mathbb{R}^{q+1}$ or $\mathbb{C}^{q+1}$. Let $K \subset S$ be a compact, proper subset of S. Then there exists an open neighbourhood U of K and, for every $x \in U$, q vectors $v_1(x), \ldots, v_q(x)$ with the following properties:*

(6.4) $v_1(x), \ldots, v_q(x)$ *are real-analytic functions on U;*

(6.5)   *for each $x \in U \cap S$, $(x, v_1(x), \ldots, v_q(x))$ is an orthonormal basis of $\mathbb{R}^{q+1}$ (or $\mathbb{C}^{q+1}$).*

By a rotation of axes, if necessary, we can suppose $(0, \ldots, 0, 1) \notin K$. We shall prove the complex case, beginning by constructing vectors $w_1(x), \ldots, w_q(x)$ which are real-analytic functions of x and such that

$(x, w_1(x), \ldots, w_q(x))$ is a basis of $\mathbb{C}^{q+1}$ in a neighbourhood of K. Then it will only remain to orthonormalize the sequence $(x, w_1(x), \ldots, w_q(x))$ by the usual Gram-Schmidt process—which doesn't change the first vector when it lies in S.

To construct $w_1(x), \ldots, w_q(x)$, we start with the following observation. The determinant

$$\begin{vmatrix} z_1 & \alpha & 0 & \cdots & 0 \\ z_2 & 0 & \alpha & \cdots & 0 \\ \vdots & \vdots & \vdots & \ddots & \cdot \\ z_q & 0 & 0 & \cdots & \alpha \\ z_{q+1} & -\bar{z}_1 & -\bar{z}_2 & \cdots & -\bar{z}_q \end{vmatrix}$$

has value

$$(-1)^{q+1}\{\alpha^{q-1}(|z_1|^2 + \cdots + |z_q|^2) - \alpha^q z_{q+1}\}.$$

Suppose now that the vector $z = (z_1, \ldots, z_{q+1})$ belongs to K. If $\alpha > 0$ is sufficiently small, the determinant does not vanish for any $z \in K$.

Here's why. The determinant vanishes precisely when $\alpha z_q = |z_1|^2 + \cdots + |z_q|^2$. Since $|z_{q+1}| \leq 1$, it follows that $|z_1|^2 + \cdots + |z_q|^2 \leq \alpha$. Moreover, $z_{q+1} > 0$ and, finally, as α tends to 0, the point $z \in K$ where the determinant vanishes tends to the point $(0, \ldots, 0, 1)$ which is not in K.

For a sufficiently small value of $\alpha > 0$, the columns of the determinant give the vectors $w_1(z), \ldots, w_q(z)$ we want.

In the case where S is the unit sphere in $\mathbb{R}^{q+1}$, Lemma 5 can be proved as follows. We suppose that the compact set K does not contain the vector $(0, \ldots, 0, -1)$ and we set $y = \frac{1}{2}(x+e)$, $x \in U$, where e denotes the vector $(0, \ldots, 0, 1)$. If σ_y denotes the operation of reflection in y, which interchanges e and x, the orthonormal frame we need is the image by σ_y of the canonical frame $(e_1, e_2, \ldots, e_q)$, $e_{q+1} = e$.

The construction of multidimensional wavelets which we have just given is due to Gröchenig ([122]).

7 Cancellation of wavelets

The cancellation (in French, "oscillations") of a wavelet is measured by the number of moments which are zero.

We return to the statement of Theorem 2 and let ψ_l, $1 \leq l \leq q$, denote functions with properties (6.1) and (6.2). Then we get

Lemma 6. $|\hat{\phi}(\xi)|^2 + |\hat{\psi}_1(\xi)|^2 + \cdots + |\hat{\psi}_q(\xi)|^2 = |\hat{\phi}(\xi/2)|^2.$

To see this, we start by replacing ξ by 2ξ and the problem is reduced to verifying that $|m_0(\xi)|^2 + |m_1(\xi)|^2 + \cdots + |m_q(\xi)|^2 = 1$. Changing the subindices leads to the identity $\sum_{\varepsilon \in R} |m_\varepsilon(\xi)|^2 = 1$. It is then enough to use Corollary 2 of Proposition 1 of Section 4.

We are in a position to prove the following important result:

Proposition 4. *The wavelets $\psi_1, \ldots, \psi_q$ which have properties (6.1) and (6.2) of Theorem 2 necessarily satisfy*

$$(7.1) \qquad \int x^\alpha \psi_l(x)\, dx = 0,$$

for $|\alpha| \le r$ and $1 \le l \le q$.

The integer r is specified by the hypothesis that V_j is an r-regular multiresolution approximation of $L^2(\mathbb{R}^n)$.

To prove (7.1), we use the fact that $|\hat{\phi}(\xi)|^2 = 1 + O(|\xi|^{2r+2})$ as ξ tends to 0 (Chapter 2, Proposition 7). Together with Lemma 6, this gives $\hat{\psi}_l(\xi) = O(|\xi|^{r+1})$, for $1 \le l \le q$. But the ψ_l decrease rapidly at infinity and their transforms are thus infinitely differentiable. Hence, $\partial^\alpha \hat{\psi}_l(\xi) = 0$ for $\xi = 0$ and $|\alpha| \le r$. Working back, we get (7.1).

8 Wavelets with compact support

We return to the case of dimension 1 to prove the following theorem:

Theorem 3. *For each integer $r \ge 1$, there exists a multiresolution approximation V_j of $L^2(\mathbb{R})$ which is r-regular and such that the associated functions ϕ and ψ have compact support.*

If $r = 0$, there is nothing to prove, because the Haar functions suffice. In that case, ϕ is the characteristic function of the interval $[0, 1]$ and

$$\psi(x) = \begin{cases} 1, & \text{on } [0, 1/2); \\ -1, & \text{on } [1/2, 1); \\ 0, & \text{elsewhere.} \end{cases}$$

As the regularity r increases, so do the supports of ϕ and ψ. I. Daubechies ([88]), to whom Theorem 3 is due, has also put a precise value on the rate of growth: there exists a constant C such that, for each $r \ge 1$, the supports of ϕ and ψ are contained in $[-Cr, Cr]$. Moreover, the linear dependence on r of the size of the supports of ϕ and ψ is optimal: there exists another constant $c > 0$ such that the diameters of the supports of ϕ and ψ are greater than cr.

Finally, the functions ϕ and ψ, that we are going to construct, are real-valued.

Before proving Theorem 3, we shall make some remarks about it. As usual, we denote by $\mathcal{I}$ the collection of all dyadic intervals $I = [k2^{-j}, (k+1)2^{-j})$, where $k, j \in \mathbb{Z}$. For $I \in \mathcal{I}$, we let ψ_I denote the function

$2^{j/2}\psi(2^j x - k)$. Let m be an integer large enough for the supports of ϕ and ψ to be contained in $[1/2 - m/2, 1/2 + m/2]$. Then the supports of ψ_I and of ϕ_I are contained in the interval mI whose centre is that of I and whose length is m times that of I.

The fundamental identity

$$(8.1) \qquad f(x) = \sum_{I \in \mathcal{I}} (f, \psi_I) \psi_I$$

taken together with the more precise identity

$$(8.2) \qquad f(x) = \sum_{|I| \le 2^{-N}} (f, \psi_I) \psi_I + \sum_{|I| = 2^{-N}} (f, \phi_I) \phi_I,$$

for any $N \ge 1$, provides an answer to the analyst's dream of finding a Fourier analysis which is local on every scale.

Let us examine (8.2) from this point of view. The right-hand side of (8.2) consists of a series of "humps" ϕ_I, such that $|I| = 2^{-N}$, and of a double series, involving those ψ_I satisfying $|I| \le 2^{-N}$.

The series in ϕ_I can be interpreted as an "image", on the scale 2^{-N} of the function f which is being analysed. This image is obtained by calculating the "smoothed means" (f, ϕ_I) of f on each interval mI whose length is $m2^{-N}$.

Once this "schematic image" has been arrived at, we want to reconstruct $f(x)$ in all its complexity. To do this, we have to "touch in the details" on finer and finer scales of dimension $m2^{-j}$, $j \ge N$. These "touchings in" all have the same structure and take place going from left to right in steps of 2^{-j}. Clearly, the "touchings in" are the terms $(f, \psi_I) \psi_I$ in the double series of (8.2).

After these remarks, we come to the proof of Theorem 3. What we are going to give is not Daubechies' original proof. Instead, we rely on Mallat's programme of constructing an r-regular multiresolution approximation by beginning with a function $m_0(\xi)$ which is infinitely differentiable, 2π-periodic and satisfies $|m_0(\xi)|^2 + |m_0(\xi + \pi)|^2 = 1$, with $m_0(0) = 1$. Naturally, we construct V_0 by looking for a $\hat{\phi}$ such that $\phi(x - k)$, $k \in \mathbf{Z}$, is an orthonormal sequence. We then let V_0 be the closed subspace generated by the sequence.

The connection between $\hat{\phi}$ and m_0 is given by $\hat{\phi}(2\xi) = m_0(\xi)\hat{\phi}(\xi)$. This means that we should set $\hat{\phi}(\xi) = \prod_1^\infty m_0(2^{-j}\xi)$. The convergence of the infinite product results from the equation $m_0(0) = 1$ and the regularity of m_0.

If, then, $\phi(x - k)$, $k \in \mathbf{Z}$, is an orthonormal sequence—and this is the first subtlety—we have constructed the space V_0 and we have

$$\mathcal{F} V_0 = \{ m(\xi)\hat{\phi}(\xi) : m(\xi) \in L^2[0, 2\pi] \text{ and is } 2\pi\text{-periodic} \}.$$

V_{-1} is defined as the set of functions $f(x/2)$, where $f \in V_0$, and we have
$\mathcal{F}V_{-1} = \{m(2\xi)\hat{\phi}(2\xi) : m(\xi) \in L^2[0, 2\pi]\}$. This implies that $\mathcal{F}V_{-1} \subset$
$\mathcal{F}V_0$, and we can build a multiresolution approximation.

There are two "delicate moments" in this programme:

(a) it is not always true that $\phi(x-k)$, $k \in \mathbf{Z}$, is an orthonormal sequence
    when ϕ is constructed from $m_0(\xi)$ as indicated above;

(b) it is not obvious that the multiresolution approximation is r-regular.

What is needed is to find conditions on $m_0(\xi)$ sufficient to deal with
these two tricky points. On the other hand, it is very easy to choose
$m_0(\xi)$ so that ϕ has compact support. We know that $m_0(\xi) = \sum \alpha_k e^{ik\xi}$,
where $\alpha_k = \frac{1}{2} \int \phi(x/2)\bar{\phi}(x+k)\,dx$. If ϕ has compact support, $\alpha_k =$
0, for $|k| > T$, and $m_0(\xi)$ is a finite trigonometric sum. Conversely,
if $m_0(\xi)$ is a finite trigonometric sum, then the object ϕ (*a priori* a
tempered distribution) defined by $\hat{\phi}(\xi) = \prod_1^\infty m_0(2^{-j}\xi)$, is a distribution
of compact support. Indeed, let σ denote the finite sum of point masses
α_k at the points $-k \in \mathbf{Z}$. Next, let σ_j denote the finite sum of the same
point masses, but placed at the points $-k2^{-j}$. Then $\phi = \sigma_1 * \sigma_2 * \cdots$ and
$\sigma_1 * \cdots * \sigma_j$ converges to ϕ in the sense of distributions. This is because
the partial products $m_0(\xi/2) \cdots m_0(\xi/2^j) = \pi_j(\xi)$ satisfy $|\pi_j(\xi)| \leq 1$
and converge uniformly on compacta to $\hat{\phi}$, and thus converge to $\hat{\phi}$ in the
sense of distributions. If the support of σ_1 is contained in the interval
$[-T/2, T/2]$, then that of $\sigma_1 * \cdots * \sigma_j$ is contained in the interval $[-T, T]$
and, on passing to the limit, the support of ϕ is contained in the same
interval $[-T, T]$.

Once ϕ has compact support and once $m_0(\xi)$ is a trigonometric polynomial, the wavelet ψ, constructed according to the recipe of Theorem 1,
has compact support. Indeed, the formula $\hat{\psi}(2\xi) = e^{-i\xi}\bar{m}_0(\xi + \pi)\hat{\phi}(\xi)$
means that $\frac{1}{2}\psi(x/2)$ is the convolution of ϕ with a finite sum of Dirac
measures situated at points of $[1 - T, 1 + T]$. It follows that the support
of ψ is compact and contained in the interval $[1/2 - T, 1/2 + T]$.

The first lemma below shows us that we still have $\|\phi\|_2 \leq 1$ (in fact,
$\|\phi\|_2 < 1$ can occur) when we define ϕ with the help of m_0.

Lemma 7. *Suppose that $m_0(\xi)$ is a 2π-periodic C^1-function such that
$|m_0(\xi)|^2 + |m_0(\xi + \pi)|^2 = 1$ and $m_0(0) = 1$. Define ϕ by the infinite
product*

$$\hat{\phi}(\xi) = m_0\left(\frac{\xi}{2}\right) m_0\left(\frac{\xi}{4}\right) \cdots$$

Then $\|\phi\|_2 \leq 1$.

To see this, it is enough to show that, for each $N \geq 1$, we have

$$I_N = \int_{-\pi 2^N}^{\pi 2^N} |\pi_N(\xi)|^2 \, dx = 2\pi \, ,$$

where $\pi_N(\xi) = m_0(\xi/2) \cdots m_0(\xi/2^N)$. Indeed, $|m_0(\xi)| \leq 1$ gives $|\hat{\phi}(\xi)| \leq |\pi_N(\xi)|$. Then

$$\int_{-\pi 2^N}^{\pi 2^N} |\hat{\phi}(\xi)|^2 \, dx \leq 2\pi$$

leads to the desired conclusion.

To calculate I_N, we note that $\pi_N(\xi)$ is $2\pi 2^N$-periodic in ξ. So we get

$$I_N = \int_0^{2\pi 2^N} |\pi_N(\xi)|^2 \, d\xi = \int_0^{\pi 2^N} + \int_{\pi 2^N}^{2\pi 2^N} \, .$$

In the last integral, we change the variable to $u = \xi - \pi 2^N$. By the periodicity of $m_0(\xi)$, all the terms of the product $\pi_N(\xi)$, except the last, remain the same. We thus get

$$I_N = \int_0^{\pi 2^N} |\pi_{N-1}(\xi)|^2 \left(|m_0(\xi 2^{-N})|^2 + |m_0(\xi 2^{-N} + \pi)|^2 \right) d\xi$$

$$= \int_0^{\pi 2^N} |\pi_{N-1}(\xi)|^2 \, d\xi$$

$$= I_{N-1} \, .$$

Working down, $I_N = \int_{-2\pi}^{2\pi} |m_0(\xi/2)|^2 \, d\xi = 2 \int_{-\pi}^{\pi} |m_0(\xi)|^2 \, d\xi = 2\pi$, since $|m_0(\xi)|^2 + |m_0(\xi + \pi)|^2 = 1$.

Consider the following two examples. If $m_0(\xi) = \cos(\xi/2)e^{i\xi/2} = (1 + e^{i\xi})/2$, then $\hat{\phi}(\xi) = e^{i\xi/2}\sin(\xi/2)/(\xi/2)$ and ϕ is the characteristic function of $[-1,0]$. This example would lead to the multiresolution approximation producing the Haar system as the orthonormal basis of wavelets.

If, however, $m_0(\xi) = (1 + e^{3i\xi})/2$, the hypotheses of Lemma 7 are still satisfied, but this choice of $m_0(\xi)$ produces $\phi(x) = 1/3$ on $[-3,0]$ and $\phi(x) = 0$ elsewhere. The sequence $\phi(x - k)$, $k \in \mathbb{Z}$, is thus not orthonormal in $L^2(\mathbb{R})$.

The next lemma explains this counter-example.

Lemma 8. *In addition to the hypotheses of Lemma 7, suppose that $m_0(\xi)$ is a finite trigonometric sum $\sum_{|k| \leq T} \alpha_k e^{ik\xi}$ and that $m_0(\xi) \neq 0$ on $[-\pi/2, \pi/2]$. Then $\phi(x - k)$, $k \in \mathbb{Z}$, is an othonormal sequence.*

Because of the preceding lemmas, we know that $\phi \in L^2(\mathbb{R})$ and has compact support and, thus, that $\hat{\phi}$ belongs to all the Sobolev spaces.

The function

$$\alpha(\xi) = \sum_{-\infty}^{\infty} |\hat{\phi}(\xi + 2k\pi)|^2$$

is therefore infinitely differentiable (Chapter 2, Lemma 7).

Showing that $\phi(x - k)$, $k \in \mathbb{Z}$ is an orthonormal sequence amounts to verifying that $\alpha = 1$. In fact, it is enough to establish that $\alpha(\xi)$ is constant. Here's why: we have $\hat{\phi}(0) = 1$ (by the infinite product expansion) and, also, $\hat{\phi}(2l\pi) = 0$ for $l \in \mathbb{Z}$, $l \neq 0$. To see this, we observe that $m_0(\pi) = 0$, because $|m_0(\xi)|^2 + |m_0(\xi + \pi)|^2 = 1$, and thus $\hat{\phi}(\xi) = 0$, as long as $2^{-j}\xi = (2m + 1)\pi$, for some $j \geq 1$. This is the case if $\xi = 2l\pi$, $l \in \mathbb{Z}$, $l \neq 0$.

We shall follow P. Tchamitchian ([227]) in showing that $\alpha(\xi)$ is constant. To simplify the notation, we set $g(\xi) = |m_0(\xi)|^2$. Then $g(\xi) + g(\xi + \pi) = 1$, $0 \leq g(\xi) \leq 1$ and $|\hat{\phi}(\xi)|^2 = g(\xi/2)g(\xi/4)\cdots$. From the relations above, we retain the following consequence:

Lemma 9.

(8.3) $$\alpha(2\xi) = \alpha(\xi)g(\xi) + \alpha(\xi + \pi)g(\xi + \pi)$$

identically in $\xi \in \mathbb{R}$.

This identity expresses $\alpha(2\xi)$ as a barycentre of $\alpha(\xi)$ and $\alpha(\xi + \pi)$ with coefficients $g(\xi)$ and $g(\xi + \pi)$. This and the periodicity of the continuous functions α and g will be enough to prove that $\alpha(\xi)$ is constant.

To prove Lemma 9, we return—for the last time—to the definition of $\alpha(\xi)$. We have $\hat{\phi}(2\xi) = m_0(\xi)\hat{\phi}(\xi)$ and

$$\alpha(2\xi) = \sum_{-\infty}^{\infty} |\hat{\phi}(2\xi + 2k\pi)|^2 = \sum_{-\infty}^{\infty} |m_0(\xi + k\pi)|^2 |\hat{\phi}(\xi + k\pi)|^2$$

$$= |m_0(\xi)|^2 \sum_{-\infty}^{\infty} |\hat{\phi}(\xi + 2k\pi)|^2 + |m_0(\xi + \pi)|^2 \sum_{-\infty}^{\infty} |\hat{\phi}(\xi + (2k + 1)\pi)|^2$$

$$= \alpha(\xi)g(\xi) + \alpha(\xi + \pi)g(\xi + \pi) \,.$$

In proving that $\alpha(\xi)$ is constant, we can forget all the above, except for (8.3) and the conditions $0 \leq g(\xi) \leq 1$, $g(\xi) + g(\xi + \pi) = 1$ and $g(\xi) > 0$ on $[-\pi/2, \pi/2]$. Let m denote the minimum and M the maximum of the continuous function $\alpha(\xi)$ on the interval $[-\pi, \pi]$; we intend to show that $M = m = \alpha(0)$.

The minimum m is attained at (at least) one point $\xi_0 \in [-\pi, \pi]$ We apply (8.3) with $\xi_1 = \xi_0/2 \in [-\pi/2, \pi/2]$. By hypothesis , we have $g(\xi_1) > 0$, so either $0 < g(\xi_1) < 1$ or $g(\xi_1) = 1$. In the first case we get $\alpha(\xi_1) = m = \alpha(\xi_1 + \pi)$; in the second we get $\alpha(\xi_1) = m$ but no information about the value of $\alpha(\xi_1 + \pi)$. Repeating the process, we get $\alpha(\xi_k) = m$, for $\xi_k = 2^{-k}\xi_0$, and, passing to the limit, it follows that $\alpha(0) = m$. Similarly, $\alpha(0) = M$. The function $\alpha(\xi)$ thus equals a constant and, as we showed above, the constant is 1.

To complete step (a) of our programme, we suppose that we already know how to choose $m_0(\xi)$ so that ϕ belongs to $L^\infty(\mathbb{R})$. Then Lemma 8 gives a multiresolution approximation V_j, $j \in \mathbb{Z}$, of $L^2(\mathbb{R})$ and the multiresolution approximation will be r-regular if, in addition, all the derivatives of ϕ up to order r belong to $L^\infty(\mathbb{R})$.

Let V_0 denote the closed subspace of $L^2(\mathbb{R})$ generated by the $\phi(x-k)$, $k \in \mathbb{Z}$. The V_j are defined, as usual, by the condition that $f(x) \in V_0$ is equivalent to $f(2^j x) \in V_j$.

To check that the subspaces V_j form a nested sequence, it is enough to show that $V_{-1} \subset V_0$. But the sequence $2^{-1/2}\phi(x/2 - k)$, $k \in \mathbb{Z}$, is an orthonormal basis of V_{-1}, and, by the construction, we have $\frac{1}{2}\phi(x/2) = \sum \alpha_k \phi(x+k)$, since $\hat{\phi}(2\xi) = m_0(\xi)\hat{\phi}(\xi)$.

Lastly, we show that $\bigcap_{-\infty}^{\infty} V_j = \{0\}$ and that $\bigcup_{-\infty}^{\infty} V_j$ is dense in $L^2(\mathbb{R})$. To do this, let $E_j : L^2(\mathbb{R}) \to V_j$ be the orthogonal projection operator. We shall show that $\lim_{j \to -\infty} E_j = 0$ and $\lim_{j \to +\infty} E_j = I$ in the strong operator topology. The kernel $E_j(x,y)$ of E_j is given by

$$(8.4) \qquad E_j(x,y) = 2^j E(2^j x, 2^j y)$$

and we have

$$(8.5) \qquad E(x,y) = \sum \phi(x-k)\bar{\phi}(y-k).$$

If we assume that $\phi \in L^\infty(\mathbb{R})$, we get $|E(x,y)| \le C$ and $E(x,y) = 0$ for $|x-y| \ge 2T$. Moreover, $\int E(x,y)\,dy = \sum \phi(x-k) = 1$, because $\hat{\phi}(0) = 1$ and $\hat{\phi}(2k\pi) = 0$, if $k \in \mathbb{Z}$, $k \ne 0$.

This all means that Lemma 13 of Chapter 2 applies and $E_j \to I$ as $j \to +\infty$. Similarly, it is clear that $E_j \to 0$ as $j \to -\infty$.

We now come to part (b) of our programme. We shall see that there are real difficulties here. The construction which we give owes much to discussions with J.-P. Kahane and Y. Katznelson, and we take this opportunity of thanking them.

The first simplifying remark is that we can forget $m_0(\xi)$ and just use $g(\xi)$. Indeed, we shall study the regularity of ϕ by trying to get $\hat{\phi}(\xi) = O(|\xi|^{-s})$ at infinity, which is clearly equivalent to $|\hat{\phi}(\xi)|^2 = O(|\xi|^{-2s})$. But the infinite product which gives $|\hat{\phi}(\xi)|^2$ only involves the function g. On the other hand, if $s > r + 1$, the condition on $\hat{\phi}$ implies that ϕ is of class C^r. The possibility of returning from g to $m_0(\xi)$ is given by the following lemma, due to F. Riesz.

Lemma 10. *Let* $g(\xi) = \sum_{-T}^{T} \gamma_k e^{ik\xi}$ *be a trigonometric polynomial which is positive or zero on the real line. Then there exists a trigonometric polynomial* $h(\xi) = \sum_{0}^{T} \alpha_k e^{ik\xi}$ *such that* $|h(\xi)|^2 = g(\xi)$. *Moreover, if the coefficients* γ_k *are real,* $h(\xi)$ *can be chosen so that the* α_k *are also real.*

In our application of Lemma 10, the coefficients γ_k will indeed be real numbers. The same will hold for the α_k, which will lead to real-valued ϕ and ψ.

For example, $g(\xi) = 1 + \cos\xi$ gives $h(\xi) = (1 + e^{i\xi})/\sqrt{2}$. The proof of Lemma 10 is left to the reader.

Finally, we show how $g(\xi)$ can be constructed. For some integer k (which will later be found to depend on s and hence on r) we let c_k denote the strictly positive constant defined by $c_k \int_0^\pi (\sin t)^{2k+1}\, dt = 1$. We thus have $c_k = O(\sqrt{k})$ as k tends to infinity. We then put $g(\xi) = 1 - c_k \int_0^\xi (\sin t)^{2k+1}\, dt$. We get $0 \le g(\xi) \le 1$ and $g(\xi) + g(\xi + \pi) = 1$ "for free". Further, $g(\xi) > 0$ on the open interval $(-\pi, \pi)$. All that remains to do is to study the behaviour at infinity of the product $G(\xi) = g(\xi/2)g(\xi/4)\cdots g(\xi/2^j)\cdots$.

(8.6) $0 \le g(t) \le 1$ for all $t \in \mathbb{R}$, $g(t)$ is continuous on $\mathbb{R}$ and $g(t+2\pi) = g(t)$;

(8.7) $g(t) \le C\sqrt{k}\left(\dfrac{3}{4}\right)^k$ for $\dfrac{2\pi}{3} \le t \le \dfrac{4\pi}{3}$ and $k \ge 1$;

(8.8) $g(t) \le |t - \pi|^{2k+2}$ for $0 \le t \le 2\pi$ and $k \ge k_0$.

To establish the latter inequalities, we can restrict ourselves to the case $0 \le t \le \pi$, since the function g is even and 2π-periodic. We then write

$$g(t) = c_k \int_t^\pi (\sin u)^{2k+1}\, du$$

and use the following observations. We have $\sin u \le \pi - u$ when $0 \le u \le \pi$, which gives an upper bound on the integral and allows us to obtain a more precise form of (8.8): in fact the constant $c_k = O(\sqrt{k})$ is absorbed by the coefficient $1/(2k + 2)$ arising from the integration.

As far as (8.7) is concerned, it is enough to replace $\sin u$ by the upper bound $\sqrt{3}/2$ when $2\pi/3 \le u \le \pi$.

It is worthwhile observing that (8.7) is more precise than (8.8), except when t is sufficiently close to π.

The next lemma will enable us to conclude our investigation into the regularity of ϕ and ψ. Its proof will follow the proof of Theorem 3.

Lemma 11. *Let* δ *and* C *be constants satisfying* $0 < \delta < 1$ *and* $C \ge 1$.

Then there exists a constant $\alpha = \alpha(\delta, C) > 0$ *such that the following property is satisfied:*

if $f(t)$ *is a function which is continuous on the whole real line, is periodic of period 1, takes its values in* $[0,1]$ *and satisfies, in addition,*

(8.9) $$0 \le f(t) \le \delta \qquad \text{if} \qquad \frac{1}{3} \le t \le \frac{2}{3}$$

and

(8.10) $$f(t) \le C|t - \frac{1}{2}|,$$

then, for each integer $j \ge 2$,

(8.11) $$\sup_{1/4 \le t \le 1/2} f(t)f(2t)\cdots f(2^{j-1}t) \le 2^{-\alpha j}.$$

In the examples we have in mind, $f(0) = 1$ so that (8.11) could not be true without some restriction on t. It is possible to establish (8.11) with the condition $a \le t \le b$, for arbitrary $0 < a < b < 1$, the exponent α then depending on the choice of a and b. But this improvement does not help us, so we restrict ourselves to the formulation of (8.9) that has been given.

For the time being, we shall assume Lemma 11 and return to the problem of the regularity of the wavelets ϕ and ψ. We shall show that, for each integer $N \ge 1$, there exists a choice of k such that, setting

$$g(x) = 1 - c_k \int_0^\xi (\sin t)^{2k+1} \, dt,$$

the infinite product

$$G(\xi) = g(\xi/2)g(\xi/4)\cdots g(\xi/2^j)\cdots$$

is $O(|\xi|^{-N})$ at infinity. This will be sufficient to show that ϕ is of class $C^{(N/2)-2}$, since $G(\xi) = |\hat{\phi}(\xi)|^2$.

To study the behaviour of $G(\xi)$ as ξ tends to infinity, we partition $|\xi| \ge \pi$ using the dyadic pairs of intervals $\pi 2^j \le |\xi| \le 2\pi 2^j$. In each pair we make the change of variable $|\xi| = 2\pi 2^{j+1}t$. This always puts us into the interval $1/4 \le t \le 1/2$. Further, on putting $f(t) = (g(2\pi t))^{1/(2k+2)}$ and remarking that $0 \le g(\xi) \le 1$, we get

$$G(\xi) \le g(\xi/2)\cdots g(\xi/2^{j+1}) = g(2\pi t)\cdots g(2\pi 2^{j-1}t)$$
$$= \left(f(t)f(2t)\cdots f(2^{j-1}t)\right)^{2k+2}.$$

Let us study the properties of the function $f(t)$. Now, if $k \ge k_0$, we have $C\sqrt{k}(\frac{3}{4})^k \le (\frac{4}{5})^{k+1} = \delta^{2k+2}$, where $\delta = 2/\sqrt{5} < 1$. Then (8.8) gives $g(2\pi t) \le (2\pi)^{2k+2}|t - \frac{1}{2}|^{2k+2}$. This shows that $f(t) = (g(2\pi t))^{1/(2k+2)}$ satisfies the hypotheses of Lemma 11, with $\delta = 2/\sqrt{5}$ and $C = 2\pi$. So, assuming the conclusion of Lemma 11, we get

$$G(\xi) \le 2^{-2\alpha(k+1)j} = O(|\xi|^{-N}) \qquad \text{where } N = 2\alpha(k+1).$$

We can now show that the growth of the diameters of the supports of the functions ϕ and ψ is (at most) linear with respect to the regularity of ϕ. Indeed, we already know that the support of ϕ is contained in $[-T, T)$ when $m_0(\xi) = \sum_{-T}^{T} \alpha_k e^{ik\xi}$. Lemma 10 shows that T is the degree of the trigonometric polynomial $g(\xi)$ The actual choice that we have made gives $T = 2k + 1$. With this choice, we have obtained $\hat{\phi}(\xi) = O(|\xi|^{-\alpha(k+1)})$ with ϕ of class $C^{\alpha k + \beta}$, for every $\beta < \alpha - 1$. The growth of the diameter of the support of ϕ is indeed what we have claimed.

We now also know that linear growth is optimal (verbal communication of I. Daubechies).

Now for the proof of Lemma 11. We begin by replacing $f(t)$ by $h(t) = f(t)f(2t)$. If we can show that $h(t)h(2t)\cdots h(2^{j-1}t) \le 2^{-\alpha j}$, it will follow that $f(t)f(2t)\cdots f(2^{j-1}t) \le 2^{-\alpha j/2}$. The point of this transformation is to preserve $0 \le h(t) \le 1$, as well as $h(t) \le C|t - \frac{1}{2}|$, while extending the range of validity of (8.9) to $1/6 \le t \le 5/6$.

If $t = 1/2$, then $h(t) = 0$, so the estimate (8.11) is trivially satisfied. We can therefore suppose that $1/4 \le t < 1/2$ and express it as a binary number:

$$t = 0.01\alpha_3\alpha_4\cdots = 0 + \frac{0}{2} + \frac{1}{4} + \frac{\alpha_3}{8} + \frac{\alpha_4}{16} + \cdots,$$

where $\alpha_j = 0$ or 1.

Since $h(t)$ is periodic, of period 1, we get

$$h(2^q t) = h(0.\alpha_{q+1}\alpha_{q+2}\ldots).$$

This suggests that, if we want to find an upper bound for the product $h(t)\ldots h(2^{j-1}t)$, we should investigate the partition modulo 1 of the sequence $2^q t$, which amounts to analysing the sequence $\alpha_1, \alpha_2 \ldots$ of 0s and 1s occurring in the binary expansion of t. Let $F \subset \{0, 1, \ldots, j-1\}$ be the set of those q such that $\alpha_{q+1} \ne \alpha_{q+2}$. If $q \in F$, we have either $h(2^q t) = h(0.10\alpha_{q+3}\ldots)$ or $h(2^q t) = h(0.01\alpha_{q+3}\ldots)$. In the first case, $\frac{1}{2} \le 0.10\alpha_{q+3}\ldots < \frac{3}{4}$, whereas in the second, $\frac{1}{4} \le 0.01\alpha_{q+3}\ldots < \frac{1}{2}$.

In both cases, $h(2^q t) \le \delta$, which is the point of the definition of F. This suggests a first upper bound of δ^N for the product $h(t)\cdots h(2^{j-1}t)$, where N is the cardinal of F. This upper bound is poor if N is much smaller than j. However, we have some further information, derived from (8.10). To use it, we divide the sequence of 0s and 1s, $0.01\alpha_3\alpha_4\ldots$, into intervals I_m consisting exclusively of 1s and J_m containing only 0s, according to the following example:

$$0.0\underbrace{11}_{I_1}\overbrace{00}^{J_1}\underbrace{111}_{I_2}\overbrace{0}^{J_2}\underbrace{1}_{I_3}\overbrace{00}^{J_3}\underbrace{1111}_{I_4}\overbrace{0}^{J_4}\cdots.$$

The boundaries between the intervals I_m and J_m are marked precisely by the elements $q \in F$. If $q \in F$, we let l_q denote the length of the new interval of 0s or 1s "announced" by q—if $q \in F$ and $q + 1 \in I_m$, then l_q is the length of J_m (which could be ∞); if $q \in F$ and $q + 1$ in J_m, then l_q is the length of I_{m+1}. Clearly, $\sum_{q \in F} l_q \geq j$. Let C be the constant of (8.10) and let A be an integer such that $2^A > C/\delta$. If $l_q \geq A$, we use (8.10) to get an upper bound for $h(2^q t)$. Indeed, $2^q t$ is written—modulo 1—as

$$0.1 \underbrace{00 \ldots 0}_{l_q \text{ terms}} 1 \ldots \qquad \text{or} \qquad 0.0 \underbrace{11 \ldots 1}_{l_q \text{ terms}} 0 \ldots$$

and, in both cases, l_q measures how close $2^q t$ is to $1/2$ (modulo 1). We thus get $h(2^q t) \leq C 2^{-l_q}$. We can then combine this with the uniform estimation $h(2^q t) \leq \delta$, used when $1 \leq l_q < A$, by writing, in both cases, $h(2^q t) \leq 2^{-\alpha l_q}$, where $\alpha > 0$ is sufficiently small—the conditions are $\delta \leq 2^{-\alpha A}$ and $C \leq 2^{A(1-\alpha)}$, which are compatible, because $\delta < 1$ and $C < 2^A$.

On multiplying these inequalities term by term, we get

$$h(t)h(2t) \ldots h(2^{j-1}t) \leq \prod_{q \in F} h(2^q t) \leq 2^{-\alpha j}$$

and Lemma 11 has been proved.

Before finishing this section, we shall give Mallat's point of view about the preceding construction, a point of view that applies equally to the dyadic interpolation algorithms of Chapter 2. Keeping the notation of Section 13 of Chapter 2, we try to turn the inclusion $\mathbf{Z} \subset \mathbf{Z}/2$ into an isometric injection

$$(8.12) \qquad\qquad J : l^2(\mathbf{Z}) \to l^2(\mathbf{Z}/2)$$

such that

$$(8.13) \qquad\qquad J \tau_k = \tau_k J \qquad \text{for all } k \in \mathbf{Z}.$$

The third condition allows us to pass from the discrete to the continuous in what follows. We try to approximate $\mathbf{R}$ by $\varepsilon \mathbf{Z}$ when ε tends to 0 and, at the same time, approximate the continuous functions $f(x)$ of the real variable x by their sampling $f(\varepsilon k)$, $k \in \mathbf{Z}$, on $\varepsilon \mathbf{Z}$. We require the operator J to be compatible with this approximation.

More precisely, we let $r \geq 1$ be an integer and say that J is r-regular if, for each function $f(x)$ belonging to the Schwartz class $S(\mathbf{R})$, we can find a family f_ε, $0 < \varepsilon \leq 1$, forming a bounded subset of $S(\mathbf{R})$, such that the image by J of the sequence $f(\varepsilon k)$, $k \in \mathbf{Z}$, is $cf(\varepsilon k) + \varepsilon^r f_\varepsilon(\varepsilon k)$, where c is a constant. Since J is a partial isometry, we get $|c| = 1/\sqrt{2}$, and we shall normalize J so that $c = 1/\sqrt{2}$. Let $\varepsilon_k \in l^2(\mathbf{Z})$ denote the sequence defined by $\varepsilon_k(l) = 0$, if $k \neq l$, and $\varepsilon_k(k) = 1$. If the image by J of ε_0 is the sum $\sum_{-\infty}^{\infty} w(j)\varepsilon_{j/2}$, then that of ε_k is $\sum_{-\infty}^{\infty} w(j)\varepsilon_{(j/2)+k}$, and the image of a sequence $(x_k)_{-\infty}^{\infty} \in l^2(\mathbb{Z})$ is $\sum_{-\infty}^{\infty} \sum_{-\infty}^{\infty} w(j)x_k\varepsilon_{(j/2)+k}$.

To simplify this description, we associate with the sequence $(x_k) \in l^2(\mathbb{Z})$ the function $F(x) = \sum_{-\infty}^{\infty} x_k e^{ikx}$ whose Fourier coefficients are the x_k. Similarly, we associate with the sequence $(y_{k/2}) \in l^2(\mathbb{Z}/2)$ the function $G(x) = \sum_{-\infty}^{\infty} y_{k/2} e^{ikx/2}$. If $(y_{k/2}) = J(x_k)$, then $G(x) = W(x/2)F(x)$, where $W(x) = \sum_{-\infty}^{\infty} w(j)e^{ijx}$. We note that $F(x)$ is 2π-periodic and belongs to $L^2[0, 2\pi]$, that the same is true for $W(x)$ and, lastly, that $G(x)$ is 4π-periodic and belongs to $L^2[0, 4\pi]$. The condition that J is a partial isometry is equivalent to

(8.14) $|W(x)|^2 + |W(x + \pi)|^2 = 2$.

We shall restrict ourselves to the case where $w(j)$, $j \in \mathbb{Z}$, decreases rapidly as $|j|$ tends to infinity. Then J is r-regular if and only if

(8.15) $|W(0)| = \sqrt{2}$, $W'(0) = \cdots = W^{(r-1)}(0) = 0$.

We now consider all the inclusions

$$\mathbb{Z} \subset \mathbb{Z}/2 \subset \mathbb{Z}/4 \subset \cdots \subset \mathbb{Z}/2^q \subset \mathbb{Z}/2^{q+1} \subset \cdots \subset \mathbb{R}.$$

We intend to associate isometric injections $J_q : l^2(\mathbb{Z}/2^q) \to l^2(\mathbb{Z}/2^{q+1})$ with those inclusions, and then to pass to the limit, looking at $\mathbb{R}$ as the limit of the sub-groups $\mathbb{Z}/2^q$.

We define J_q from J (which will be denoted by J_0 in what follows), by a simple change of scale, but without altering the sequence $w(j)$, $j \in \mathbb{Z}$.

Starting from the sequence ε_k, $k \in \mathbb{Z}$, which is the canonical *orthonormal basis* of $l^2(\mathbb{Z})$, we apply J_0, then $J_1, \ldots$, then J_{q-1}. This gives an *orthonormal sequence* $\phi_q(x - k)$, $x \in \mathbb{Z}/2^q$, in $l^2(\mathbb{Z}/2^q)$, and Mallat's heuristic idea is that, in a natural sense, this orthonormal sequence will converge to $\phi(x - k)$, $k \in \mathbb{Z}$, as q tends to infinity.

We thus would like to "pass to the limit", thinking of $L^2(\mathbb{R})$ as the "limit" of the Hilbert spaces $l^2(\mathbb{Z}/2^q)$. To do the calculations more conveniently, we let the discrete measure $\mu_q = \sum f(k/2^q)\delta_{k/2^q}$ correspond to the sequence $f_q(x)$, $x \in \mathbb{Z}/2^q$, and we hope that "convergence" of the sequences $f_q \in l^2(\mathbb{Z}/2^q)$ will coincide with weak convergence of the corresponding measures μ_q. This requirement means we have to renormalize the Hilbert spaces $l^2(\mathbb{Z}/2^q)$ by multiplying the natural norm by $2^{-q/2}$. Naturally, the limit of the measures μ_q will be calculated by taking Fourier transforms, defined here by $\mathcal{F}(\mu_q)(\xi) = \int e^{ix\xi} d\mu_q(\xi)$.

Before renormalization, we thus have

(8.16) $\mathcal{F}(\mu_q)(\xi) = W(\xi/2^q)\mathcal{F}(\mu_{q-1})(\xi)$.

This identity clearly gives $\mathcal{F}(\mu_q) = W(\xi/2^q) \cdots W(\xi/2)\mathcal{F}(\mu_0)(\xi)$. The renormalization required for passing to the limit amounts to multiplying

μ_q by $2^{-q/2}$. To achieve this, it is enough to replace $W(\xi)$ systematically by $(1/\sqrt{2})W(\xi)$, that is, by the function we have called $m_0(\xi)$.

Finally, we take Dirac measure at 0, denoted by $\sigma_0 = \delta_0$, as our starting point, and we let σ_q denote the corresponding measures $2^{-q/2}\mu_q$. We then have

(8.17) $$\mathcal{F}\sigma_q(\xi) = m_0(\xi/2)\ldots m_0(\xi/2^q).$$

If J is r-regular and if the constant which is involved in the regularity is $c = 1/\sqrt{2}$, then $m_0(0) = 1$. If $w(j)$, $j \in \mathbf{Z}$, decreases rapidly, then the measures σ_q converge weakly to $\phi(x)\,dx$ and $\phi(x)$ belongs to $L^2(\mathbf{R})$.

Having got this far, we are nonetheless obliged to return to the subtle part of the proof of Daubechies' theorem. In fact, there is no reason to suppose that the sequence $\phi(x-k)$, $k \in \mathbf{Z}$ is orthonormal "in the limit". A counter-example is given by choosing $w(0) = 1/\sqrt{2}$, $w(3) = 1/\sqrt{2}$, and $w(j) = 0$ otherwise.

The heuristic description we have given has thus only allowed us to get as far as the equation $I_N = 2\pi$ of the proof of Lemma 7. But it has an indisputable aesthetic quality, which is why we have chosen to present it here. It also gives a striking geometric description of formula (2.8). Indeed, returning to the notation of Chapter 2 and keeping the heuristic approach above, V_0 is the "limit" of the spaces $J_{q-1}\cdots J_0(l^2(\mathbf{Z}))$, V_1 is the "limit" of the spaces $J_{q-1}\cdots J_1(l^2(\mathbf{Z}))$, and so on.

Thus, to construct the space W_0, which is the orthogonal complement of V_0 in V_1, we start with $W_0^\sharp$, the orthogonal complement of $J_0(l^2(\mathbf{Z})$ in $l^2(\mathbf{Z}/2)$, and then get W_0 as the "limit" of the spaces $J_{q-1}\cdots J_1(W_0^\sharp)$. To construct an orthonormal basis of W_0, of the form $\psi(x-k)$, $k \in \mathbf{Z}$, it is thus enough to take an orthonormal basis $\psi^\sharp(x-k)$, $x \in \mathbf{Z}/2$, $k \in \mathbf{Z}$, of $W_0^\sharp$, to "shift it isometrically" using $J_{q-1}\cdots J_1$, and then proceed to the limit. To construct $\psi^\sharp$, we put $\psi^\sharp(k/2) = (-1)^k\bar{w}(k+1)$, $k \in \mathbf{Z}$, which leads to (2.8).

The same remarks apply in dimension n, on consideration of the inclusions $\mathbf{Z}^n \subset \mathbf{Z}^n/2 \subset \cdots \subset \mathbf{Z}^n/2^j \subset \cdots$, which we try to turn into isometric injections $J_q : l^2(\mathbf{Z}^n/2^q) \to l^2(\mathbf{Z}^n/2^{q+1})$, all derived from the first by a change of scale. The problem is still how to "pass to the limit" in the sequence $J_{q-1}\cdots J_0(l^2(\mathbf{Z}^n))$, as q tends to infinity. But we have only a few results in this direction.

9 Wavelets with compact support in higher dimensions

We do not know any direct method for constructing an r-regular multiresolution approximation V_j, $j \in \mathbf{Z}$, of $L^2(\mathbf{R}^n)$ which produces wavelets of compact support. The difficulty is twofold. Even if we could manage to construct V_j, $j \in \mathbf{Z}$, such that ϕ has compact support, Gröchenig's algorithm would not produce wavelets of compact support as in dimension 1. These difficulties disappear if we are satisfied with constructing multidimensional wavelets by the tensor product method. This gives us wavelets of compact support, which we shall describe explicitly for the convenience of the reader.

Let $\mathcal{Q}$ be the collection of dyadic cubes

$$Q = \{x \in \mathbf{R}^n : 2^j x - k \in [0,1)^n\}, \qquad k \in \mathbf{Z}^n,\ j \in \mathbf{Z},$$

in $\mathbf{R}^n$. Let E denote the set of $2^n - 1$ sequences $(\varepsilon_1, \ldots, \varepsilon_n)$ of 0s and 1s, excluding the sequence $(0, \ldots, 0)$.

For any integer $r \in \mathbf{N}$, we let ϕ and ψ be the functions of compact support and of regularity r, constructed, as in the last section, from an r-regular multiresolution approximation.

We then define, for $\varepsilon \in E$ and $Q \in \mathcal{Q}$,

$$\psi_Q^\varepsilon(x) = 2^{nj/2} \psi^{\varepsilon_1}(2^j x_1 - k_1) \ldots \psi^{\varepsilon_n}(2^j x_n - k_n)$$

with the convention that $\psi^0 = \phi$ and $\psi^1 = \psi$.

If $m \geq 1$ is defined by the condition that the supports of ϕ and ψ are contained in the interval $m[0,1] = [1/2 - m/2, 1/2 + m/2]$, then the supports of the n-dimensional wavelets ψ_Q^ε are contained in the cube mQ, where the cube mQ has the same centre as Q and an edge of mQ is m times as long as one of Q.

The wavelets constructed in this way have the four fundamental properties of a wavelet basis which enable us to use this basis in other spaces in functional analysis as well as in the reference space $L^2(\mathbf{R}^n)$.

The four properties are:

(a) being an orthonormal basis of $L^2(\mathbf{R}^n)$;

(b) localization: support $\psi_q^\varepsilon \subset mQ$;

(c) regularity: $|\partial^\alpha \psi_Q^\varepsilon| \leq C 2^{j|\alpha|} 2^{nj/2}$, for every multi-index $\alpha \in \mathbf{N}$ of order $|\alpha| \leq r$;

(d) cancellation: $\int x^\alpha \psi_Q^\varepsilon(x)\, dx = 0$, when $|\alpha| \leq r$.

The advantage of wavelets with compact support lies in enabling us to decompose arbitrary distributions $S \in \mathcal{D}'(\mathbf{R}^n)$, of order less than r, without worrying about growth at infinity. As we shall see in more detail in the next section, for such a distribution we have

$$(9.1) \quad S = \sum_{Q \in \mathcal{Q}_0} (S, \phi_Q)\phi_Q + \sum_{j \geq 0} S_j \quad \text{where} \quad S_j = \sum_{Q \in \mathcal{Q}_j} \sum_{\varepsilon \in E} (S, \psi_Q^\varepsilon)\psi_Q^\varepsilon.$$

Here, $\mathcal{Q}_j$ is the collection of dyadic cubes of edge 2^{-j} and $(S, u) = \langle S, \bar{u} \rangle$, for each distribution S and test function u (as before, $\langle \cdot, \cdot \rangle$ denotes the bilinear form implementing the duality between distributions and test functions).

Identity (9.1) has the following intuitive meaning. $\sum_{Q\in\mathcal{Q}_0}(S,\phi_Q)\phi_Q$ provides a preliminary sketch of the distribution S. The sketch gives only rough information (on the scale 1) about S. This rough information is essentially given by the means of S about each $k\in\mathbb{Z}^n$, obtained from the integrals $\int S(x)\bar{\phi}(x-k)\,dx$. The means in question are the coefficients c_k, $k\in\mathbb{Z}^n$, which are used to put together the sketch $\sum c_k\phi(x-k)$ of S.

Then follows the process of touching in. As in the theory of martingales, these touchings in must be compatible with the description we have given of the mean values. But that is guaranteed by the orthogonality relations between the ψ_Q, $Q\in\mathcal{Q}_j$, $j\geq 0$ and the ϕ_Q, $Q\in\mathcal{Q}_0$. The touching in of details gets more and more precise and is done on ever finer scales. The infinite series of touchings in is the series $S_0 + S_1 + S_2 + \cdots$ which is added to the sketch of S to reproduce S in all its complexity.

10 Wavelets and spaces of functions and distributions

Throughout this chapter, we have been giving procedures for constructing families of wavelets ψ_λ, $\lambda\in\Lambda$. By hypothesis, these wavelets form an orthonormal basis of $L^2(\mathbb{R}^n)$ so that we can write

$$(10.1) \qquad f(x) = \sum_{\lambda\in\Lambda}(f,\psi_\lambda)\psi_\lambda(x)$$

for every function $f\in L^2(\mathbb{R}^n)$. But if we had wanted to limit ourselves to functional analysis in L^2, the Haar system would have been good enough. The advantage of wavelets compared to the Haar system is twofold: numerically, wavelet coefficients are imperceptible where the function $f(x)$ is regular and are significant only near singularities of f. In other words, wavelet series of standard functions, which have only isolated singularities (or surfaces of singularities...) are "sparse" series. This property is the more significant because a greater number of the moments of ψ_λ are zero—in the Haar system, only the integrals of the functions themselves are zero.

The second advantage of wavelet bases is the ease with which they adapt to various commonly used functional analysis norms. That is, if f belongs to one of the classical spaces (Sobolev spaces, Besov spaces, Hardy spaces...) the corresponding wavelet series will converge automatically to f in the appropriate norm. These properties will be the content of later chapters, but we shall give the reader a taste of what is to come by illustrating our remarks with the examples of Sobolev and Hölder spaces.

But there are difficulties about using (10.1) in situations which are more general than just $L^2(\mathbb{R}^n)$. Here is one obstacle: if $f(x)$ is identically equal to 1, each of the scalar products (f, ψ_λ) is zero—because the integral of each wavelet is zero—and (10.1) would give $1 = 0$. So we cannot replace $L^2(\mathbb{R}^n)$ by $L^\infty(\mathbb{R}^n)$ and still believe in (10.1).

Another obstacle, in some sense the dual of the preceding, is the case of a function $f(x) \in \mathcal{D}(\mathbb{R}^n)$ whose integral equals 1. Then (10.1) provides a decomposition of this function into a series of terms of integral zero. So the series (10.1) cannot converge in $L^1(\mathbb{R})$ norm: if it did, we could integrate term by term. It seems, then, that the use of (10.1) must essentially be reserved for the spaces $L^p(\mathbb{R}^n)$, when $1 < p < \infty$.

However, the difficulties we have described disappear once we put the wavelets back into the context of the r-regular multiresolution approximations from which they have been constructed. Put $\Gamma_0 = \mathbb{Z}^n$, $\Gamma_j = 2^{-j}\Gamma_0$, and $\Lambda_j = \Gamma_{j+1}\backslash\Gamma_j$, $j \in \mathbb{N}$. Let ϕ denote the function which is centrally involved in the construction of the wavelets, and put $\phi_\lambda(x) = \phi(x - \lambda)$, $\lambda \in \Gamma_0$. Then identity (10.1) can be abandoned in favour of

$$(10.2) \qquad f(x) = \sum_{\lambda \in \Gamma_0} (f, \phi_\lambda)\phi_\lambda(x) + \sum_0^\infty \sum_{\lambda \in \Lambda_j} (f, \psi_\lambda)\psi_\lambda(x).$$

Both paradoxes we have alluded to disappear immediately when this new algorithm is used. Indeed, if $f(x)$ is identically equal to 1, each scalar product (f, ϕ_λ) equals 1 and (10.2) becomes the identity $1 = \sum_{\lambda \in \Lambda_0} \phi_\lambda(x)$.

More generally, if $f(x)$ is a polynomial of degree not greater than r, (10.2) is satisfied because $E_0(x^\alpha) = x^\alpha$, for $|\alpha| \le r$. By contrast, a naive use of (10.1) would lead to $x^\alpha = 0$: a contradiction.

So we shall use identity (10.2) when the function space to which f belongs is characterized by a local regularity—or irregularity—condition, together with a global property, of a different nature, describing "growth at infinity". An example of such a situation is the case of non-homogeneous Hölder spaces $C^s = \dot{C}^s \cap L^\infty$, $0 < s < 1$: the local property is $|f(y) - f(x)| \le C|y - x|^s$ for every $x, y \in \mathbb{R}^n$ (we can clearly restrict ourselves to the case $|y - x| \le 1$, given the global property $\|f\|_\infty \le C_0$).

On the other hand, (10.1) is appropriate for the homogeneous Hölder space $\dot{C}^s$ (where the condition $f \in L^\infty$ is omitted). This homogeneous Hölder space is, thus, a space of functions modulo the constant functions, and so 1 does equal 0 in the space $\dot{C}^s$.

The second paradox also disappears when (10.2) is used. Indeed, if $f \in \mathcal{D}(\mathbb{R}^n)$, the series $\sum_{\lambda \in \Gamma_0}(f, \phi_\lambda)\phi_\lambda(x)$ converges in $L^1(\mathbb{R}^n)$ norm to a function $g \in L^1(\mathbb{R}^n)$ whose integral equals that of f. The double series $\sum_0^\infty \sum_{\lambda \in \Lambda_j} (f, \psi_\lambda) \psi_\lambda(x)$ converges, in L^1 norm, to a function h of zero integral, and we get $f = g + h$.

We illustrate the use of series (10.1) and (10.2) by giving characterizations of the Hölder spaces C^s and the Sobolev spaces H^s in terms of wavelet coefficients.

To obtain such characterizations, we must start with a mathematical object f whose wavelet coefficients we can calculate. That object will be a distribution of order less than r, for wavelets coming from an r-regular multiresolution approximation of $L^2(\mathbb{R}^n)$. But that is not all. In order not to get the contradiction $1 = 0$, we further suppose that $f(x) = \sum_{\lambda \in \Lambda}(f, \psi_\lambda) \psi_\lambda(x)$, in the sense of numerical identity whenever we multiply both sides by a function of compact support in C^r and integrate term by term.

If we have taken all these precautions, we get

Theorem 4. *The distribution f belongs to $H^s(\mathbb{R}^n)$, where $-r < s < r$, if and only if its wavelet coefficients $\alpha(\lambda) = (f, \psi_\lambda)$ satisfy*

$$(10.3) \qquad \sum_{j<0} \sum_{\lambda \in \Lambda_j} |\alpha(\lambda)|^2 + \sum_{j \geq 0} \sum_{\lambda \in \Lambda_j} 4^{js}|\alpha(\lambda)|^2 < \infty.$$

Similarly, f belongs to the (non-homogeneous) Hölder space C^s, where $0 < s < r$, if and only if there are constants C_0 and C such that the coefficients $\beta(\lambda) = (f, \phi_\lambda)$, $\lambda \in \Gamma_0$, and $\alpha(\lambda) = (f, \psi_\lambda)$, $\lambda \in \Lambda_j$, $j \geq 0$, satisfy

$$(10.4) \qquad |\beta(\lambda)| \leq C_0 \qquad \text{and} \qquad |\alpha(\lambda)| \leq C 2^{-nj/2} 2^{-js}.$$

Before proving these two statements, some comments may be of use. There exists no characterization of the (non-homogeneous) Hölder space C^s purely in terms of the moduli of wavelet coefficients, and it is easy to see that the reason derives from the absence of such a characterization of $L^\infty(\mathbb{R}^n)$.

The fact that $L^\infty(\mathbb{R}^n)$ cannot be thus characterized has already been mentioned in Section 2. The moduli of the wavelet coefficients of $\log|x|$ and of $x/|x|$ have the same asymptotic behaviour when $n = 1$ and we use the Littlewood-Paley wavelets. As we have seen, this anomaly is explained by the role played by the Hilbert transform, H: H changes $x/|x|$ to $\log|x|$ and changes a wavelet basis ψ_I into an "indistinguishable" basis $\tilde{\psi}_I$, when the former is derived from the Littlewood-Paley multiresolution approximation.

To end these preliminary remarks, we may observe that the Sobolev spaces H^s may also be characterized using the series (10.2). In fact, $f \in H^s$, where $-r < s < r$, if and only if $\sum_{\lambda \in \Gamma_0} |\beta(\lambda)|^2 < \infty$ and

$\sum_0^\infty \sum_{\lambda \in \Lambda_j} 4^{js} |\alpha(\lambda)|^2 < \infty$. Indeed, we have

$$\sum_{\lambda \in \Gamma_0} |\beta(\lambda)|^2 = \sum_{-\infty}^{-1} \sum_{\lambda \in \Lambda_j} |\alpha(\lambda)|^2,$$

because V_0 is the Hilbert direct sum of the spaces W_j, $j < 0$.

It remains to prove the theorem. As far as H^s is concerned, this is really easy. Invoking Theorem 8 of Chapter 2, we get $f \in H^s$ if and only if $E_0(f)$ belongs to $L^2(\mathbb{R}^n)$ and there is a sequence $\varepsilon_j \in l^2(\mathbb{N})$ such that $\|D_j(f)\|_2 \le \varepsilon_j 2^{-js}$. Now, the wavelets ψ_λ, $\lambda \in \Lambda_j$, form an orthonormal basis of W_j. So we have $\|D_j(f)\|_2^2 = \sum_{\lambda \in \Lambda_j} |(f, \psi_\lambda)|^2$. Similarly, we get $\|E_0(f)\|_2^2 = \sum_{\lambda \in \Gamma_0} |(f, \phi_\lambda)|^2$, so that (10.3) coincides with the characterization given in Chapter 2.

For the Hölder space C^s we use the following lemma:

Lemma 12. *The norms $\|f\|_\infty$ and $\sup_{\lambda \in \Lambda_0} |(f, \psi_\lambda)|$ are equivalent on the subspace W_0 of $L^2(\mathbb{R}^n)$.*

In one direction, we have $|(f, \psi_\lambda)| \le \|\psi_\lambda\|_1 \|f\|_\infty = C \|f\|_\infty$, and in the other, if $\sup_{\lambda \in \Lambda_0} |(f, \psi_\lambda)| \le 1$, we get $|f(x)| \le \sum_{\lambda \in \Lambda_0} |\psi_\lambda(x)| \le C'$. The last inequality is a consequence of the localization of the wavelets, and the lemma is proved. An immediate corollary is that the norms $\|f\|_\infty$ and $2^{nj/2} \sup_{\lambda \in \Lambda_j} |(f, \psi_\lambda)|$ are equivalent on W_j. This can be seen by a simple change of scale.

Since the conditions $E_0(f) \in L^\infty$ and $\|D_j(f)\|_\infty \le C 2^{-js}$ characterize C^s, the characterization in terms of wavelet series follows immediately. Exactly as we indicated in Chapter 2, the spaces C^s we are dealing with must not become the usual C^m spaces when $s = m$, a positive integer, but should be obtained from the Zygmund class. Thus, if $s = 1$, C^s should be replaced by the space of functions $f : \mathbb{R}^n \to \mathbb{C}$ which are continuous, bounded, and satisfy $|f(x + y) + f(x - y) - 2f(x)| \le C|y|$, for a certain constant C. If $s = 2$, we replace C^s by the space of functions whose first order partial derivatives belong to the Zygmund class, and similarly for integer $s > 2$.

For the time being, we shall leave the characterization of function spaces by the order of magnitude of their wavelet coefficients, but this theme will be resumed in Chapters 5 and 6.

11 Wavelet series and Fourier series

In this section, we are going to examine the competing merits of wavelet series and Fourier series. At first sight there is no contest, because Fourier series represent periodic functions, whereas wavelet series represent functions that vanish at infinity (in the mean). In order to make a fair comparison, we shall, therefore, first construct periodic wavelets: apart from the function identically equal to 1, periodic wavelets will be the periodifications, of period 1, of wavelets

$$2^{j/2}\psi(2^j x - k), \qquad k \in \mathbb{Z}, j \geq 0.$$

The most astonishing result we obtain will be the remarkable fact that "full" wavelet series (those having plenty of non-zero coefficients) represent really pathological functions, whereas "normal" functions have "sparse" or "lacunary" wavelet series. On the other hand, Fourier series of the usual functions are "full", whereas lacunary Fourier series represent pathological functions.

This phenomenon has a simple explanation. Analysis by wavelets is a local Fourier analysis which takes place at every scale. It has the advantage of being concentrated near the singular support of the function analysed. In other words, away from the singular support, the function analysed is infinitely differentiable and the corresponding wavelet coefficients are negligible.

Enough of these generalities! We must get to the meat of the construction of periodic wavelets.

Let V_j, $j \in \mathbb{Z}$, be an r-regular multiresolution approximation of $L^2(\mathbb{R})$. (We shall deal with the n-dimensional case later.) We suppose that $r \geq 1$ in what follows. If $1 \leq p \leq 2$ we let V_j^p denote $V_j \cap L^p(\mathbb{R})$, whereas, if $2 \leq p < \infty$, V_j^p denotes the completion of V_j in $L^p(\mathbb{R})$-norm. Finally, if $p = \infty$, we give $L^\infty(\mathbb{R})$ the $\sigma(L^\infty, L^1)$-topology, as usual, and let V_j^∞ denote the completion of V_j in that topology. It can be verified immediately that, if $\phi(x)$ is the function canonically associated with the multiresolution approximation V_j, then the space V_0^∞ is described by the equivalence

$$(11.1) \quad f \in V_0^\infty \iff f(x) = \sum_{-\infty}^{\infty} c_k \phi(x - k) \quad \text{with} \quad c_k \in l^\infty(\mathbb{Z}).$$

Further,

$$f(x) \in V_0^\infty \iff f(2^j x) \in V_j^\infty.$$

Let $P_j \subset V_j^\infty$ be the subspace consisting of functions of period 1. Then we have

Lemma 13. *If $j \leq 0$, the P_j are identical and consist of just the constant functions. If $j > 0$, the dimension of P_j is 2^j.*

We first note that the constants are in every V_j^∞, since $\sum_{-\infty}^{\infty} \phi(x - k) = 1$. Moreover, we have $V_j^\infty \subset V_{j+1}^\infty$, for every $j \in \mathbb{Z}$, and the first statement of the lemma will be proved if we can show that every function $f \in P_0$ is constant. To do this, we write $f(x) = \sum_{-\infty}^{\infty} c_k \phi(x - k)$, where

$c_k = \int_{-\infty}^{\infty} f(x)\bar{\phi}(x-k)\,dx$. The sequence c_k, $k \in \mathbb{Z}$, is constant, so the same holds for $f(x)$.

Now consider the case $j > 0$. This time, we shall write $f(x) = \sum_{-\infty}^{\infty} c(k)\phi(2^j x - k)$, where $c(k) = \int_{-\infty}^{\infty} f(x)\bar{\phi}(2^j x - k)$. The identity $f(x+1) = f(x)$ now gives $c(k+2^j) = c(k)$. The converse is immediate, and the dimension of the vector space P_j is thus 2^j.

Lemma 14. *The union of the P_j is dense in the Banach space of functions which are continuous on the real line and periodic, of period 1.*

The orthogonal projection operator, $E_j : L^2(\mathbb{R}) \to V_j$ is given by

$$E_j f(x) = \int E_j(x,y)f(y)\,dy,$$

where $E_j(x,y) = 2^j E(2^j x, 2^j y)$ and

$$E(x,y) = \sum_{-\infty}^{\infty} \phi(x-k)\bar{\phi}(y-k).$$

We know that $|E(x,y)| \le C(1+|x-y|)^{-2}$ and that $\int E(x,y)\,dy = 1$. It follows that, if f is bounded and uniformly continuous on $\mathbb{R}$, then $\|f - E_j(f)\|_\infty \to 0$ as $j \to \infty$. Moreover, if f is periodic, of period 1, we get $E_j(f) \in P_j$, for $j \in \mathbb{N}$. Lemma 14 is therefore proved.

Definition 1. *A nested sequence P_j, $j \in \mathbb{N}$, constructed by the procedure above is called an r-regular multiresolution approximation of $L^2(\mathbb{T})$.*

Let Γ_j denote the finite group of residues $k2^{-j}$ modulo 1. Then P_j is a vector subspace of $L^2(\mathbb{T})$, invariant under the action of Γ_j.

Let $\phi_j \in P_j$ be the function defined by

(11.2) $$\phi_j(x) = \sum_{k \in \mathbb{Z}} 2^{j/2} \phi(2^j(x-k)).$$

Lemma 15. *The orbit of ϕ_j under the action of Γ_j is an orthonormal basis of P_j.*

Since the dimension of P_j equals the cardinality of Γ_j, it is enough to verify that the functions $\phi_j(x-2^{-j}m)$, $0 \le m < 2^j$, form an orthonormal sequence in $L^2[0,1]$. This leads us to calculate

$$\sum_k \sum_l \int_0^1 \phi(2^j x - 2^j k - m)\bar{\phi}(2^j x - 2^j l - m')\,dx.$$

We apply the change of variable $x - k = 2^{-j}t$, where $0 \le x < 1$, $k \in \mathbb{Z}$, and $t \in \mathbb{R}$. This leads to

$$\sum_q \int_{-\infty}^{\infty} \phi(t-m)\bar{\phi}(t-2^j q - m')\,dt.$$

If $0 \leq m < 2^j$, $0 \leq m' < 2^j$ and $m \neq m'$, each of the integrals vanishes, and the sum over $q \in \mathbb{Z}$ yields 0. If $m = m'$, the only non-zero integral is when $q = 0$ and then it equals 1. Lemma 15 is proved.

We now let Q_j denote the orthogonal complement of P_j in P_{j+1} and set

$$(11.3) \qquad \psi_j(x) = 2^{j/2} \sum_{-\infty}^{\infty} \psi(2^j(x-k)).$$

Lemma 16. *For each $j \in \mathbb{N}$, the functions $\psi_j(x - k2^{-j})$, $0 \leq k < 2^j$, form an orthonormal basis of Q_j.*

The proof is the same as that of Lemma 15. We first check that the functions $\psi_j(x - k2^{-j})$ belong to Q_j, that is, that they are functions in P_{j+1} (obviously) which are orthogonal to P_j. Then we verify that the functions $\psi_j(x - k2^{-j})$ are mutually orthogonal. We conclude by observing that the dimension of Q_j is 2^j (because that of P_j is 2^j and the dimension of P_{j+1} is 2^{j+1}). 2^j is the number of the vectors $\psi_j(x - k2^{-j})$ forming an orthonormal set in Q_j. So that set is a basis of Q_j.

Now

$$L^2(\mathbb{T}) = P_0 \oplus Q_0 \oplus Q_1 \oplus Q_2 \oplus \cdots,$$

so it follows that the constant function 1 together with the collection $\{\psi_j(x - k2^{-j}) : 0 \leq k < 2^j, j \in \mathbb{N}\}$, constitute an orthonormal basis of $L^2(\mathbb{T})$. We shall order this basis lexicographically to get a sequence. That is, we define the sequence $g_m(x)$, $m \in \mathbb{N}$ of functions in $L^2(\mathbb{T})$ by

$$(11.4) \qquad g_0(x) = 1$$

$$g_1(x) = \psi_0(x)$$

$$g_2(x) = \psi_1(x) \qquad g_3(x) = \psi_1(x - \tfrac{1}{2})$$

$$g_4(x) = \psi_2(x) \; g_5(x) = \psi_2(x - \tfrac{1}{4}) \; g_6(x) = \psi_2(x - \tfrac{1}{2}) \; g_7(x) = \psi_2(x - \tfrac{3}{4})$$

$$g_8(x) = \psi_3(x) \ldots$$

The above sequence is an orthonormal basis of $L^2(\mathbb{T})$, and it has other remarkable properties which we shall spell out in the theorem below. But first we need to fix some notation.

Let E be a Banach space, with norm denoted by $\| \cdot \|$, and let

$$e_0, e_1, \ldots, e_k, \ldots$$

be a sequence of elements of E. The sequence is a Schauder basis of E if, for each $x \in E$, there exists a unique sequence of coefficients $\alpha_0, \alpha_1, \ldots, \alpha_k, \ldots$ such that

$$\lim_{m \uparrow \infty} \|x - (\alpha_0 e_0 + \cdots + \alpha_m e_m)\| = 0.$$

We can then write $x = \alpha_0 e_0 + \alpha_1 e_1 + \cdots$ and we say that the Schauder basis is *unconditional* if the series is a summable family (that is, if it still converges to x after an arbitrary permutation of its terms). We shall come back to this idea throughout the present volume, and we shall give several equivalent definitions.

We can now state the theorem announced above.

Theorem 5. *Let V_j be an r-regular multiresolution approximation of $L^2(\mathbb{R})$, with $r \geq 1$. Then the sequence $g_m(x)$, $m \in \mathbb{N}$, of periodic wavelets constructed out of the V_j is a Schauder basis of the spaces $C(\mathbb{T})$ and $L^1(\mathbb{T})$. It is also a Schauder basis of the (usual) $C^k(\mathbb{T})$ spaces for $0 \leq k < r$. Further, the same sequence $g_m(x)$ is an unconditional basis of the Hölder space C^α, when $0 < \alpha < 1$, of the Zygmund class Λ_*, if $r \geq 2$, of the L^p spaces, when $1 < p < \infty$, for the Hardy space $H^1(\mathbb{T})$, for its dual $BMO(\mathbb{T})$, etc.*

This means that we have a *"universal identity"*

$$(11.5) \qquad f = \sum_{0}^{\infty} (f, g_m) g_m .$$

To begin with, if f belongs to $L^2(\mathbb{T})$, the partial sums

$$\sigma_l(f) = \sum_{0}^{l-1} (f, g_m) g_m$$

converge to f in $L^2(\mathbb{T})$ norm. If f is in $C(\mathbb{T})$ as well, that is, if f is continuous and 1-periodic, the partial sums $\sigma_l(f)$ converge uniformly to f. Such a property does not hold in the case of Fourier series. Again, if f belongs to $L^1(\mathbb{T})$, the partial sums $\sigma_l(f)$ converge to f in $L^1(\mathbb{T})$ norm. This property, too, is not shared by Fourier series.

In the case of the spaces $L^p(\mathbb{T})$, the characters $e^{2k\pi ix}$, $k \in \mathbb{Z}$, do form a Schauder basis: the partial sums $\sum_{-l}^{l} c(k)e^{2k\pi ix}$ of the Fourier series of a function $f \in L^p(\mathbb{T})$ converge to f in $L^p(\mathbb{T})$ norm. For wavelet series, the position is even better, since the wavelet series of a function $f \in L^p(\mathbb{T})$ converges unconditionally to f.

The only standard function spaces where Fourier series work as well as wavelet series are the space $L^2(\mathbb{T})$ and the Sobolev spaces.

There is one case, however, that of the Wiener algebra, where Fourier series score a clear victory over wavelet series. The Wiener algebra, $A(\mathbb{T})$ is the subalgebra of $C(T)$ consisting of the functions whose Fourier series are absolutely convergent. Wavelets cannot even form a Schauder basis of the Wiener algebra.

Before moving to the proof of Theorem 5, let us look at the particular

case of the periodic wavelets arising from the Littlewood-Paley multiresolution approximation. It is easy to show that the functions $g_m(x)$ are
finite trigonometric series

$$\sum_{l_m \leq |l| \leq L_m} c(k,l)e^{2l\pi ix} \qquad \text{where} \qquad \frac{m}{3} \leq l_m < L_m \leq \frac{4m}{3}.$$

The characters $e^{2l\pi ix}$, $l \in \mathbb{Z}$, of a standard Fourier series have thus,
in the case of these wavelet series, become "wave-packets" obtained by
regrouping characters with approximately the same frequency on a log-
arithmic scale. We shall not try to give a more precise mathematical
sense to the last remark.

Among the assertions of Theorem 5, certain apply only to periodic
wavelet series, because of properties where the lexicographic order plays
a part. We shall give a proof of these. But we refer the reader to
Chapters 5 and 6 for all that concerns those function spaces for which
the wavelets form an unconditional basis: there the periodic case does
not differ from the general case.

We start by studying wavelet series of functions which are continuous
on the real line and periodic, of period 1. The outline of the proof is as
follows. To show that $\|f - \sigma_l(f)\|_\infty$ tends to 0, as l tends to infinity, for
every function f, it is enough to verify this property for f belonging to a
dense linear subspace of $C(\mathbb{T})$ and to show that there exists a constant
C such that, for each $l \in \mathbb{N}$, we have $\|\sigma_l(f)\|_\infty \leq C\|f\|_\infty$.

The dense linear subspace we use is the union of the spaces P_j of
Definition 1. If f belongs to P_j, we have $\sigma_l(f) = f$ once $l > 2^j$. All that
remains to prove is that the uniform upper bound holds. To do this, we
identify f with a continuous periodic function on $\mathbb{R}$, of period 1, and
we suppose at first that $l = 2^j$. Then $\sigma_l(f) = E_j(f)$, and we know that
the operators E_j are uniformly bounded on $L^\infty(\mathbb{R})$. This fact reduces,
after a change of scale, to knowing that $E_0 : L^\infty(\mathbb{R}) \to L^\infty(\mathbb{R})$ is a
continuous operator—the continuity follows from $C(1 + |x - y|)^{-2}$ being
an upper bound for the kernel of E_0.

Suppose now that $2^j \leq l < 2^{j+1}$. We write $f_l(x) = \sigma_l(f) = E_j(f) +
r_l(x)$, where $r_l(x) = \sum_{2^j \leq k < l}(f, g_k)g_k(x)$.

Since we have already controlled $\|E_j(f)\|_\infty$, we need only work with
$\|r_l\|_\infty$. Now,

$$\|r_l\|_\infty \leq \|f\|_\infty \left(\sup_{2^j \leq k < 2^{j+1}} \|g_k\|_1 \right) \left\| \sum_{2^j \leq k < 2^{j+1}} |g_k(x)| \right\|_\infty.$$

In fact, $\|g_k\|_1 \leq C2^{-j/2}$, when $2^j \leq k < 2^{j+1}$, and

$$\sum_{2^j \leq k < 2^{j+1}} |\psi_j(x - k2^{-j})| \leq 2^{j/2} \left\| \sum_{-\infty}^{\infty} |\psi(x - k)| \right\|_{\infty} = C2^{j/2}.$$

All this gives $\|r_l\|_\infty \leq C\|f\|_\infty$, and the uniform convergence of the functions f_l to f has been completely established.

The case of the spaces $C^k(\mathbb{T})$, where $0 \leq k < r$, is essentially similar and is left to the reader. The one extra fact needed is Theorem 5 of Chapter 2.

The proof of the convergence of $\sigma_l(f)$ to f in $L^1(\mathbb{T})$ norm, when f belongs to $L^1(\mathbb{T})$, is a corollary of the preceding result. In fact, the essential point is the proof of the upper bound $\|\sigma_l(f)\|_1 \leq C\|f\|_1$, uniformly in l. Now, this bound is obtained directly by duality from the inequality $\|\sigma_l(f)\|_\infty \leq \|f\|_\infty$. Indeed, we have

$$\|\sigma_l(f)\|_1 = \sup\{(|\sigma_l(f), g)| : \|g\|_\infty \leq 1\}$$

and we can suppose that the functions $g(x)$ are continuous and periodic, of period 1. But $(\sigma_l(f), g) = (f, \sigma_l(g))$, and $C\|f\|_1$ is an upper bound for the scalar product. This gives the required inequality.

Similarly, if μ is a Borel measure on $\mathbb{T}$, then the $\sigma_l(\mu)$ form a bounded sequence of functions in $L^1(\mathbb{T})$, and this sequence converges to μ in the $\sigma(M(\mathbb{T}), C(\mathbb{T}))$ topology. This remark will allow us to show that certain distributions S are not Borel measures: it will be enough to verify that $\sup_{l \in \mathbb{N}} \|\sigma_l(S)\|_1 = \infty$.

In the results which we have just described, the order of the terms plays a fundamental part. Anticipating a discussion which is to be resumed in Chapter 6, we know that the space $C(\mathbb{T})$ and the space $L^1(\mathbb{T})$ do not have unconditional bases—we cannot expect the series (11.5) to converge if we disturb the order of the terms.

The case of the $L^p(\mathbb{T})$ spaces, $1 < p < \infty$ is entirely different. Here is a statement which sharpens Theorem 5, and whose proof will be given in Chapter 6. *A series $\sum_0^\infty \alpha_k g_k(x)$ converges in $L^p(\mathbb{T})$, (we suppose that $1 < p < \infty$) if and only if $\left(\sum_0^\infty |\alpha_k|^2 |g_k(x)|^2 \right)^{1/2}$ belongs to $L^p(\mathbb{T})$.* The condition therefore involves only the moduli of the α_k and is monotone, in the sense that, once it is satisfied for a sequence $|\alpha_k|$, it holds for any sequence β_k satisfying $|\beta_k| \leq |\alpha_k|$, for all $k \in \mathbb{N}$.

The case of the Hölder spaces C^α also leads to a remarkable explicit characterization. *The series $\sum_0^\infty \alpha_k g_k(x)$ defines a Hölder function of exponent $s > 0$ if and only if $\alpha_k = O(k^{-s-1/2})$.* It is necessary to suppose that $0 < s < r$, and to replace the space C^1 by the Zygmund class Λ_*, the space C^2 by the periodic primitives, of period 1, of functions in the

Zygmund class, and so forth. This characterization of Hölder functions
is an immediate consequence of the results of Section 10. There is no
analogue in the case of Fourier series, except in the case of a lacunary
series whose local and global properties can be read off from the order
of magnitude of the coefficients.

We shall continue the comparison between wavelet series and Fourier
series. But first we shall indulge in a digression.

Weierstrass constructed a continuous, nowhere differentiable function.
G. Freud generalized the construction to show that if $q > 1$, $\lambda_0 > 0$, and
$\lambda_{k+1} > q\lambda_k$, for every $k \in \mathbb{N}$, then each series $\sum_0^\infty \alpha_k \cos \lambda_k x$, such that
$\sum_0^\infty |\alpha_k| < \infty$ but $\alpha_k \lambda_k$ does not tend to 0, defines a function which is
continuous on the real line but is nowhere differentiable.

To prove this, Freud considered a function $\psi \in L^1(\mathbb{R})$ whose Fourier
transform $\hat{\psi}(\xi)$ was of class C^4, say, was 1 at the point 1 and 0 outside
the interval $[1/q, q]$. It is clear that $\psi(x)$ is continuous and $O(|x|^4)$ at
infinty.

Suppose that the sum $f(x) = \sum_0^\infty \alpha_k \cos \lambda_k x$ is differentiable at x_0.
Then Freud calculated the integral $I_k = \int_{-\infty}^\infty f(x)\psi(\lambda_k(x - x_0))\, dx$ in
two different ways. On the one hand, if $f(x)$ is differentiable at x_0,
replacing $f(x)$ by $f(x) - f(x_0) - (x - x_0)f'(x_0) = o(x - x_0)$ does not
change I_k. This gives $I_k = o(\lambda_k^{-2})$. But the direct calculation gives
$I_k = \frac{1}{2}\alpha_k \lambda_k^{-1} e^{i\lambda_k x_0}$. Hence the assumption of differentiability at a point
leads to the conclusion that the sequence $\alpha_k \lambda_k$ tends to 0.

Wavelet technology was already present, in an embryonic state, in that
proof. The functions $\lambda_k^{1/2}\psi(\lambda_k(x-x_0))$ are very similar to our wavelets—
they have, at any rate, analogous properties of regularity, localization
and cancellation.

Freud's remarkable proof would obviously not work for general non-
lacunary Fourier series, whereas it will extend to all wavelet series, as
we shall demonstrate.

Still restricting ourselves to periodic wavelet series $\sum_0^\infty \alpha_m g_m(x)$, we
set $m = 2^j + k$, where $j \in \mathbb{N}$ and $0 \le k < 2^j$ (as long as $m \ne 0$). We let
I_m denote the interval $[k/2^j, (k + 1)/2^j)$. The interval qI_m $(q \ge 1)$ will
denote the interval with the same centre as I_m and of length $q/2^j$.

We then have

Proposition 5. *Let the sum $f(x)$ of the series $\sum_0^\infty \alpha_m g_m(x)$ be a continuous function on the real line which is differentiable at a point x_0.
Then, for each fixed $q \ge 1$, we have $\alpha_m = o(m^{-3/2})$, as m tends to infinity through a sequence of numbers such that the intervals qI_m contain*
x_0.

The proof poses no problems, as $\alpha_m = \int_0^1 f(x)\bar{g}_m(x)\,dx$. Thus we can repeat Freud's argument, word for word.

Corollary 1. *Let α_m be a sequence of real or complex numbers such that, for two constants $C_2 \geq C_1 > 0$ and for all $m \geq 1$, we have $C_1 m^{-3/2} \leq |\alpha_m| \leq C_2 m^{-3/2}$. Then the sum $f(x)$ of the wavelet series $\sum_0^\infty \alpha_m g_m(x)$ belongs to the Zygmund class Λ_* (and, in particular, to each Hölder space C^α for $0 < \alpha < 1$) but is nowhere differentiable.*

We shall show that the historic Weierstrass series (or a related series) appears as a special case of the general result of Corollary 1.

To do that, we consider the particular wavelet $\psi(x)$ arising from the Littlewood-Paley multiresolution approximation—this wavelet is described in Section 2. Then we have $\sum_{-\infty}^\infty \psi(x-k) = -\sqrt{2}\cos 2\pi x$, as a straightforward calculation shows.

Now take a periodic wavelet series

$$f(x) = \sum_{j \in \mathbf{N}} \sum_{k=0}^{2^j - 1} \alpha(j,k)\psi_j\left(x - \frac{k}{2^j}\right),$$

such that $\alpha(j,k) = \alpha(j)$ does not depend on k. Then

$$f(x) = -\sqrt{2}\sum_0^\infty 2^{j/2}\alpha(j)\cos 2\pi 2^j x,$$

a lacunary Fourier series.

Lacunary Fourier series, which are very special, are thus found to be swimming in a shoal of wavelet series, the moduli of whose coefficients tend regularly to 0.

Let us consider Takagi's fractal function ([226]), the definition of which we now recall. We begin with the continuous periodic function $\Delta(x)$, of period 1 on $\mathbf{R}$, defined by

$$\Delta(x) = \begin{cases} 2x, & \text{if } 0 \leq x \leq \frac{1}{2}; \\ 2 - 2x, & \text{if } \frac{1}{2} \leq x \leq 1. \end{cases}$$

Takagi's function is then defined by

$$T(x) = \sum_1^\infty 2^{-j}\Delta(2^{j-1}x).$$

This function has a quite remarkable structure. Consider the affine transformations of $\mathbf{R}^2$ given by $A(x,y) = (x/2, (x+y)/2)$ and $B(x,y) = ((x+1)/2, (1+y-x)/2)$. We let each word $A^{m_1}B^{m_1}\cdots A^{m_q}B^{m_q}$, where $m_j, n_j \in \mathbf{N}$, $1 \leq j \leq q$, act on the point $(0,0)$. Then the closure of the set thereby obtained is the graph of $T(x)$ over the domain $[0,1]$

We next take the piecewise linear wavelet $\psi(x)$ of Figure 1. It satisfies

$\sum_{-\infty}^{\infty} \psi(x-k) = \sqrt{3}(1-2T(x))$, so that Takagi's function can be written

$$T(x) = \frac{1}{2} - \frac{1}{2\sqrt{3}} \sum_{1}^{\infty} \sum_{-\infty}^{\infty} 2^{-j}\psi(2^j x - k).$$

Takagi's function is nowhere differentiable, as can be seen by applying Proposition 5. Indeed, we have $\int_{-\infty}^{\infty} \psi(x)\,dx = \int_{-\infty}^{\infty} x\psi(x)\,dx = 0$. Takagi's function belongs to the Hölder space C^α for each $\alpha < 1$. On the other hand, it does not belong to the Zygmund class. This can be seen by remarking that $T(\pm 2^{-j}) \geq j2^{-j}$, which is inconsistent with having $|T(2^{-j}) + T(-2^{-j}) - 2T(0)| \leq C2^{-j}$. This shows that Theorem 5 becomes inexact when $\alpha = r$ (the regularity of the multiresolution approximation).

Corollary 1 of Proposition 5 leads directly to the following result.

Corollary 2. *There exists a dense open subset of the Zygmund class Λ_* which is composed of nowhere differentiable functions.*

Indeed, let Ω_ε, $\varepsilon > 0$ denote the set of functions $f(x) = \sum \alpha_m g_m(x)$ such that $\alpha_m = O(m^{-3/2})$ and $|\alpha_m| \geq \varepsilon m^{-3/2}$, for every $m \geq 1$. Corollary 1 shows that $f(x)$ belongs to Λ_* and is nowhere differentiable. The union Ω of the Ω_ε, $\varepsilon > 0$, is an open subset of Λ_*. Indeed, if f belongs to Ω_ε and if $\|f - g\| \leq \delta$ (in Λ_* norm), then $|\alpha_m - \beta_m| \leq C\delta m^{-3/2}$, where the β_m are the wavelet coefficients of g. Choosing $\delta > 0$ sufficiently small for $C\delta < \varepsilon/2$ to hold, we get $g \in \Omega_{\varepsilon/2}$.

Finally, we show that this open set is dense in Λ_*. Indeed, if $h(x) = \sum \gamma_m g_m(x)$ belongs to Λ_*, then $|\gamma_m| \leq Cm^{-3/2}$ for some constant $C > 0$ and every $m \geq 1$. For each $\varepsilon > 0$, we carry out the following operation on the coefficients γ_m: if $|\gamma_m| \geq \varepsilon m^{-3/2}$, we leave γ_m unchanged, but if $|\gamma_m| < \varepsilon m^{-3/2}$, we replace γ_m by $\varepsilon m^{-3/2}$. Calling the new coefficients $\tilde{\gamma}_m$, we set $\tilde{h}(x) = \sum \tilde{\gamma}_m g_m(x)$. Clearly $\tilde{h} \in \Omega_\varepsilon$ and $\|h - \tilde{h}\| \leq C\varepsilon$: the union of the Ω_ε is indeed dense in Λ_*.

We continue our comparison of periodic wavelet series with Fourier series by examining a special case.

Consider, for $0 < \alpha < 1$, the function $|\sin \pi x|^{-\alpha}$, which is certainly periodic, of period 1. Its Fourier coefficients $c(k)$, $k \in \mathbb{Z}$, have an asymptotic expansion $c(k) = \gamma(\alpha)|k|^{-1+\alpha} + O(k^{-3+\alpha})$, which can be found in [239], Chapter 5. The constant $\gamma(\alpha)$ is non-zero, so the singularity at 0 of the function $|\sin \pi x|^{-\alpha}$ affects *all* the Fourier coefficients. On the other hand, its wavelet coefficients do not seem in the least affected by the singularity of the function $|\sin \pi x|^{-\alpha}$, except where the singularity is too close to the interval I_m which defines the localization of the wavelet $g_m(x)$. More precisely, if $2^j \leq m < 2^{j+1}$ and if $l = \min(m-2^j, 2^{j+1}-m)$,

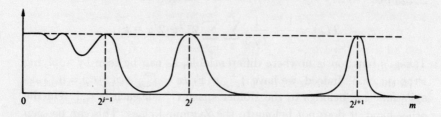

Figure 4

we get $|\alpha_m| \le C_N 2^{j(\alpha-1/2)}/(1+l)^N$, for each integer $N \ge 1$, when the periodic wavelets $g_m(x)$ arise from the Littlewood-Paley multiresolution approximation. For example, if $\alpha = 1/2$, the following graph gives the order of magnitude of the wavelet coefficients.

The wavelet series of the function $|\sin \pi x|^{-\alpha}$ is thus a "sparse" series.

Let us go a little further and consider the "lacunary" wavelet series $\sum_0^\infty a_j g_{2^j}(x)$, where the $g_m(x)$ are still our periodic wavelets. Suppose that $a_j = O(2^{j(\alpha-1/2)})$. Then the sum $f(x)$ of the lacunary wavelet series is a periodic function, of period 1, which is infinitely differentiable except at $0, \pm1, \pm2, \dots$ and whose singularities are of the same type as those of $|\sin \pi x|^{-\alpha}$. In other words, we have

$$f(x) = O(|\sin \pi x|^{-\alpha}), \qquad f'(x) = O(|\sin \pi x|^{-\alpha-1}), \qquad \cdots$$

Conversely, "full" wavelet series resemble lacunary Fourier series. Let us systematically consider all 1-periodic wavelet series

$$f(x) = \sum_0^\infty \alpha_m g_m(x)$$

such that the sequence of amplitudes $|\alpha_m|$ is decreasing. With this condition, we have the following remarkable statement.

Theorem 6. *Let the sequence $|\alpha_m|$ be decreasing. If $\sum_0^\infty |\alpha_m|^2 = \infty$, then $\sum_0^\infty \alpha_m g_m(x)$ is never the wavelet series of a bounded Borel measure. Such series define genuine distributions. If, on the other hand, $\sum_0^\infty |\alpha_m|^2 < \infty$, the wavelet series belongs to every $L^p(\mathbb{T})$ space and even to $BMO(\mathbb{T})$.*

John and Nirenberg's space BMO will be studied in Chapter 5. We remark here that $L^p(\mathbb{T}) \subset L^q(\mathbb{T})$ when $p \ge q$ and that $BMO(\mathbb{T})$ is a subspace of all the $L^p(\mathbb{T})$ spaces.

The dichotomy described by Theorem 6 can be found in two other situations.

Suppose that $f(x) = \sum_0^\infty (a_k \cos \lambda_k x + b_k \sin \lambda_k x)$ is a Hadamard-lacunary Fourier series (that is, $\lambda_0 > 0$ and $\lambda_{k+1} \geq q\lambda_k$ for a fixed $q > 1$). Then either $\sum_0^\infty (|a_k|^2 + |b_k|^2) < \infty$ and $f(x)$ belongs to BMO and all the L^p spaces, or $\sum_0^\infty (|a_k|^2 + |b_k|^2) = \infty$ and $f(x)$ is necessarily a true distribution—it cannot be a Borel measure.

In the case of random Fourier series ([149]) there is also a dichotomy theorem, similar in every point except that the "ultimate" space, BMO, is not reached. But the random series $f(\omega, x) = \sum_0^\infty (\pm a_k \cos kx \pm b_k \sin kx)$, where the $\pm$ represent independent equidistributed Bernoulli random variables, almost surely defines a true distribution (which is not a Borel measure) when $\sum_0^\infty (|a_k|^2 + |b_k|^2) = \infty$ and, conversely, belongs to all the L^p spaces when the latter series converges.

To prove Theorem 6, suppose at first that $\sum |\alpha_m|^2 < \infty$. We want to show that $f(x)$ belongs to BMO. We shall feel free to make use of Carleson's criterion for BMO, not given until Theorem 4 of Chapter 5.

Returning to dyadic form, we write $m = 2^j + k$, for $0 \leq k < 2^j$, and get $g_m(x) = \psi_j(x - 2^{-j}k)$. We put $c(j,k) = \alpha_m$ and let $I(j,k)$ denote the dyadic interval $[2^{-j}k, 2^{-j}(k+1))$. Then "Carleson's condition" is the existence of a constant C such that, for every dyadic interval $I \subset [0,1)$, we have

$$(11.6) \qquad \sum_{I(j,k) \subset I} |c(j,k)|^2 \leq C|I|.$$

To check that this condition is satisfied, we note that the fact that the $|\alpha_m|$ are decreasing implies that $|c(j,k)| \leq |c(j,0)| = \omega_j$, say, for $0 \leq k < 2^j$. Further, $\sum |\alpha_m|^2 < \infty$ gives $\sum 2^j \omega_j^2 < \infty$. We now sum the left-hand side of (11.6) with j fixed (and $2^{-j} \leq |I|$). The number of values of k which appear is exactly $2^j |I|$, and the partial sum we are calculating has $2^j \omega_j^2 |I|$ as an upper bound. Summing over j now gives (11.6).

Suppose now that $\sum_0^\infty |\alpha_m|^2 = \infty$. We consider the partial sums $f_m(x) = \alpha_0 + \cdots + \alpha_m g_m(x)$. To show that f is not a bounded Borel measure, it is enough to apply Theorem 5 and then to show that $\|f_m\|_1$ tends to infinity. We shall do even better and prove that there exists a constant $c > 0$ such that

$$(11.7) \qquad c \left(\sum_0^m |\alpha_q|^2 \right)^{1/2} \leq \|f_m\|_1.$$

To get (11.7), we use the first part of the theorem. We have $\|f_m\|_4 \leq C\|f_m\|_{\text{BMO}} \leq C' \left(\sum_0^m |\alpha_q|^2 \right)^{1/2}$. But Hölder's inequality implies that $\|f_m\|_1 \geq \|f_m\|_2^3 \|f_m\|_4^{-2} \geq c \left(\sum_0^m |\alpha_q|^2 \right)^{1/2}$. This completes the proof.

As we mentioned at the beginning of this section, the construction of periodic wavelets generalizes easily to the n-dimensional case. In order not to embroil the reader in a welter of stifling notation, we restrict ourselves to the case of dimension 2. We use the functions of a real variable ϕ_j and ψ_j defined by (11.2) and (11.3). They enable us to define functions $g_k(x, y)$, $k \in \mathbb{N}$, which will be the periodic two-dimensional wavelets, of period 1 in both x and y. Those functions are given by $g_0(x, y) = 1$, $g_1(x, y) = \psi_0(x)\phi_0(y)$ (in fact, $\phi_0(y) = 1$), $g_2(x, y) = \phi_0(x)\psi_0(y)$ and $g_3(x, y) = \psi_0(x)\psi_0(y)$, then

$$g_4(x, y) = \psi(x)\phi_1(y), \ldots, g_{15}(x, y) = \psi_1(x - \frac{1}{2})\psi_1(y - \frac{1}{2}),$$

and then

$$g_{16}(x, y) = \psi_2(x)\phi_2(y), \ldots, g_{63}(x, y) = \psi_2(x - \frac{3}{4})\psi_2(y - \frac{3}{4}),$$

and so forth. As can be seen, this sequence falls into blocks, and the $g_k(x, y)$ such that $4^j \leq k < 4^{j+1}$ form an orthonormal basis of the orthogonal complement of $P_j \otimes P_j$ in $P_{j+1} \otimes P_{j+1}$.

12 Notes and comments

As we indicated in the introduction, the first orthonormal wavelet basis was constructed by J.O. Strömberg in [223]. This is how he did it. Let V_j, $j \in \mathbb{Z}$, denote the multiresolution approximation of $L^2(\mathbb{R})$ given by the spline functions of order $r \geq 1$. To construct a function ψ belonging to W_0 and such that $\psi(x - k)$, $k \in \mathbb{Z}$, is an orthonormal basis of W_0, Strömberg introduced an intermediate space between V_0 and V_1, which we denote by $V_{1/2}$. A function f lies in $V_{1/2}$ if and only if f belongs to V_1 and the restriction of f to $[0, \infty)$ coincides with the restriction of a function of V_0 to $[0, \infty)$: in other words, the points of discontinuity of the rth derivative of f lie in $\mathbb{N} \cup -\mathbb{N}/2$. Let $T : L^2(\mathbb{R}) \to L^2(\mathbb{R})$ be the translation operator defined by $Tf(x) = f(x-1)$. Then $V_{1/2} \subset T(V_{1/2})$, the $T^k(V_{1/2})$ form a nested sequence, their union is dense in V_1 and their intersection is V_0. Lastly, $V_{1/2}$ has codimension 1 in $T(V_{1/2})$, in fact, $V_{1/2}$ is the kernel of the linear form λ defined on $T(V_{1/2})$ by

$$\lambda(f) = \left(\frac{d}{dx}\right)^r f(\frac{1}{2} + 0) - \left(\frac{d}{dx}\right)^r f(\frac{1}{2} - 0).$$

Strömberg then considered the function $\psi \in T(V_{1/2})$, normalized by $\|\psi\|_2 = 1$, which is orthogonal to $V_{1/2}$ in the Hilbert space $L^2(\mathbb{R})$. The function ψ is unique up to multiplication by a constant of modulus 1. Clearly, $\psi(x - k)$, $k \in \mathbb{Z}$, is an orthonormal basis of the orthogonal complement W_0 of V_0 in V_1. Indeed, W_0 is the direct sum of the orthogonal complements R_k of $T^k(V_{1/2})$ in $T^{k+1}(V_{1/2})$, and we have $R_k = T^k(R_0)$.

We must still calculate ψ. We depart from Strömberg's method and indicate that the calculation follows from the general algorithms of this chapter, as long as we make the right choice of function $\phi \in V_0$ such that $\phi(x-k)$, $k \in \mathbb{Z}$, is an orthonormal basis of V_0. Any two choices ϕ_1 and ϕ_2 are related by the identity $\hat{\phi}(\xi) = \chi(\xi)\hat{\phi}_1(\xi)$, where $\chi(\xi)$ is a 2π-periodic, unimodular, real-analytic function (if ϕ_1 and ϕ_2 decrease exponentially). All we need do is to require that ϕ have support in $(-\infty, 0]$: then ϕ is unique up to multiplication by a constant of modulus 1. Having made the right choice of ϕ, we define the 2π-periodic, real-analytic function $m_0(\xi)$ by $\hat{\phi}(2\xi) = m_0(\xi)\hat{\phi}(\xi)$, next $m_1(\xi)$ by $m_1(\xi) = e^{-i\xi}\bar{m}_0(\xi + \pi)$ and, finally, ψ by $\hat{\psi}(2\xi) = m_1(\xi)\hat{\phi}(\xi)$. Then ψ is Strömberg's wavelet (up to multiplication by a constant of modulus 1). It follows that $\psi(x)$ is exponential at infinity: this exponential decrease is the same as that of the wavelets symmetric about $1/2$ which we have singled out in this chapter.

Strömberg's discovery did not get the publicity it deserved. His wavelets could have been used straightaway in the work of G. Battle and P. Federbush ([8], [9] and [10]) on renormalization in quantum field theory.

The next orthonormal wavelet basis was discovered by chance during the summer of 1985. I then believed that no function ψ in the Schwartz class $S(\mathbb{R})$ could be such that the sequence $2^{j/2}\psi(2^j x - k)$, $j, k \in \mathbb{Z}$, would form an orthonormal basis of $L^2(\mathbb{R})$. But this initial point of view had to change in the light of the bizarre calculations and manipulations leading to the construction of the Littlewood-Paley wavelet.

The calculation of the corresponding multi-dimensional wavelets was done in collaboration with P.G. Lemarié ([166]).

A year later, and with Mallat's collaboration, the notion of r-regular multiresolution approximation saw the light of day. All that had appeared miraculous in our previous calculations now became very simple and quite natural.

At this point, it is appropriate to recall Calderón's well-known identity ([35]) which is a continuous version of wavelet series expansions. Let $\psi \in L^1(\mathbb{R}^n)$ be a function whose integral is zero and whose Fourier transform $\hat{\psi}(\xi)$ satisfies

$$\int_0^\infty |\hat{\psi}(t\xi)|^2 \frac{dt}{t} = 1,$$

for each $\xi \neq 0$. We put $\tilde{\psi}(x) = \bar{\psi}(-x)$, $\psi_t(x) = t^{-n}\psi(x/t)$ and similarly for $\tilde{\psi}_t$.

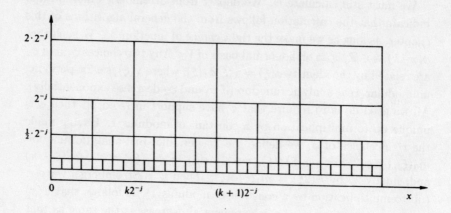

Figure 5

Then, for every function $f \in L^2(\mathbb{R}^n)$, we have

$$(12.1) \qquad f = \int_0^\infty f * \tilde{\psi}_t * \psi_t \frac{dt}{t} .$$

To see this, it is enough to take the Fourier transforms of both sides.
This leads to the straightforward formula

$$\hat{f}(\xi) = \int_0^\infty \hat{f}(\xi)|\hat{\psi}(t\xi)|^2 \frac{dt}{t} ,$$

which results from the condition on ψ.

Grossmann and Morlet give Calderón's identity the following remark-
able interpretation. We use ψ to define wavelets $\psi_{(a,b)}$, $a > 0$, $b \in \mathbb{R}^n$,
by $\psi_{(a,b)}(x) = a^{-n/2}\psi((x - b)/a)$: b is the centre of the wavelet and a
is its width. We then let the "wavelet transformation" of $f \in L^2(\mathbb{R}^n)$
be the function defined on $\mathbb{R}^{n+1}_+$ by $F(a, b) = (f, \psi_{(a,b)})$. Once these
wavelet coefficients have been calculated, we can recover f completely
by the formula

$$(12.2) \qquad f(x) = \int_0^\infty \int_{\mathbb{R}^n} F(a, b)\psi_{(a,b)}(x)\, db \frac{da}{a^{1+n}} .$$

All now follows as if the $\psi_{(a,b)}$ formed an orthonormal wavelet basis: after calculating the wavelet coefficients as if we were dealing with
coordinates of f, we use the coordinates as coefficients to recover f.

It goes without saying that (12.2) is just a paraphrase of (12.1), but
its geometrical language is more striking.

Finally—and we shall come back to this in the next chapter—we can
replace (12.2) by a discrete version, obtained by paving $\mathbb{R}^{n+1}$ by the
"Whitney cubes" of Figure 5.

We summarily replace a by 2^{-j} and b by $k2^{-j}$, $k \in \mathbb{Z}^n$, then $da\,db$ by the volume of the Whitney cube, that is, $2^{-(n+1)j}$. This volume exactly compensates for a^{1+n}, and (12.2) becomes

$$(12.3) \qquad f(x) \cong \sum_j \sum_k \alpha(j,k) 2^{nj/2} \psi(2^j x - k) \,.$$

Of course, (12.3) is just an approximation. The mathematical theory of this type of approximation—which gives an exact formula on iteration—has been developed by Daubechies and forms the material of the next chapter.

We should not close this chapter without saying a few words about the application of wavelets to renormalization calculations in constructive field theory. After several stages, which we shall not mention here, the problem becomes that of defining particular path integrals for spaces of functions of 2, 3 or 4 real variables. We follow P. Federbush and restrict ourselves to the case of dimension 2.

We must try to interpret and then calculate integrals of the form

$$I(p) = \int_\Omega p(\phi) e^{-S(\phi)} \, d\phi \,,$$

where Ω is the vector space of real-valued functions—actually, and unfortunately, distributions—defined on $\mathbb{R}^2$ such that

$$S(\phi) = \frac{1}{2} \|\nabla\phi\|_2^2 + \frac{M^2}{2} \|\phi\|_2^2 + \lambda\|\phi\|_4^4 \,,$$

where $M > 0$ is the mass and $\lambda > 0$ a small parameter. We put

$$S_0(\phi) = \frac{1}{2} \|\nabla\phi\|_2^2 + \frac{M^2}{2} \|\phi\|_2^2 \qquad \text{and} \qquad S_I(\phi) = \lambda\|\phi\|_4^4 \,,$$

where I stands for "non-linear interaction".

We ought to make the meaning of the measure $d\phi$ more precise. We decide to choose the measure which works for $\lambda = 0$ and to imitate the definition of a Wiener integral in that case.

To this end, we want to make a "change of variables" in which ϕ becomes a sequence $\alpha = (\alpha_0, \alpha_1, \ldots)$ while $d\phi$ becomes the normalized Gaussian measure $\prod_0^\infty \mu_j$, where $d\mu_j = (2\pi)^{-1/2} e^{-\alpha_j^2/2} \, d\alpha_j$. We suppose that our change of variables is linear and bijective. So $\phi = \alpha_0 u_0 + \cdots + \alpha_k u_k + \cdots$ and this series expansion is such that the measure $\exp((-\|\nabla\phi\|_2^2 - M^2\|\phi\|_2^2)/2) \, d\phi$ becomes $\prod_0^\infty \mu_j$. This implies that $\|\nabla\phi\|_2^2 + M^2\|\phi\|_2^2 = ((M^2 - \Delta)\phi, \phi) = \sum_0^\infty \alpha_k^2$. In other words, the functions $(-\Delta + M^2)^{1/2} u_k$ must form an orthonormal basis of $L^2(\mathbb{R}^2)$. G. Battle and Federbush choose $(-\Delta + M^2)^{1/2} u_k$ to be precisely an orthonormal wavelet basis. In fact, these authors consider only dyadic cubes of side ≤ 1, that is, they use the orthonormal basis composed of the functions $\phi(x - \lambda)$, $\lambda \in \mathbb{R}^n$, and the wavelets ψ_λ, $\lambda \in \Lambda_j$, $j \geq 0$. To be consistent with Battle and Federbush's notation, we let ψ_k, $k \geq 0$, denote this orthonormal basis and obtain the remarkable identity $\phi(x) = (-\Delta + M^2)^{-1/2} \sum_0^\infty \alpha_k \psi_k$. This means that $\phi(x)$ appears as a random series where the coefficients are independant normal variables of the same law. The situation is analogous to that of the Wiener integral, where the "variable of integration" describes the set of trajectories of Brownian motion.

In fact, a representation of Brownian motion can be obtained precisely by calculating the primitives of the random series $\sum \alpha_k \psi_k(t)$, where, this time, the $\psi_k(t)$ are one-dimensional wavelets and the α_k have the same meaning as above.

Let us return to the calculation of the integral

$$I(p, \lambda) = \int_\Omega p(\phi) e^{-S(\phi)} \, d\phi = \int_\Omega p(\phi) e^{-\lambda \|\phi\|_4^4 - S_0(\phi)} \, d\phi.$$

We must make the meaning of $p(\phi)$ precise. It is a polynomial, in the usual sense, in a finite number of variables α_k. So the integrals $I(p, \lambda)$ arise as the moments of a probability measure (whose existence for $\lambda > 0$ is exactly the problem posed by constructive field theory).

Now we use the characterization of classical spaces of functions or distributions by the order of magnitude of wavelet coefficients to be found in Chapter 6. We want to know whether $\|\phi\|_4$ is almost surely finite, with respect to the measure $e^{-S_0(\phi)} \, d\phi$. This amounts to knowing whether $\sum \alpha_k \psi_k(x)$ belongs almost surely to the space $W^{-1,4}$ whose elements are defined to be the distributions $S = f_0 + \partial f_1/\partial x + \partial f_2/\partial y$, where f_0, f_1 and f_2 belong to $L^4(\mathbb{R}^2)$. Applying the criterion to be given in Chapter 6 (which involves only the moduli of the wavelet coefficients) we find that $\|\phi\|_4^4$ is almost surely infinite! The integral $I(p, \lambda)$ behaves like an infinite product diverging to 0. To get some sense out of this, we must *renormalize*. The first idea which comes to mind is that the manner of divergence of our integral does not depend greatly on the choice of the polynomial p and thus that the quotient $I(p_2, \lambda)/I(p_1, \lambda)$, where p_1 and p_2 are polynomials, has a better chance of converging than $I(p, \lambda)$. This point of view leads to passing to the limit, as in the case of an infinite product, by restricting the indices k of α_k to finite sets A whose union is $\mathbb{N}$. In $S_0(\phi)$ and $S_I(\phi)$, we substitute 0 for all the α_k with $k \notin A$. The integrals then involve a finite set of real variables and the first renormalization consists—in the obvious notation—of looking for the limit of the quotient $I^A(p, \lambda)/I^A(1, \lambda)$ as A describes an increasing sequence of finite subsets of $\mathbb{N}$ whose union is $\mathbb{N}$.

But this is not enough, and we have to proceed to a second renormalization applied at the same time as the first (called "cutoff" in Feder-bush's text). This second renormalization consists of taming the term $\|\phi\|_4^4$ (the cause of the divergence to 0) by replacing it by an algebraic expression having greater cancellation. We refer the reader to [107] and to the references in that article.

4

Non-orthogonal wavelets

1 Introduction

Well before orthonormal wavelet bases existed, wavelets had been used by J. Morlet (a geophysical engineer with O.R.I.C., Elf-Aquitaine) for the numerical processing of seismic signals recorded during oil prospecting expeditions.

Morlet's methods were mathematically justified, *post facto*, by Daubechies ([87]) and this chapter is dedicated to the statement and proof of the L^2 convergence of Morlet's iterative algorithm.

Unlike the case of orthogonal wavelets, L^2 convergence does not necessarily imply that "Morlet's wavelets" can be used in any function space other than the reference space L^2. In fact, results by P. Tchamitchian and then by P.G. Lemarié have enabled the following to be established: for every exponent $p > 2$, there exists a function $\theta(x)$, of the real variable x, belonging to the Schwartz class $S(\mathbb{R})$, all of whose moments are zero and which satisfies two apparently contradictory properties as follows:

(a) the collection of functions $2^{j/2}\theta(2^j x - k)$, $j, k \in \mathbb{Z}$, is a Riesz basis of $L^2(\mathbb{R})$;

(b) the above collection is not complete in $L^p(\mathbb{R})$.

These properties are not due to any special pathology of the spaces $L^p(\mathbb{R})$, $2 < p < \infty$, which are, in any case, not in the least pathological. The same happens if we try to decompose the Hölder spaces C^α using non-orthogonal wavelets. There is, in fact, a similar statement, namely that, for every $m \geq 1$ and every α in the interval $(0, m)$, there exist two C^m functions ϕ and ψ of compact support, such that the collection of functions $\phi(x - k)$, $k \in \mathbf{Z}$, and $2^{j/2}\psi(2^j x - k)$, $j \in \mathbf{N}$, $k \in \mathbf{Z}$, is a Riesz basis of $L^2(\mathbf{R})$, while the self-same collection is not complete in C^α (with the weak topology, in which the test functions are dense in C^α).

Similarly, the above collection is not complete in the space of continuous functions with the topology of uniform convergence on compacta.

The use of non-orthogonal wavelets in function spaces other than L^2 poses tough questions which we do not yet know how to resolve.

2 Frames

Let H be a Hilbert space with inner product denoted by $\langle \cdot, \cdot \rangle$. Let Λ be a discrete set, let $l^2(\Lambda)$ be the space of square-summable families x_λ, $\lambda \in \Lambda$, and let ε_λ, $\lambda \in \Lambda$ be the canonical basis of $l^2(\Lambda)$ given by $\varepsilon(\lambda') = 1$, if $\lambda' = \lambda$, and 0 otherwise.

A family of vectors e_λ, $\lambda \in \Lambda$, in the Hilbert space H is a *frame* if the linear operator T, defined by $T(\varepsilon_\lambda) = e_\lambda$, $\lambda \in \Lambda$, on finite linear combinations of the ε_λ, can be extended to a continuous linear mapping of $l^2(\Lambda)$ to H and if the extended operator T is a surjection.

This means that there exists a constant C_1 such that

$$(2.1) \qquad \left\| \sum \alpha_\lambda e_\lambda \right\|_H \leq C_1 \left(\sum |\alpha_\lambda|^2 \right)^{1/2}$$

and a constant C_2 (given by Banach's Open Mapping Theorem) such that, for each $x \in H$, there exists a sequence $\beta_\lambda \in l^2(\Lambda)$ such that

$$(2.2) \qquad x = \sum \beta_\lambda e_\lambda \quad \text{and} \quad \left(\sum |\beta_\lambda|^2 \right)^{1/2} \leq C_2 \|x\|.$$

We then form the adjoint operator $T^* : H \to l^2(\Lambda)$ of T. This operator is continuous and we get $\|T^*(x)\| \geq (C_2)^{-1}\|x\|$, for every $x \in H$. Explicitly, the two properties of T^* give

$$(2.3) \qquad c_1\|x\| \leq \left(\sum_{\lambda \in \Lambda} |\langle x, e_\lambda \rangle|^2 \right)^{1/2} \leq c_2\|x\|,$$

for two constants $c_2 \geq c_1 > 0$, which is more useful in practice.

We can transform (2.3) again, by introducing a new continuous linear operator $L : H \to H$, defined by

$$(2.4) \qquad L(x) = \sum_{\lambda \in \Lambda} \langle x, e_\lambda \rangle e_\lambda.$$

Then the meaning of (2.3) is precisely that, in the sense of self-adjoint operators,

$$(2.5) \qquad c_1 I \leq L \leq c_2 I.$$

Renormalizing, if necessary, by multiplying each e_λ by the same constant $\delta > 0$, we may suppose that $c_2 < 2$ in (2.5).

The existence of the two constants c_1 and c_2, such that $0 < c_1 \le c_2 < 2$ and (2.5) holds, is equivalent to the existence of one constant $r < 1$ such that

$$(2.6) \qquad \qquad \|I - L\| \le r.$$

Here is how engineers read (2.6). The family e_λ, $\lambda \in \Lambda$, is treated as if it were an orthonormal basis. So one proceeds as if x could be written in the form $x = \sum_{\lambda \in \Lambda} \langle x, e_\lambda \rangle e_\lambda + x_1$, where the error term x_1 is small. In fact $\|x_1\| \le r\|x\|$, so this algorithm can be iterated to obtain, successively

$$x_1 = \sum_{\lambda \in \Lambda} \langle x_1, e_\lambda \rangle e_\lambda + x_2$$

$$\dots$$

$$x_k = \sum_{\lambda \in \Lambda} \langle x_k, e_\lambda \rangle e_\lambda + x_{k+1}$$

$$\dots$$

Clearly, $\|x_k\| \le r^k \|x\|$, and by putting all these identities together, an exact decomposition

$$(2.7) \qquad \qquad x = \sum_{\lambda \in \Lambda} \langle y, e_\lambda \rangle e_\lambda$$

is obtained, where y can be effectively calculated from x. Setting $L = I + R$ gives $x_k = (-R)^k x$ and thus $y = L^{-1}(x)$.

Supposing only that every vector $x \in H$ can be written robustly and stably (in the sense of (2.1) and (2.2)) as a linear combination of the e_λ, $\lambda \in \Lambda$, we have finally got a simple and effective algorithm for giving a special decomposition.

An important special class of frames is given by Riesz bases, where, in addition to (2.1) and (2.2), we have the inverse inequality to (2.1), namely

$$(2.8) \qquad \qquad \left\| \sum \alpha_\lambda e_\lambda \right\| \ge \delta \left(\sum |\alpha|^2 \right)^{1/2},$$

for a certain constant $\delta > 0$.

In other words, the operator $T : l^2(\Lambda) \to H$ is an isomorphism of Hilbert spaces.

3 Ingrid Daubechies' criterion

Let $\psi \in L^1(\mathbb{R})$ be a function of zero mean. It defines a sequence $\beta(k)$,

$k \in \mathbb{Z}$, by the formula

(3.1) $$\beta(k) = \sup_{1 \le |\xi| \le 2} \sum_{-\infty}^{\infty} |\hat{\psi}(2^j \xi)||\hat{\psi}(2^j \xi + 2k\pi)|.$$

We now suppose that

$$\sup_{1 \le |\xi| \le 2} \sum_{-\infty}^{\infty} |\hat{\psi}(2^j \xi)|^2 \le C < \infty$$

and introduce the constant

(3.2) $$\gamma = \inf_{1 \le |\xi| \le 2} \sum_{-\infty}^{\infty} |\hat{\psi}(2^j \xi)|^2.$$

Theorem 1. *With the above notation and under the above hypotheses let*

(3.3) $$\gamma > 2 \sum_{1}^{\infty} \Big(\beta(k)\beta(-k) \Big)^{1/2}.$$

Then the collection $2^{j/2}\psi(2^j x - k)$, $j, k \in \mathbb{Z}$, is a frame for $L^2(\mathbb{R})$.

To see this, we use criterion (2.3), which leads to evaluating the sum

(3.4) $$S(f) = \sum_{j} \sum_{k} |(f, \psi_{j,k})|^2,$$

where

$$\psi_{j,k}(x) = 2^{j/2}\psi(2^j x - k).$$

Parseval's identity gives

$$(f, \psi_{j,k}) = 2^{-j/2} \frac{1}{2\pi} \int_{-\infty}^{\infty} \hat{f}(\xi)\bar{\hat{\psi}}(2^{-j}\xi)e^{ik2^{-j}\xi} d\xi.$$

We use the following remark to calculate $\sum_k |(f, \psi_{j,k})|^2$, for fixed $j \in \mathbb{Z}$:

Lemma 1. *Let $g(t)$ be a function in $L^1 \cap L^2$ and, for any $\delta > 0$, let $G(t)$ be the $2\pi\delta^{-1}$-periodic function defined by $G(t) = \sum_{-\infty}^{\infty} g(t + 2k\pi\delta^{-1})$. Then*

$$\sum_{-\infty}^{\infty} \left| \int_{-\infty}^{\infty} g(t)e^{ik\delta t} dt \right|^2 = \frac{2\pi}{\delta} \int_{0}^{2\pi/\delta} |G(t)|^2 dt.$$

This lemma is just Plancherel's identity for the $2\pi\delta^{-1}$-periodic function $G(t)$.

In our case, $\delta = 2^{-j}$ and $g(\xi) = \hat{f}(\xi)\bar{\hat{\psi}}(2^{-j}\xi)$, so that $G(\xi) = G_j(\xi) = \sum_{-\infty}^{\infty} \hat{f}(\xi + 2\pi l 2^j)\bar{\hat{\psi}}(2^{-j}\xi + 2\pi l)$. So, in the end, we get

(3.5) $$S(f) = \frac{1}{2\pi} \sum_{-\infty}^{\infty} \int_{0}^{2\pi 2^j} |G_j(\xi)|^2 d\xi.$$

We then expand $|G_j(\xi)|^2$, using the double inequality

$$\sum |z_l|^2 - 2\sum\sum_{l<m} |z_l||z_m| \le |\sum z_l|^2 \le \sum |z_l|^2 + 2\sum\sum_{l<m} |z_l||z_m|.$$

After integrating with respect to ξ, we get

(3.6) $$S_0(f) - 2S_1(f) \le S(f) \le S_0(f) + 2S_1(f).$$

The plan of the proof is to check that

(3.7) $$\gamma\|f\|_2^2 \le S_0(f) \le C\|f\|_2^2$$

and also that

(3.8) $$S_1(f) \le \sum_1^\infty \left(\beta(k)\beta(-k)\right)^{1/2}\|f\|_2^2.$$

The principal term $S_0(f)$ is easily estimated. We first observe that the intervals $[2\pi l 2^j, 2\pi(l+1)2^j)$ form a partition of $\mathbb{R}$. Summing over $l \in \mathbb{Z}$, we get $S_0(f) = (1/2\pi)\int_{-\infty}^\infty |\hat{f}(\xi)|^2 \sum_{-\infty}^\infty |\hat{\psi}(2^{-j}\xi)|^2\, d\xi$. We then use the hypotheses on ψ, and (3.7) follows.

To estimate $S_1(f)$, we write

$$S_1(f) = \frac{1}{\pi}\sum_{-\infty}^\infty \sum\sum_{l<m} I(j,l,m),$$

where

$$I(j,l,m) =$$

$$\int_0^{2\pi 2^j} |\hat{f}(\xi+2\pi l 2^j)||\hat{\psi}(2^{-j}\xi+2\pi l)||\hat{f}(\xi+2\pi m 2^j)||\hat{\psi}(2^{-j}\xi+2\pi m)|\, d\xi.$$

The first step in simplifying the calculation of $S_1(f)$ is to set $m = l+k$, $k \ge 1$, and then to put $t = \xi + 2\pi l 2^j$. The two variables $l \in \mathbb{Z}$ and $\xi \in [0, 2\pi 2^j)$ are thus replaced by $t \in \mathbb{R}$, so we are left with $k \ge 1$ and $t \in \mathbb{R}$. This gives

$$S_1(f) = \frac{1}{\pi}\sum_1^\infty \int_{-\infty}^\infty \sigma_k(t)|\hat{f}(t)|\, dt,$$

where

$$\sigma_k(t) = \sum_{-\infty}^\infty |\hat{\psi}(2^{-j}t)||\hat{f}(t+2\pi k 2^j)||\hat{\psi}(2^{-j}t+2\pi k)|.$$

We then use the definition of $\beta(k)$ and the Cauchy-Schwarz inequality to get

$$\sigma_k(t) \le \left(\beta(k)\right)^{1/2}\left(\sum_{-\infty}^\infty |\hat{\psi}(2^{-j}t)||\hat{\psi}(2^{-j}t+2\pi k)||\hat{f}(t+2\pi k 2^j)|^2\right)^{1/2}.$$

Another application of Cauchy-Schwarz gives

$$\int_{-\infty}^\infty \sigma_k(t)|\hat{f}(t)|\, dt \le \|f\|_2\|\sigma_k\|_2 = \sqrt{2\pi}\|f\|_2\|\sigma_k\|_2.$$

This leads to an evaluation of the quantities

$$I_k = \sum_{-\infty}^{\infty} \int_{-\infty}^{\infty} |\hat{\psi}(2^{-j}t)||\hat{\psi}(2^{-j}t + 2\pi k)||\hat{f}(t + 2\pi k 2^j)|^2 \, dt.$$

To do this, we make the change of variable $s = t + 2\pi k 2^j$ and get

$$I_k = \sum_{-\infty}^{\infty} \int_{-\infty}^{\infty} |\hat{\psi}(2^{-j}s)||\hat{\psi}(2^{-j}s - 2\pi k)||\hat{f}(s)|^2 \, ds \le \beta(-k)\|\hat{f}\|_2^2$$

$$= 2\pi\beta(-k)\|f\|_2^2.$$

This gives

$$S_1(f) \le 2\sum_1^{\infty} \left(\beta(k)\beta(-k)\right)^{1/2} \|f\|_2^2,$$

which concludes the proof of Theorem 1.

Armed with this criterion and a small computer, Daubechies took the wavelet $\psi(x) = (2/\sqrt{3})\pi^{-1/4}(1-x^2)e^{-x^2/2}$, that is, the second derivative of the Gaussian $e^{-x^2/2}$, normalized so that $\|\psi\|_2 = 1$. She showed that $2^{j/2}\psi(2^j x - k)$, $j, k \in \mathbb{Z}$, was a frame. The method used also enabled the calculation of the rate of convergence of the iterative algorithm of Section 2. This example is important because of its use by Morlet in the numerical processing of seismic signals ([121]).

In practice, we "oversample" by a long way and, with the same choice of ψ, form the frame

$$\psi_{j,k}^{\sharp}(x) = \gamma 2^{j/8}\psi(2^{j/4}x - k/2), \qquad j, k \in \mathbb{Z}, \gamma > 0.$$

Then, with a convenient choice of $\gamma > 0$, we get $r < 10^{-2}$ in (2.6). As a first approximation, we can then decompose f using the $\psi_{j,k}^{\sharp}$ as if they formed an orthonormal basis. If the precision is not satisfactory, then one or two iterations suffice.

It is worth stressing that these calculations of rates of convergence only apply to the norm in $L^2(\mathbb{R}^n)$. We shall shortly examine the problems encountered in other function spaces.

A further remarkable application of Daubechies' criterion is the improvement of the algorithms for atomic decomposition of functions in the complex Hardy space $\mathbb{H}^2(\mathbb{R})$, whose definition we now recall.

Let Ω denote the open upper half-plane $y > 0$ in the complex plane. Then $\mathbb{H}^2$ is the Hilbert space of functions $f(z)$ which are holomorphic on Ω and satisfy $\sup_{y>0} \int_{-\infty}^{\infty} |f(x + iy)|^2 \, dx < \infty$. If this condition is satisfied, $f(x)$ has a trace on the real axis, denoted by $f(x + i0)$, and defined by $f(x + i0) = \lim_{y \downarrow 0} f(x + iy)$. This limit exists in $L^2(\mathbb{R})$ and also almost everywhere. Lastly, the Hilbert space $\mathbb{H}^2$ embeds isometrically in $L^2(\mathbb{R})$ and is defined by the Paley-Wiener theorem: $f \in \mathbb{H}^2$ if and only if $f(x + i0) \in L^2(\mathbb{R})$ and $\hat{f}(\xi) = 0$ for $\xi < 0$.

We do not yet know whether there exists a function ψ belonging to the Schwartz class $\mathcal{S}(\mathbb{R})$ and to $\mathbb{H}^2$ such that the functions $2^{j/2}\psi(2^j x - k)$, $j, k \in \mathbb{Z}$, form a Hilbert basis of $\mathbb{H}^2$ or even a Riesz basis of $\mathbb{H}^2$.

On the other hand, Daubechies' criterion easily provides frames for $\mathbb{H}^2(\mathbb{R})$. The existence of such frames was a consequence of results of G. Weiss and his coworkers ([73]). The example below is the first to have given explicit numerical constants.

We take $\psi(x) = (x + i)^{-2}$, which clearly belongs to $\mathbb{H}^2(\mathbb{R})$. Then Daubechies' criterion shows that the functions $2^{j/2}\psi(2^j x - k)$, $j, k \in \mathbb{Z}$, constitute a frame for $\mathbb{H}^2(\mathbb{R})$. We can even restrict ourselves to the functions with $k \in 2\mathbb{Z}$: taking only one in every two functions, we still get a frame for $\mathbb{H}^2$. But, in the other direction, it is easy to show that there is a real constant $q_0 > 2$ such that, when $q > q_0$, the functions $2^{j/2}(2^j x - kq + i)^{-2}$, $j, k \in \mathbb{Z}$, form an unconditional basis of a proper subspace $F \subset \mathbb{H}^2$, $F \neq \mathbb{H}^2$.

Is there a critical value q_0 such that, when $1 \leq q < q_0$, the functions $2^{j/2}(2^j x - kq + i)^{-2}$, $j, k \in \mathbb{Z}$ constitute a frame for $\mathbb{H}^2$ and when $q > q_0$ the functions $2^{j/2}(2^j x - kq + i)^{-2}$ form a Riesz basis of a proper subspace of $\mathbb{H}^2$? We do not know the answer to this question.

4 Riesz Bases and L^p convergence

We start with a wavelet $\psi \in \mathcal{S}(\mathbb{R})$ such that the collection $2^{j/2}\psi(2^j x - k)$, $j, k \in \mathbb{Z}$, is an orthonormal basis of $L^2(\mathbb{R})$. We know that all the moments of ψ are then zero. We can even choose ψ so that its Fourier transform has support in $[2\pi/3, 8\pi/3]$.

Let $0 \leq r \leq 1$. With the help of this parameter, we construct a perturbation of ψ, denoted by θ and defined by $\theta(x) = \psi(x) - r\sqrt{2}\psi(2x)$. Then

$$2^{j/2}\theta(2^j x - k) = \theta_{j,k}(x) = 2^{j/2}\psi(2^j x - k) - r2^{(j+1)/2}\psi(2^{j+1}x - 2k).$$

In other words, with the obvious notation, $\theta_{j,k} = (1 - rU)\psi_{j,k}$, where $U : L^2(\mathbb{R}) \to L^2(\mathbb{R})$ is the partial isometry defined by $U(\psi_{j,k}) = \psi_{j+1,2k}$.

Theorem 2. If $2 < p < \infty$ and if $2^{-(1/2-1/p)} < r < 1$, then the set of functions $\theta_{j,k}$ is a Riesz basis of $L^2(\mathbb{R})$ but is not total in $L^p(\mathbb{R})$.

More precisely, we show that the closure in $L^p(\mathbb{R})$ of the vector space of linear combinations of the $\theta_{j,k}$ has infinite codimension.

We first verify that the $\theta_{j,k}$ form a Riesz basis of $L^2(\mathbb{R})$: it is enough to observe that the operator $1 - rU : L^2(\mathbb{R}) \to L^2(\mathbb{R})$ is an isomorphism, as long as $r \in [0, 1)$.

For the other half of the proof of Theorem 2, it is enough to construct a function $g \in L^q(\mathbb{R})$ which is orthogonal to all the functions $\theta_{j,k}$ without being identically zero. We therefore require $(g, \psi_{j,k}) = r(g, \psi_{j+1,2k})$ for all $j, k \in \mathbb{Z}$. This gives an infinity of choices, of which we describe one example. We suppose g is decomposed according to the wavelet basis $\psi_{j,k}$. Then $g(x) = \sum_j \sum_k \alpha(j,k)\psi_{j,k}$, and we take $\alpha(j,k) = 0$, if $j = 0$ and $k \neq 1$, whereas $\alpha(0,1) = 1$. This, together with the equation $\alpha(j,k) = r\alpha(j+1, 2k)$, implies that $\alpha(j,k) = 0$ if $j \leq -1$. Further, if $j \geq 1$, $\alpha(j, 2^j) = 1/r^j$ and $\alpha(j,k) = 0$ if $k \neq 2^j$.

Then $g(x) = \sum_0^\infty g_j(x)$ and $\|g_j\|_q = r^{-j} 2^{j/2} 2^{-j/q} \|\psi\|_q$. It follows that $g \in L^q(\mathbb{R})$ as long as $2^{1/2-1/q} < r$. Theorem 2 is fully proven.

Let us now show how to adapt the proof of Theorem 2 to the case of Hölder spaces C^α, for $\alpha > 0$. We begin by noting that C^α is the dual of the Besov space $B_1^{-\alpha,1}$ and that the functions of the Schwartz space $\mathcal{D}(\mathbb{R})$ form a dense linear subspace of C^α in the $\sigma(C^\alpha, B_1^{-\alpha,1})$ topology. So the natural question is to ask whether the collection of functions $\theta_{j,k}$, $j, k \in \mathbb{Z}$, is complete in C^α with the $\sigma(C^\alpha, B_1^{-\alpha,1})$ topology. To show that this is not the case, it is enough to construct a function $g \in B_1^{-\alpha,1}$ which is orthogonal to all the functions $\theta_{j,k}$, without being identically zero.

We repeat the same construction as in the $L^p(\mathbb{R})$ case. We get $g(x) = \sum_0^\infty g_j(x)$ and the norm of $g(x)$ in $B_1^{-\alpha,1}$ is $O(r^{-j} 2^{-j(\alpha+1/2)})$. The conclusion is that the series converges if $r > 2^{-(\alpha+1/2)}$.

The significance of this is that, even when r is very close to 0, not all functions $f \in C^\alpha$ can be approximated by linear combinations of the functions $\theta_{j,k}$, $j, k \in \mathbb{Z}$, when

$$\alpha + \frac{1}{2} > \log_2\left(\frac{1}{r}\right).$$

Before concluding this chapter, we mention that it is not known whether this pathology still occurs for Morlet's "natural" frames. To be precise, if we use the wavelet $\psi(x) = (1 - x^2)e^{-x^2/2}$, we do not know whether the functions $2^{j/2}\psi(2^j x - k)$, $j, k \in \mathbb{Z}$, form a complete set in $L^p(\mathbb{R})$ for $1 < p < \infty$.

5

Wavelets, the Hardy space H^1 and its dual BMO

1 Introduction

This chapter is devoted to a particularly elegant application of wavelet theory: the study of the (real) Hardy space $H^1(\mathbb{R}^n)$ and its dual BMO($\mathbb{R}^n$).

Let us start with some general definitions for series of vectors in a Banach space.

Let B be a Banach space and $\sum_0^\infty x_k$ a series of elements of B. We say that this series *converges unconditionally* to an element $x \in B$ if, for each $\varepsilon > 0$, there exists a finite set $F(\varepsilon) \subset \mathbb{N}$ such that, for every finite set $F \subset \mathbb{N}$ containing $F(\varepsilon)$,

$$\left\| x - \sum_{k \in F} x_k \right\| \leq \varepsilon.$$

We say that a series $\sum_0^\infty x_k$ is *unconditionally convergent* if there exists an x such that the series converges unconditionally to x. Thus, if a series $\sum_0^\infty x_k$ is unconditionally convergent, there exists a constant C such that, for every integer $n \geq 1$ and every sequence $\alpha_0, \ldots, \alpha_n$ of real or complex numbers satisfying $|\alpha_0| \leq 1, \ldots, |\alpha_n| \leq 1$, we have

$$\| \alpha_0 x_0 + \cdots + \alpha_n x_n \| \leq C.$$

For a finite-dimensional space, this noteworthy inequality implies that the series converges absolutely. That is clearly not the case for infinite-dimensional spaces.

Another way to express the unconditional convergence of a series of vectors $\sum_0^\infty x_k$ of B is to write that the series is *commutatively* convergent: for every permutation of the integers $\sigma : \mathbb{N} \to \mathbb{N}$, the series $\sum_0^\infty x_{\sigma(k)}$ converges in norm to $x = \sum_0^\infty x_k$.

The notion of unconditional convergence is thus appropriate for any situation where the order of the x_k does not play a special role. We can then use the words *summable family* to underline that the set of indices is no longer ordered.

Let us proceed to the definitions of a Schauder basis and an unconditional basis of a Banach space.

A set of vectors $e_0, e_1, \ldots, e_k, \ldots$ in a Banach space B is called a *Schauder basis* of B if every vector $x \in B$ can be written

(1.1) $$x = \alpha_0 e_0 + \alpha_1 e_1 + \cdots + \alpha_k e_k + \cdots,$$

where the $\alpha_0, \alpha_1, \ldots, \alpha_k, \ldots$ are scalar coefficients and the convergence of (1.1) is defined by

(1.2) $$\lim_{n \to \infty} \| x - \alpha_0 e_0 - \alpha_1 e_1 - \cdots - \alpha_n e_n \| = 0 .$$

Further, we require the coefficients $\alpha_0, \alpha_1, \ldots, \alpha_k \ldots$ to be uniquely defined by (1.1) and (1.2).

A Schauder basis is an *unconditional basis* if the following property is satisfied: for all $x \in B$, the series (1.1) converges unconditionally to x.

An equivalent form of this definition is to require that the sequence of e_k, $k \in \mathbb{N}$, be total in B—the finite linear combinations of the e_k, $k \in \mathbb{Z}$, form a dense subset of B—and that the following property be satisfied: there exists a constant $C \geq 1$ such that, for every integer n, every sequence of scalars α_k such that $\sup |\alpha_k| \leq 1$, and for any sequence of scalars β_k

$$\| \sum_0^n \alpha_k \beta_k e_k \| \leq C \| \sum_0^n \beta_k e_k \| .$$

Yet another way to express the fact that a Schauder basis is an unconditional basis is to ask the following question: is there a condition on the moduli $|\alpha_k|$ of the coefficients α_k, which is necessary and sufficient for a formal series $\sum_0^\infty \alpha_k e_k$ to converge in B? To say that the basis e_k, $k \in \mathbb{N}$, is unconditional means that necessary and sufficient conditions for a series $\sum_0^\infty \alpha_k e_k$ to converge in B can only apply to the moduli $|\alpha_k|$ and that, if they are satisfied for the sequence α_k, then they are automatically satisfied for any sequence β_k satisfying $|\beta_k| \leq |\alpha_k|$, for every $k \in \mathbb{N}$.

A final point of view is to study the continuous linear operators $T : B \to B$ which have the e_k, $k \in \mathbb{N}$, as eigenvectors. Let λ_k be the corresponding eigenvalues. Then $|\lambda_k| \leq \|T\|$, because we suppose that $e_k \neq 0$. To say that the sequence e_k, $k \in \mathbb{N}$, is an unconditional basis of the Banach space B means that, in the opposite direction, every sequence $\lambda_k \in l^\infty(\mathbb{N})$ is the sequence of eigenvalues of a certain bounded linear operator $T : B \to B$, completely defined by $T(e_k) = \lambda_k e_k$, $k \in \mathbb{N}$.

When the sequence e_k, $k \in \mathbb{N}$, is the usual trigonometric system, the operators which have the functions e^{ikx}, $k \in \mathbb{Z}$, as eigenvectors are called *multipliers*. The functions $(2\pi)^{-1/2}e^{ikx}$ form an orthonormal basis of $L^2(0, 2\pi)$ and, in an obvious way, every sequence $\lambda_k \in l^\infty(\mathbb{Z})$ is a multiplier (of the Fourier coefficients of the functions in $L^2(0, 2\pi)$). As far as the $L^p(0, 2\pi)$, $p \neq 2$, are concerned, the theory of multipliers is subtle. At the time of writing, we know only sufficient conditions, which relate this theory to that of Calderón-Zygmund operators.

If, however, we replace the trigonometric system by the periodic wavelets $g_k(x)$ of Section 11 of Chapter 3, then the spaces $L^p(0, 2\pi)$ are no problem, as long as $1 < p < \infty$. Indeed, for every bounded sequence $\lambda_k \in l^\infty(\mathbb{N})$, the operator T, defined by $T(g_k) = \lambda_k g_k$, is always a Calderón-Zygmund operator. Moreover, as we shall see in Chapter 7, all these operators are bounded on L^p.

The non-periodic case is similar and the wavelets form an unconditional basis of $L^p(\mathbb{R}^n)$ for $1 < p < \infty$.

In the course of this chapter, we shall consider the limiting case where $p = 1$. The wavelets do not form an unconditional basis of $L^1(\mathbb{R}^n)$. Only certain functions $f \in L^1(\mathbb{R}^n)$ have the special property that their wavelet series converge unconditionally to f. The purpose of this chapter is to characterize such functions.

As may be guessed from the title of this chapter, the solution to the enigma is the $H^1(\mathbb{R}^n)$ space of Stein and Weiss. Moreover, the wavelet series which converge unconditionally in $L^1(\mathbb{R}^n)$ also converge unconditionally in the Banach space $H^1(\mathbb{R}^n) \subset L^1(\mathbb{R}^n)$ to a function $f \in H^1(\mathbb{R}^n)$.

Naturally, this account of the space H^1 differs from the traditional approach, which we shall mention for the convenience of the reader. In this chapter, we shall prove that the atomic definition of H^1 (as it appears in [75]) is equivalent to the definition using wavelets. We shall describe other possible definitions of H^1 in the last section.

The dual space of H^1 is the space BMO of John and Nirenberg. We shall give a proof of this duality—in fact it is a tautology, given the point of view adopted. The wavelets form an unconditional basis of the space BMO too—the convergence of the formal series takes place in the $\sigma(\text{BMO}, H^1)$ topology. The wavelet coefficients of a function in BMO satisfy Carleson's well-known quadratic conditions. This remarkable result confirms the fact that wavelets enable us to perform Fourier analysis locally in a way which is very convenient for the definition of BMO using local L^2 norms.

2 Equivalent definitions of the space $H^1(\mathbb{R}^n)$

We start from the problem posed in the introduction and say that $f \in H^1(\mathbb{R}^n)$ if $f \in L^1(\mathbb{R}^n)$ and the wavelet series $\sum_{\lambda \in \Lambda} (f, \psi_\lambda) \psi_\lambda(x)$ converges unconditionally to f.

As we implied in the introduction, this imposes the existence of a constant $C \geq 1$ such that, for each finite subset $F \subset \Lambda$ and every sequence $\varepsilon(\lambda)$, $\lambda \in F$, taking the values ± 1, we have

$$(2.1) \qquad \| \sum_{\lambda \in F} \varepsilon(\lambda)(f, \psi_\lambda) \psi_\lambda(x) \|_1 \leq C.$$

The upper bound, over all F and sequences $\varepsilon(\lambda)$, $\lambda \in F$, of the left-hand side of (2.1) will be the first norm we consider on the space $H^1(\mathbb{R}^n)$.

We clearly need to specify which particular wavelets ψ_λ are being used. Throughout this section, we shall keep to the regular wavelets of compact support of Sections 8 and 9 of Chapter 3. That the definition is independent of the particular wavelet basis used (as long as the corresponding multiresolution approximation is r-regular for $r \geq 1$) is easy to establish from the arguments to follow. We shall, however, delay the proof until Chapter 7, where it will be a striking consequence of the general theory of Calderón-Zygmund operators.

We shall now transform (2.1), using Khinchin's well-known inequality: there exists a constant $C > 1$ such that, for every integer $n \geq 1$ and every sequence $a_1, \ldots, a_n$ of complex numbers,

$$(2.2) \qquad \left(\sum_1^n |a_j|^2\right)^{1/2} \leq C 2^{-n} \sum_{\varepsilon_1} \cdots \sum_{\varepsilon_n} |\varepsilon_1 a_1 + \cdots \varepsilon_n a_n|,$$

where, on the right-hand side, the sum is taken over all sequences $\varepsilon = (\varepsilon_1, \ldots, \varepsilon_n)$ of 1s and -1s.

The inequality implies the existence of a constant C such that

$$\int_{\mathbb{R}^n} \left(\sum_{\lambda \in F} |(f, \psi_\lambda)|^2 |\psi_\lambda(x)|^2\right)^{1/2} dx \leq C,$$

for every finite subset $F \subset \Lambda$.

This leads to the second definition of $H^1(\mathbb{R}^n)$, of the form

$$(2.3) \qquad \int_{\mathbb{R}^n} \left(\sum_{\lambda \in F} |(f, \psi_\lambda)|^2 |\psi_\lambda(x)|^2\right)^{1/2} dx < \infty.$$

The left-hand side of (2.3) will be the second norm that we use on $H^1(\mathbb{R}^n)$. It is not obvious that this second norm is equivalent to the first, because, in passing from (2.1) to (2.3), we have lost information—we have replaced an inequality which is uniform, with respect to sequences of ± 1, by an inequality in the mean. One of the surprises of the theory is that these two norms are equivalent.

To show this, we go via three other norms. Firstly, we write the wavelets $\psi_\lambda(x)$ explicitly as $2^{nj/2}\psi^\varepsilon(2^j x - k)$, where $j \in \mathbb{Z}$, $k \in \mathbb{Z}^n$, and where $\varepsilon \in E$, the finite set $\{0,1\}^n \backslash \{(0,0,\ldots,0)\}$, of cardinal $2^n - 1$.

Replacing, if necessary, the basic wavelets by appropriate translates by $l \in \mathbb{Z}^n$, we may suppose (as we shall do in all that follows) that $\int_0^1 \cdots \int_0^1 |\psi^\varepsilon(x)|^2\, dx \neq 0$. It clearly follows that $|\psi^\varepsilon(x)| \geq c > 0$ when x is in a some cube $A^\varepsilon \subset [0,1)^n$ satisfying $|A^\varepsilon| \geq \gamma > 0$.

For each $\lambda = k2^{-j} + \varepsilon 2^{-j-1}$ belonging to Λ, we denote by $Q(\lambda)$ the dyadic cube defined by $2^j x - k \in [0,1)^n$ and by $R(\lambda)$ the subcube defined by $2^j x - k \in A^\varepsilon$. We always have

$$(2.4) \qquad\qquad |R(\lambda)| \geq \gamma |Q(\lambda)|.$$

Then the third norm to be used will be

$$(2.5) \qquad \left\| \left(\sum_{\lambda \in \Lambda} |\alpha(\lambda)|^2 |Q(\lambda)|^{-1} \chi_{R(\lambda)}(x) \right)^{1/2} \right\|_1,$$

where $\chi_{R(\lambda)}$ is the characteristic function of $R(\lambda) \subset Q(\lambda)$.

The only relevant property of $R(\lambda)$ is the inequality (2.4) and not the precise geometrical definition of $R(\lambda)$

The fourth norm we use is given by (2.5), but with $R(\lambda) = Q(\lambda)$.

The last norm in our investigation of $H^1(\mathbb{R}^n)$ does not involve wavelet series. It is the well-known atomic definition of $H^1(\mathbb{R}^n)$.

Definition 1. *An atom of $H^1(\mathbb{R}^n)$ is a function $a(x)$, belonging to $L^2(\mathbb{R}^n)$, such that there exists a ball B in $\mathbb{R}^n$, whose volume is denoted by $|B|$, and for which the three following properties hold:*

$$(2.6) \qquad\qquad a(x) = 0 \quad if \quad x \notin B;$$

$$(2.7) \qquad\qquad \|a\|_2 \leq |B|^{-1/2};$$

and

$$(2.8) \qquad\qquad \int_B a(x)\, dx = 0.$$

Applying the Cauchy-Schwarz inequality, we see that $a(x)$ is integrable and that $\|a\|_1 \leq 1$, so that property (2.8) makes sense.

We shall say that a function $f \in L^1(\mathbb{R}^n)$ belongs to *atomic H^1* if there exists a sequence $a_j(x)$ of atoms and a sequence λ_j of scalar coefficients such that

(2.9) $\qquad \sum_{0}^{\infty} |\lambda_j| < \infty \qquad$ and $\qquad f(x) = \sum_{0}^{\infty} \lambda_j a_j(x)$.

The right-hand series clearly converges in $L^1(\mathbb{R}^n)$.

There is no uniqueness about the above decomposition, and this absence of uniqueness is one of the trump cards of this theory, as we shall soon see.

The norm $\|f\|_{\text{at}}$ is defined as the infimum of the quantities $\sum_{0}^{\infty} |\lambda_j|$ corresponding to all the possible atomic decompositions of f.

The main result of this chapter is the following theorem.

Theorem 1. *The above five definitions of* $H^1(\mathbb{R}^n)$ *are equivalent.*

To clarify the proof, let us denote the following properties—whose equivalence we have to prove—by A,B,C, D and E.

(A) $\qquad \sup_{F \subset \Lambda} \sup_{\varepsilon(\lambda)=\pm 1} \| \sum_{\lambda \in F} \varepsilon(\lambda)\alpha(\lambda)\psi_\lambda(x)\|_1 < \infty;$

(B) $\qquad \left(\sum_{\lambda \in \Lambda} |\alpha(\lambda)|^2 |\psi_\lambda(x)|^2 \right)^{1/2} \in L^1(\mathbb{R}^n);$

(C) $\qquad \left(\sum_{\lambda \in \Lambda} |\alpha(\lambda)|^2 |Q(\lambda)|^{-1} \chi_{R(\lambda)}(x) \right)^{1/2} \in L^1(\mathbb{R}^n),$

where $R(\lambda) \subset Q(\lambda)$ satisfies $|R(\lambda)| \geq \gamma |Q(\lambda)|$, for a constant $\gamma > 0$;

(D) $\qquad \left(\sum_{\lambda \in \Lambda} |\alpha(\lambda)|^2 |Q(\lambda)|^{-1} \chi_{Q(\lambda)}(x) \right)^{1/2} \in L^1(\mathbb{R}^n);$

and, lastly,

(E) $\qquad f(x) = \sum \alpha(\lambda)\psi_\lambda(x)$ has an atomic decomposition.

We already know that (A)$\Rightarrow$(B), and it is clear that (B)$\Rightarrow$(C). The main part of the proof is the implication (C)$\Rightarrow$(E). The method is to take the series $f(x) = \sum_{\lambda \in \Lambda} \alpha(\lambda)\psi_\lambda(x)$ and to group the terms according to a partition of Λ into disjoint sets $\Lambda(k,l)$. The atoms (together with their coefficients) $\lambda_j a_j(x)$ will then be given by the partial sums $\sum_{\lambda \in \Lambda(k,l)} \alpha(\lambda)\psi_\lambda(x)$. The grouping will be done directly on the coefficients $\alpha(\lambda)$, using certain level sets which we shall define in due course.

On the way to establishing the implication (C)$\Rightarrow$(E), we shall prove (C)$\Rightarrow$(D) as well as the inequality

(2.10) $\qquad \|f\|_1 \leq C \|(\sum_{\lambda \in \Lambda} |\alpha(\lambda)|^2 |Q(\lambda)|^{-1} \chi_{R(\lambda)}(x))^{1/2}\|_1,$

for every series $f(x) = \sum \alpha(\lambda)\psi_\lambda(x)$.

The verification of (E)$\Rightarrow$(D) is an easy exercise, because it is sufficient to do it for each atom taken separately. After that, (E)$\Rightarrow$(A) is obvious.

A further remark might be useful. To apply one of the five criteria, we must have a mathematical object f to which we apply the criterion. This object will be a distribution of order less than r, where r is the regularity of the wavelets. But not just any distribution will do. For example, if $f(x) = 1$, all the wavelet coefficients of f are zero, and so all the criteria are satisfied. However, the function 1 does not belong to $H^1(\mathbb{R}^n)$.

Such problems are avoided by requiring the series $\sum_{\lambda \in \Lambda}(f, \psi_\lambda)\psi_\lambda$ to converge to f in the sense of distributions.

We now begin the proof of the implication (C)$\Rightarrow$(E) of Theorem 1.

3 Atomic decomposition at the coefficient level

For each $j \in \mathbb{Z}$ and $k \in \mathbb{Z}^n$, let $Q(j,k)$ denote the dyadic cube defined by $2^j x - k \in [0,1)^n$. The collection of all these dyadic cubes is denoted by $\mathcal{Q}$ and we shall study certain spaces of (real or complex) sequences $\alpha(Q)$, $Q \in \mathcal{Q}$. For such a sequence α, the *support* of α is the set of dyadic cubes $Q \in \mathcal{Q}$ such that $\alpha(Q) \neq 0$, while the *socle* of α is the union of the dyadic cubes Q in the support of α.

If Q and Q' are dyadic cubes and $Q \cap Q'$ is non-empty, then one of the dyadic cubes is contained in the other.

Let $\mathcal{C}$ be a non-empty collection of dyadic cubes Q whose volumes $|Q|$ do not exceed a certain finite upper bound. Then every cube $Q \in \mathcal{C}$ is contained in a cube $\tilde{Q} \in \mathcal{C}$ which is maximal in $\mathcal{C}$. The maximal cubes $\tilde{Q}$ are pairwise disjoint. Let the union Ω of all the cubes $Q \in \mathcal{C}$ be called the *socle* of $\mathcal{C}$. Then the maximal dyadic cubes in $\mathcal{C}$ form a partition of the socle of $\mathcal{C}$.

We now come to the definition of the "tent space" T_1 (which is a Banach space). This space is the discrete version of the space of [71].

To each sequence $\alpha(Q)$, $Q \in \mathcal{Q}$, we assign the measurable function $S(\alpha) : \mathbb{R}^n \to [0, \infty]$ defined by

$$S(\alpha)(x) = \Big(\sum_{Q \ni x} |\alpha(Q)|^2 |Q|^{-1} \Big)^{1/2}.$$

Definition 2. *We say that a sequence of complex numbers* $\alpha(Q)$, $Q \in \mathcal{Q}$, *belongs to* T_1 *if the function* $S(\alpha)$ *belongs to* $L^1(\mathbb{R}^n)$, *and we put* $\|\alpha\| = \int_{\mathbb{R}^n} S(\alpha)\, dx$.

Using the inequality $|\alpha(Q)||Q|^{1/2} \leq \|\alpha\|$, it is easy to verify that T_1 is a Banach space.

But where do the "tents" come in?

Let R be a dyadic cube. The "tent over R" is the collection $\mathcal{C}$ of all dyadic cubes $Q \subset R$. To understand this terminology, we use the following picture. With $n = 1$, we represent the dyadic interval $[k/2^j, (k+1)/2^j)$, $j, k \in \mathbb{Z}$, by the complex number $(k+\frac{1}{2})2^{-j} + i2^{-j-1}$. Then we can sketch the complex numbers associated with a tent and verify (with a bit of imagination) that this set of points looks like a tent whose base is the interval R with which we started.

Definition 3. *We say that a sequence $\alpha = (\alpha(Q))_{Q \in \mathcal{Q}}$ is an atom if there exists a dyadic cube R containing the socle of α such that*

$$(3.1) \qquad \sum |\alpha(Q)|^2 \leq \frac{1}{|R|}.$$

We can then choose R to be the smallest dyadic cube containing the socle of α and this choice of R will be called the *base* of the atom α. The support of the atom α is thus contained in the "tent above R".

Let us verify that α belongs to T_1 and that, further, $\|\alpha\| \leq 1$. To do this, we note that the support of $S(\alpha)$ is contained in the base R of the atom α and that, by the Cauchy-Schwarz inequality, we have $\int S(\alpha) \, dx = \int_R S(\alpha) \, dx \leq |R|^{1/2} \left(\int (S(\alpha))^2 \, dx \right)^{1/2} \leq 1$. The last inequality follows from writing $S(\alpha)$ in the form

$$S(\alpha)(x) = \left(\sum |\alpha(Q)|^2 |Q|^{-1} \chi_Q(x) \right)^{1/2},$$

where χ_Q is the characteristic function of Q.

Theorem 2. *The following three properties of a sequence*

$$\alpha = (\alpha(Q))_{Q \in \mathcal{Q}}$$

are equivalent:

(3.2)   *there exists a constant $\gamma \in (0, 1]$ such that, for each cube $Q \in \mathcal{Q}$, there is a measurable subset $R \subset Q$ with $|R| \geq \gamma|Q|$ and such that*

$$\left(\sum_{Q \in \mathcal{Q}} |\alpha(Q)|^2 |Q|^{-1} \chi_R(x) \right)^{1/2} \in L^1(\mathbb{R}^n);$$

(3.3) $\alpha \in T_1$;

(3.4)   *there exists a sequence of atoms α_j in T_1 and a sequence λ_j of scalars such that $\sum_0^\infty |\lambda_j| < \infty$ and $\alpha = \sum_0^\infty \lambda_j \alpha_j$.*

Before beginning the proof, we define a magnification $E^\star$ of each measurable subset $E \subset \mathbb{R}^n$ with measure $|E| < \infty$. We choose a coefficient

$\beta \in (0,1)$ (which will be strictly less than γ in what follows). Then $E^\star$ is the union of the cubes $Q \in \mathcal{C}$, where $\mathcal{C}$ is the collection of all cubes $Q \in \mathcal{Q}$ such that $|Q \cap E| \geq \beta |Q|$.

Now, each cube $Q \in \mathcal{C}$ is contained in a maximal cube (maximal with respect to inclusion) $\tilde{Q} \in \mathcal{C}$, which exists because $Q \in \mathcal{C}$ implies that $|Q| \leq \beta^{-1} |E|$. The maximal cubes are pairwise disjoint and form a partition of $E^\star$. So we get

$$|E^\star| = \sum |\tilde{Q}| \leq \beta^{-1} \sum |\tilde{Q} \cap E| \leq \beta^{-1} |E|.$$

Further, $E^\star$ contains E up to a set of measure zero. Indeed, by Lebesgue's theorem

$$\lim_{Q \ni x} \frac{|Q \cap E|}{|Q|} = 1 \qquad \text{for almost all } x \in E,$$

the limit being taken as the diameters of the dyadic cubes containing x tend to 0. So, for almost all $x \in E$, there exists at least one dyadic cube Q containing x such that $|Q \cap E| \geq \beta |Q|$, which implies that $Q \subset E^\star$ and so $x \in E^\star$.

Clearly, if E and F are measurable sets of finite measure with $E \subset F$, then $E^\star \subset F^\star$.

We return to Theorem 2 and put

$$\sigma(x) = \Big(\sum_{Q \in \mathcal{Q}} |\alpha(Q)|^2 |Q|^{-1} \chi_R(x) \Big)^{1/2}.$$

Let E_k denote the set of x satisfying $\sigma(x) > 2^k$. We have $E_k \supset E_{k+1}$ and

$$(3.5) \qquad \sum_{-\infty}^{\infty} 2^k |E_k| \leq 2 \int_{\mathbb{R}^n} \sigma(x) \, dx.$$

This inequality will give us $\sum_0^\infty |\lambda_j| \leq C \|\sigma\|_1$ in (3.4).

We then consider the collection $\mathcal{C}_k$, $k \in \mathbb{Z}$, of the dyadic cubes Q satisfying $|Q \cap E_k| \geq \beta |Q|$, where $\beta \in (0, \gamma)$ will, from now on, be fixed. The union of the cubes $Q \in \mathcal{C}_k$ is thus $E_k^\star$ and we have $|E_k^\star| \leq \beta^{-1} |E_k|$, which gives

$$(3.6) \qquad \sum_{-\infty}^{\infty} 2^k |E_k^\star| \leq 2\beta^{-1} \|\sigma\|_1.$$

Let $Q(k,l)$ denote the dyadic cubes which are maximal in $\mathcal{C}_k$. By what has gone before, they form a partition of $E_k^\star$.

We now claim that the support $\mathcal{C}$ of α is contained in the union of the $\mathcal{C}_k$. Indeed, if $\alpha(Q) \neq 0$, then, for a certain integer k, $|\alpha(Q)||Q|^{-1/2} > 2^k$, so that $\sigma(x) > 2^k$ for all $x \in R \subset Q$. In conclusion, by the definition of E_k, we get $E_k \cap Q \supset R$ and $|E_k \cap Q| \geq |R| \geq \gamma |Q| > \beta |Q|$, which gives $Q \in \mathcal{C}_k$.

Now we let Δ_k denote the collection of $Q \in \mathcal{C}_k$ which are not in $\mathcal{C}_{k+1}$ ($\Delta_k = \mathcal{C}_k \backslash \mathcal{C}_{k+1}$). Finally, we write $Q \in \Delta(k,l)$ if $Q \in \Delta_k$ and $Q \subset Q(k,l)$. The sets $\Delta(k,l)$ thus form a partition of Δ_k (some of them may be empty and will then be omitted). So the support $\mathcal{C}$ of the sequence α is the disjoint union of the sets $\Delta(k,l)$, $k \in \mathbb{Z}$, $l \in F(k) = \{l : \Delta(k,l) \neq \emptyset\}$.

Lemma 1. *With the above notation, we restrict the sequence $\alpha(Q)$ to each of the disjoint subsets $\Delta(k,l)$ of the support of α to get (after an obvious renormalization) the atomic decomposition, as claimed, in the form*

$$(3.7) \qquad \alpha = \sum \sum \lambda(k,l) \alpha_{k,l},$$

where the $\alpha_{k,l}$ are atoms and the base of $\alpha_{k,l}$ is contained in $Q(k,l)$.

To prove this, we start by establishing the following inequality:

$$(3.8) \qquad \sum_{Q \in \Delta(k,l)} |\alpha(Q)|^2 \leq \frac{1}{\gamma - \beta} \int_{Q(k,l) \backslash E_{k+1}} \sigma^2(x)\,dx \leq \frac{1}{\gamma - \beta} 4^{k+1} |Q(k,l)|.$$

Now for the proof of (3.8). The dyadic cube Q belongs to $\Delta(k,l)$, if, on the one hand, $Q \subset Q(k,l)$ but, on the other, Q is not in $\mathcal{C}_{k+1}$. So $|Q \cap E_{k+1}| < \beta|Q|$ and, as a result,

$$(3.9) \qquad |R \cap E_{k+1}| < \beta|Q| \leq \frac{\beta}{\gamma}|R|.$$

Rewriting, this gives $|R \backslash E_{k+1}| \geq (1 - \beta/\gamma)|R| \geq (\gamma - \beta)|Q|$. But $\sigma^2(x) \geq \sum_{Q \in \Delta(k,l)} |\alpha(Q)|^2 |Q|^{-1} \chi_R(x)$, so we get

$$\int_{Q(k,l) \backslash E_{k+1}} \sigma^2(x)\,dx \geq \sum_{Q \in \Delta(k,l)} |\alpha(Q)|^2 |Q|^{-1} |R \backslash E_{k+1}|$$

$$\geq (\gamma - \beta) \sum_{Q \in \Delta(k,l)} |\alpha(Q)|^2,$$

which gives the first inequality of (3.8). The second is a consequence of $\sigma(x) \leq 2^{k+1}$ when $x \notin E_{k+1}$.

We now put

$$\lambda(k,l) = |Q(k,l)|^{1/2} \left(\sum_{Q \in \Delta(k,l)} |\alpha(Q)|^2 \right)^{1/2},$$

and define the atoms $\alpha_{k,l}(Q)$ by $\alpha_{k,l}(Q) = (\lambda(k,l))^{-1}\alpha(Q)$, when $Q \in \Delta(k,l)$, and $\alpha_{k,l} = 0$ otherwise.

By the construction, the $\alpha_{k,l}$ are atoms and the base of each $\alpha_{k,l}$ is contained in $Q(k,l)$.

All that remains is to verify that $\sum \sum \lambda(k,l) \leq C\|\sigma\|_1$ But, by (3.8), this series is bounded by

$$\frac{2}{(\gamma - \beta)^{1/2}} \sum \sum 2^k |Q(k,l)| = \frac{2}{(\gamma - \beta)^{1/2}} \sum 2^k |E_k^{\star}|$$

$$\leq \frac{2}{\beta(\gamma - \beta)^{1/2}} \sum 2^k |E_k|$$

$$\leq \frac{4}{\beta(\gamma - \beta)^{1/2}} \|\sigma\|_1 .$$

4 Back to earth

We now are in a position to prove the implications (B)⇒(C)⇒(E) and (D)⇒(C)⇒(E) of Theorem 1.

So let us assume that $\left(\sum_{\lambda \in \Lambda} |\alpha(\lambda)|^2 |\psi_\lambda(x)|^2\right)^{1/2} \in L^1(\mathbb{R}^n)$. We have already noted that, if $\lambda = 2^{-j}k + 2^{-j-1}\varepsilon$, with $\varepsilon \in E$, then $|\psi_\lambda(x)| \geq c|Q|^{-1/2}$ for each $x \in R$, where $R \subset Q$, $|R| \geq \gamma|Q|$, and $c, \gamma > 0$ are constants. Thus, with the notation of the proof of Theorem 2, $\sigma(x) \in L^1(\mathbb{R}^n)$. Now the sequence $\alpha(\lambda)$, $\lambda \in \Lambda$, is composed of the subsequences $\alpha^\varepsilon(Q)$, $Q \in \mathcal{Q}$. By Theorem 2, for each $\varepsilon \in E$, the sequence $\alpha^\varepsilon(Q)$ belongs to the tent space T_1 and thus has an atomic decomposition $\alpha^\varepsilon(Q) = \sum \lambda_i^\varepsilon \alpha_i^\varepsilon(Q)$.

We conclude the proof of the implication by showing that the functions $a_i^\varepsilon(x) = \sum_{Q \in \mathcal{Q}} \alpha_i^\varepsilon(Q)\psi_Q^\varepsilon(x)$ are atoms of $H^1(\mathbb{R}^n)$. This is particularly pleasant if the basic wavelets ψ^ε have compact support. In that case, the support of ψ_Q^ε is contained in mQ, for some constant $m \geq 1$. Let Q_i^ε denote the base of the atom $\alpha_i^\varepsilon(Q)$ of the space T_1. Then the support of $a_i^\varepsilon(x)$ is contained in mQ_i^ε and

$$\|a_i^\varepsilon\|_2 = \left(\Sigma_Q |\alpha_i^\varepsilon(Q)|^2\right)^{1/2} \leq \frac{1}{\sqrt{|Q_i^\varepsilon|}} .$$

Hence the series defining $a_i^\varepsilon(x)$ also converges in $L^1(\mathbb{R}^n)$ and we can integrate term by term to get $\int a_i^\varepsilon(x)\, dx = 0$.

We have now shown that the functions $m^{-n/2}a_i^\varepsilon(x)$ are atoms of $H^1(\mathbb{R}^n)$ and so the atomic decomposition of the sequences in T_1 yields, via the wavelets, an atomic decomposition of functions. (The case where the wavelets do not have compact support will be dealt with in the next section.)

We have, *a fortiori*, proved the fundamental inequality

$$(4.1) \quad \left\|\sum \alpha(\lambda)\psi_\lambda(x)\right\|_1 \leq C \left\|\left(\sum_{\lambda \in \Lambda} |\alpha(\lambda)|^2 |Q(\lambda)|^{-1} \chi_{R(\lambda)}(x)\right)^{1/2}\right\|_1 .$$

It is now easy to prove that (E) implies (B). We set

$$S(f)(x) = \Big(\sum_{\lambda \in \Lambda} |(f, \psi_\lambda)|^2 |\psi_\lambda(x)|^2\Big)^{1/2}.$$

By convexity, to show that $S(f)$ is integrable when $\|f\|_{\mathrm{at}}$ is finite, it is enough to show it for each atom taken separately.

Lemma 2. *There is a constant C such that $\int_{\mathbf{R}^n} S(f)(x)\,dx \leq C$, for every atom f of $H^1(\mathbf{R}^n)$ whose associated ball B has centre x_0 and radius $r > 0$.*

To see this, we take $C > 1$ to be a constant, depending on m, whose value will be fixed in the course of the proof. We decompose $\mathbf{R}^n$ into a central ball $|x - x_0| \leq Cr$, denoted by $\tilde{B}$, surrounded by dyadic shells R_k, $k \in \mathbf{N}$, defined by $2^k Cr \leq |x - x_0| < 2^{k+1} Cr$. Then

$$\int_{\mathbf{R}^n} S(f)\,dx = \int_{\tilde{B}} S(f)\,dx + \sum_0^\infty \int_{R_k} S(f)\,dx.$$

To estimate the first integral, we simply use the Cauchy-Schwarz inequality and get

$$\int_{\tilde{B}} S(f)\,dx \leq |\tilde{B}|^{1/2} \|S(f)\|_2 = C^{n/2} |B|^{1/2} \|f\|_2 \leq C^{n/2}.$$

For the integrals $\int_{R_k} S(f)\,dx$, we first take into account the wavelets' good localization and replace the sum over $\lambda \in \Lambda$ in $S(f)$ by a partial sum over $\lambda \in \Lambda^k$. The definition of Λ^k will be given shortly. Then we shall estimate the wavelet coefficients (f, ψ_λ), using the fact that $\int f(x)\,dx = 0$ and that the wavelets ψ_λ are "flat enough" when $\lambda \in \Lambda^k$.

Here are the very simple details of the estimates. We have $\psi_\lambda(x) = \psi_Q^\varepsilon(x)$, so the support of ψ_λ is contained in mQ. If mQ does not intersect B, then $(f, \psi_\lambda) = 0$, and if mQ does not meet R_k, we need not involve λ in the estimate of $\int_{R_k} S(f)\,dx$. So we define Λ^k by the condition that mQ meets both B and the shell R_k. This means that $2^{-j} \geq cr2^k$, where 2^{-j} is the length of the side of Q, and $c \geq 0$ is a geometric constant, which is strictly positive when C is large enough.

To get an upper bound for (f, ψ_λ) we take account of the regularity of the wavelet ψ_λ and use the fact that $\int f(x)\,dx = 0$. We get $|(f, \psi_\lambda)| \leq C2^{nj/2}2^j r$. This then gives, for $x \in R_k$, $S(f)(x) \leq C'2^{-k(n+1)}r^{-n}$ and $\sum_0^\infty \int_{R_k} S(f)\,dx \leq C''$.

The proof of Theorem 1 ends with the straightforward verification of the implication (C)$\Rightarrow$(A), by returning to the bound given by (4.1) and observing that the right-hand side of (4.1) does not change if the $\alpha(\lambda)$ are replaced by $\pm\alpha(\lambda)$.

5 Atoms and molecules

When the wavelets used do not have compact support, the atomic decomposition at the coefficient level no longer leads to an atomic decomposition of the function $f \in H^1(\mathbb{R}^n)$, but gives instead a *molecular* decomposition.

Molecules belong to the weighted spaces $L^2(\mathbb{R}^n, \omega(x)\, dx)$ whose definition we now recall. Let $\omega(x) > 0$ be a measurable function on $\mathbb{R}^n$ such that $\int_{\mathbb{R}^n} (\omega(x))^{-1}\, dx < \infty$. Then $L^2(\mathbb{R}^n, \omega(x)\, dx)$ denotes the space of measurable functions such that $\int |f(x)|^2 \omega(x)\, dx < \infty$.

Under these conditions, the Cauchy-Schwarz inequality gives

$$L^2(\mathbb{R}^n, \omega(x)\, dx) \subset L^1(\mathbb{R}^n).$$

We consider the special case where $\omega_s(x) = (1 + |x|)^s$, $s > n$, and let M^s be the subspace of $L^2(\mathbb{R}^n, \omega_s\, dx)$ consisting of functions whose integral is zero.

Proposition 1. *For every $s > n$, M^s is a dense linear subspace of $H^1(\mathbb{R}^n)$.*

For f belonging to M^s, we define the functions f_j by

$$f_0(x) = \begin{cases} f(x), & \text{if } |x| \le 1, \\ 0, & \text{if } |x| > 1, \end{cases}$$

and, for $j > 0$,

$$f_j(x) = \begin{cases} f(x), & \text{if } 2^{j-1} < |x| \le 2^j, \\ 0, & \text{otherwise.} \end{cases}$$

The hypothesis $f \in M^s$ then becomes $\|f_j\|_2 \le \varepsilon_j 2^{-js/2}$, with $\sum_0^\infty |\varepsilon_j|^2 < \infty$, and $\sum_0^\infty I_j = 0$, where $I_j = \int f_j(x)\, dx$. By the Cauchy-Schwarz inequality, we get $|I_j| \le c(n)\varepsilon_j 2^{-j(s-n)/2}$. Then, putting $\sigma_j = I_j + I_{j+1} + \cdots$, we get $|\sigma_j| \le \eta_j 2^{-j(s-n)/2}$, where

$$\eta_j = c(n) \sum_{k \ge j} \varepsilon_k 2^{-(s-n)(k-j)/2}$$

belongs to $l^2(\mathbb{N})$.

We can change f_j into an atom , by replacing it by

$$a_j(x) = f_j(x) + \sigma_{j+1}\theta_{j+1}(x) - \sigma_j\theta_j(x),$$

where $\theta_j(x) = \gamma(n)2^{-nj}$, if $|x| \le 2^j$, and $\theta_j(x) = 0$, otherwise. The constant $\gamma(n)$ is adjusted so that the integral of $\theta_j(x)$ equals 1, which guarantees that the integral of $a_j(x)$ is 0. Since the integral of f is zero, $\sigma_0 = 0$ and $\sum_0^\infty a_j(x) = \sum_0^\infty f_j(x) = f(x)$.

The support of $a_j(x)$ is contained in the ball $|x| \le 2^{j+1}$, so we arrive at $\|a_j\|_2 \le \varepsilon_j' 2^{-js}$, where $\varepsilon_j' \in l^2(\mathbb{N})$. We have thus obtained an atomic decomposition of f.

The fact that M^s is dense in $H^1(\mathbb{R}^n)$ is a simple consequence of the fact that the atoms of H^1 are in M^s.

Armed with Proposition 1, we are in a position to define the molecules of $H^1(\mathbb{R}^n)$.

Definition 4. *Let $s > n$ be a real number. A molecule $f \in M^s$, centred on x_0 and of width $d > 0$, is defined to be a function belonging to M^s which is normalized by*

$$(5.1) \qquad \left(\int_{\mathbb{R}^n} |f(x)|^2 \left(1 + \frac{|x - x_0|}{d} \right)^s dx \right)^{1/2} \le d^{-n/2}.$$

It is an immediate consequence of the proof of Proposition 1 that the norm of such a molecule in $H^1(\mathbb{R}^n)$ is no larger than a constant $C(s, n)$ independent of x_0 and of $d > 0$.

The atoms of $H^1(\mathbb{R}^n)$ are molecules (after multiplication by a constant $\gamma(s, n) > 0$). Replacing the atoms of (3.4) systematically by molecules would thus not change the definition of $H^1(\mathbb{R}^n)$.

As G. Weiss pertinently remarks in [75], this freedom in the definition of $H^1(\mathbb{R}^n)$ is essential when we want to show that a continuous linear operator $T : L^2(\mathbb{R}^n) \to L^2(\mathbb{R}^n)$ extends to a continuous linear operator $T : H^1(\mathbb{R}^n) \to H^1(\mathbb{R}^n)$. It is enough to establish that T takes the atoms of $H^1(\mathbb{R}^n)$ to molecules of $H^1(\mathbb{R}^n)$. We shall see this remark applied, in Chapter 7, to the Calderón-Zygmund operators.

6 The space BMO of John and Nirenberg

The letters BMO stand for "Bounded Mean Oscillation", the meaning of which will be explained in the definitions to follow.

We start by looking at continuous linear functionals on $H^1(\mathbb{R}^n)$. If λ is such a functional, it is completely determined by its restriction to the dense linear subspace $M^s \subset H^1(\mathbb{R}^n)$. What is more, M^s is of codimension 1 in $L^2(\mathbb{R}^n, (1 + |x|)^s dx)$. Denoting by $\theta(x)$ a function whose integral is 1 and which belongs to the latter space, we extend λ to the whole of $L^2(\mathbb{R}^n, (1 + |x|)^s dx)$, by fixing the value of $\lambda(\theta)$. Thus λ has become an element of the dual of $L^2(\mathbb{R}^n, (1 + |x|)^s dx)$, that is, a function $b(x)$ in $L^2(\mathbb{R}^n, (1 + |x|)^{-s} dx)$.

The atoms $a(x)$ of the atomic decomposition of $H^1(\mathbb{R}^n)$ clearly belong to M^s and, since $b(x)$ defines a continuous linear functional on $H^1(\mathbb{R}^n)$, there must necessarily be a constant C such that, for every atom $a(x)$, $\left| \int a(x)b(x) dx \right| \le C$. In other words, for every ball B in $\mathbb{R}^n$ and for every square-summable function $a(x)$ with support in B whose integral is 0,

$$\left| \int a(x)b(x)\, dx \right| \leq C|B|^{1/2}\|a\|_2 .$$

This condition is equivalent to the existence of a constant γ_B such that

(6.1) $$\left(\int_B |b(x) - \gamma_B|^2\, dx \right)^{1/2} \leq C|B|^{1/2}.$$

The constant γ_B which minimizes the left-hand side of (6.1) is precisely the mean of $b(x)$ on the ball B, denoted by $m_B b$. The inequality (6.1) has become

(6.2) $$\left(\int_B |b(x) - m_B b|^2\, dx \right)^{1/2} \leq C|B|^{1/2}.$$

The quantity

$$\left(\frac{1}{|B|} \int_B |b(x) - \gamma_B|^2\, dx \right)^{1/2}$$

is the standard deviation of the random variable $b(x)$ on the ball B with respect to the probability measure $|B|^{-1}\, dx$. Such standard deviations are the "mean oscillations" referred to by the letters BMO.

We can now state the following result.

Theorem 3. *Let $b(x) \in L^2_{\mathrm{loc}}(\mathbb{R}^n)$ be a function such that, with the notation defined above,*

(6.3) $$\sup_B \left(\frac{1}{|B|} \int_B |b(x) - m_B b|^2\, dx \right)^{1/2} = \|b\|_{\mathrm{BMO}} < \infty ,$$

where the upper bound is taken over all balls B in $\mathbb{R}^n$.

Then $b(x)$ defines a continuous linear functional λ on $H^1(\mathbb{R}^n)$ by

(6.4) $$\lambda(f) = \sum_0^\infty \lambda_j \int b(x)a_j(x)\, dx ,$$

whenever $a_j(x)$ are atoms, $\sum_0^\infty |\lambda_j| < \infty$ and $f(x) = \sum_0^\infty \lambda_j a_j(x)$.

Conversely, every continuous linear functional on $H^1(\mathbb{R}^n)$ is defined in this way.

This statement would seem to be obvious, given the preceding discussion. The only subtlety is the following problem: does the definition of $\lambda(f)$ given by (6.4) depend on the representation of f?

If b belongs to $L^\infty(\mathbb{R}^n)$, the answer to the question is trivial. Indeed, if $0 = \sum_0^\infty \lambda_j a_j(x)$, the partial sums $\sum_0^\infty \lambda_j a_j(x)$ converge to 0 in $L^1(\mathbb{R}^n)$ and we can integrate, after multiplication by $b(x)$, and pass to the limit. But condition (6.3) does not imply $b(x) \in L^\infty(\mathbb{R}^n)$. We deal with the problem by first noting that BMO is a lattice. This means that, if f and g are real-valued functions in BMO, then the same is true for $\sup(f, g)$ and $\inf(f, g)$. This follows from the following lemma:

Lemma 3. *If $f(x)$ belongs to* BMO$(\mathbb{R}^n)$, *the same is true for $|f(x)|$.*

Indeed, we put $\alpha(B) = |m_B f|$, and the triangle inequality gives

$$(6.5) \qquad \Big| |f(x)| - \alpha(B) \Big| \le \Big| f(x) - m_B f \Big|.$$

Squaring (6.5) and taking the mean over B gives the result.

Returning to Theorem 3, we suppose that $b(x)$ satisfies (6.3) and consider the real and imaginary parts separately. So we may suppose that $b(x)$ is real-valued.

We now observe that the constant functions belong to BMO and have zero norm. For each $N \ge 1$, we define $b_N(x)$ by $b_N(x) = N$, if $b(x) \ge N$, $b_N(x) = -N$, if $b(x) \le -N$, and $b_N(x) = b(x)$, if $= N < b(x) < N$. By Lemma 3, the functions $b_N(x)$ are uniformly bounded in BMO and $\|b_N\|_{\text{BMO}} \le 3\|b\|_{\text{BMO}}$.

Suppose that $\sum_0^\infty \lambda_j a_j(x) = 0$, where $\sum_0^\infty |\lambda_j| < \infty$ and the $a_j(x)$ are atoms. Since the functions $b_N(x)$ are now in $L^\infty(\mathbb{R}^n)$, we have $\sum_0^\infty \lambda_j \int a_j(x) b_N(x)\, dx = 0$. Further, the functions $b_N(x)$ are uniformly bounded in BMO, so $\big| \int a_j(x) b_N(x)\, dx \big| \le C\|b\|_{\text{BMO}}$. Lastly,

$$\lim_{N \to \infty} \int a_j(x) b_N(x)\, dx = \int a_j(x) b(x)\, dx,$$

since, if B_j is the ball corresponding to the atom $a_j(x)$, then $b(x) \in L^2(B_j)$ and $b_N(x)$ converges to $b(x)$ in $L^2(B_j)$.

To conclude, we apply the dominated convergence theorem (for $l^1(\mathbb{N})$) to get

$$0 = \lim_{N \to \infty} \sum_0^\infty \lambda_j \int a_j(x) b_N(x)\, dx = \sum_0^\infty \lambda_j \int a_j(x) b(x)\, dx,$$

as claimed.

It will be useful to give an apparently more precise form of the definition (6.3). Instead of using atoms to generate H^1, we use molecules and get, once again by duality,

$$(6.6) \qquad \left(\frac{1}{|B|} \int_{\mathbb{R}^n} |b(x) - m_B b|^2 \left(1 + \frac{|x - x_0|}{d} \right)^{-s} dx \right)^{1/2} \le C.$$

We now complete our account of the space BMO by giving its description in terms of wavelet series. To simplify matters a little, we shall use the wavelets of compact support from Sections 8 and 9 of Chapter 3. As an exercise, the reader is invited to rewrite the proof of the following

theorem using a different wavelet basis. The only essential point is to
begin with a multiresolution approximation of regularity $r \geq 1$.

Theorem 4. *Let $b(x)$ be a function belonging to $\mathrm{BMO}(\mathbb{R}^n)$. Then
its wavelet coefficients $\alpha(\lambda) = (b, \psi_\lambda)$ satisfy Carleson's condition, as
follows:*

*For $\lambda = 2^{-j}k + 2^{-j-1}\varepsilon$, let $Q(\lambda)$ denote the cube defined by $2^j x - k \in
[0,1)^n$. Then there exists a constant C such that, for each dyadic cube
$Q \in \mathcal{Q}$,*

$$(6.7) \qquad \sum_{Q(\lambda) \subset Q} |\alpha(\lambda)|^2 \leq C|Q|.$$

*Conversely, if the coefficients $\alpha(\lambda)$, $\lambda \in \Lambda$, satisfy (6.7), then the series
$\sum \alpha(\lambda)\psi_\lambda(x)$ converges, in the $\sigma(\mathrm{BMO}, H^1)$-topology, to a function of
BMO.*

We first show why (6.7) holds when $b(x)$ belongs to $\mathrm{BMO}(\mathbb{R}^n)$.

The support of the wavelet ψ_λ is contained in the cube $mQ(\lambda)$, which
has the same centre as the cube $Q(\lambda)$ but is m times as big. We write
$b(x)$ as $b(x) = b_1(x) + b_2(x) + c(Q)$, where $c(Q)$ is the mean of $b(x)$ on
mQ and where $b_1(x) = b(x) - c(Q)$, if $x \in mQ$, and $b_1(x) = 0$ otherwise.
By the geometric remark about the support of the wavelets, $(b_2, \psi_\lambda) = 0$
if $Q(\lambda) \subset Q$. On the other hand, the integral of each wavelet ψ_λ is zero.
So $(b, \psi_\lambda) = (b_1, \psi_\lambda)$ and we get

$$\sum_{Q(\lambda) \subset Q} |(b, \psi_\lambda)|^2 \leq \sum_{\lambda \in \Lambda} |(b_1, \psi_\lambda)|^2 = \|b_1\|_2^2 \leq m^n \|b\|_{\mathrm{BMO}}^2 |Q|.$$

The wavelets' regularity has played no part in the reasoning above.
However, the regularity is essential for the converse implication.

Suppose, then, that (6.7) is satisfied. Let $\tilde{B}$ be a ball with centre x_0
and of radius $r > 0$. We define the integer $q \in \mathbb{Z}$ by $2^{-q} \leq r < 2^{-q+1}$
and we split the sum $\sum_{\lambda \in \Lambda} \alpha(\lambda)\psi_\lambda(x)$ into two parts, each of which
is itself divided into two parts. At first, we consider "small" cubes of
side $2^{-j} \leq 2^{-q}$ and then "large" cubes for which $j < q$. The wavelets
corresponding to the small cubes are themselves of two kinds—their
supports either meet B or don't meet B. If a "small" cube $Q(\lambda)$ has
the property that $mQ(\lambda)$ intersects with B, then $Q(\lambda)$ is necessarily
contained in MB, where $M > 1$ is a constant depending only on m. If
$b_1 + b_2$ denotes the decomposition of $b = \sum \alpha(\lambda)\psi_\lambda(x)$ according to the
small and large cubes, then b_1 splits into $b_{1,1} + b_{1,2}$, and $b_{1,2}(x) = 0$ on
B, whereas $b_{1,1}$ involves the small cubes $Q(\lambda)$ contained in MB. Then

$$\|b_{1,1}\|_2^2 \leq \sum_{Q(\lambda) \subset MB} |\alpha(\lambda)|^2 \leq C|B|.$$

We turn our attention to the large cubes and the corresponding sub-series b_2 of $\sum \alpha(\lambda)\psi_\lambda(x)$. We now use the fact that the wavelets are "flat" and that, moreover, for a given size 2^{-j} of the large cube $Q(\lambda)$, only M^n wavelets $\psi_\lambda(x)$ are not identically zero on B (again, because the support of $\psi_\lambda(x)$ is contained in $mQ(\lambda)$). For each of the remaining M^n wavelets, we have $|\psi_\lambda(x) - \psi_\lambda(x_0)| \leq C2^j 2^{nj/2}|x - x_0|$, because of the wavelets' regularity. Now the modulus of the corresponding coeffi-cient $|\alpha(\lambda)|$ is bounded above by $C|Q(\lambda)|^{1/2} = C2^{-nj/2}$. We still have to perform the summation: $\sum_{j<q} 2^j |x - x_0| = 2^q |x - x_0| \leq 2$, since $|x - x_0| \leq r < 2^{q+1}$, where x_0 is the centre of B and $r > 0$ is its radius.

We have proved Theorem 4. In Chapter 11, we shall need the following remark, which is a "corollary of the proof" of Theorem 4.

Let $w_\lambda(x)$, $\lambda \in \Lambda$, *be functions in* $L^2(\mathbb{R}^n)$ *with the following three properties:*

$$(6.8) \qquad |w_\lambda(x)| \leq \frac{C2^{nj/2}}{\left(1 + |2^j x - k|\right)^{n+1}} \qquad \text{for } \lambda = 2^{-j}k + 2^{-j-1}\varepsilon;$$

$$(6.9) \qquad \int w_\lambda(x)\, dx = 0;$$

there exists a constant C' *such that, for every sequence* $\alpha(\lambda) \in l^2(\Lambda)$,

$$(6.10) \qquad \Big\| \sum \alpha_\lambda w_\lambda(x) \Big\|_2 \leq C' \big(\sum |\alpha_\lambda|^2\big)^{1/2}.$$

Then, for each $b \in \text{BMO}(\mathbb{R}^n)$, *the coefficients* $\alpha(\lambda) = \int b(x)w_\lambda(x)\, dx$ *satisfy Carleson's condition* (6.7).

To see this, we begin by observing that the dual form of (6.10) is the following inequality:

$$(6.11) \qquad \Big(\sum_{\lambda \in \Lambda} |(f, w_\lambda)|^2\Big)^{1/2} \leq C'\|f\|_2,$$

for every $f \in L^2(\mathbb{R}^n)$.

Then, by an obvious change of variable, we reduce the proof of (6.7) to the case where $Q = [0,1)^n$. We let B be the cube defined by $-2 \leq x_1 \leq 2, \ldots, -2 \leq x_n \leq 2$ and, as above, we decompose b as $b(x) = b_0 + b_1(x) + b_2(x)$, where b_0 is the mean of $b(x)$ over B, $b_1(x) = b(x) - b_0$, if $x \in B$, and 0, elsewhere, and $b_2(x)$ is zero on B.

The coefficients (b_0, w_λ) are zero. The coefficients (b_1, w_λ) can be dealt with by using (6.11), so we just need to find a bound for the coefficients (b_2, w_λ). We use (6.6), which gives

$$\int_{B^c} |x|^{-n-s}|b_2(x)|^2 \, dx \leq C\|b\|_{\text{BMO}}^2,$$

where $B^c = \mathbb{R}^n \backslash B$. If $Q(\lambda)$ is contained in $[0,1)^n$, then $2^{-j}k$ is in $[0,1)^n$, and if x is not in B, then $|x - 2^{-j}k| \geq c|x|$, where $c > 0$ is a constant. This simple remark gives

$$|(b_2, w_\lambda)| \leq C2^{nj/2} \int_{B^c} |2^j x - k|^{-n-1} |b_2(x)| \, dx$$

$$\leq C2^{-j}2^{-nj/2} \int_{B^c} |x|^{-n-1} |b_2(x)| \, dx$$

$$\leq C'2^{-j}2^{-nj/2} \|b\|_{\text{BMO}} ,$$

by the Cauchy-Schwarz inequality and the factorization

$$|x|^{-n-1} = |x|^{-(n/2)-((1+s)/2)} |x|^{-(n/2)-((1-s)/2)} ,$$

for $0 < s < 1$.

All that remains is to calculate $\sum_{Q(\lambda) \subset [0,1)^n} 4^{-j} 2^{-nj}$, which is easy, because, for each $j \geq 0$, the number of λs involved is $(2^n - 1)2^{nj}$.

7 Maurey's theorem

Dyadic H^1 space is defined as follows. We imitate the atomic decomposition of the usual H^1 space but require the atoms of dyadic H^1 to satisfy a more restrictive condition, namely, that an atom $a(x)$ of dyadic H^1 has support in a dyadic cube Q such that $\|a\|_2 \leq |Q|^{-1/2}$ and $\int a(x) \, dx = 0$.

An immediate consequence of the definition is that dyadic H^1 is contained in the usual H^1 space.

They are not the same. For example, a function of a real variable x can only belong to dyadic H^1 if the two integrals $\int_0^\infty f(x) \, dx$ and $\int_{-\infty}^0 f(x) \, dx$ are zero. This is because a dyadic interval $I \in \mathcal{I}$ is entirely contained in either $[0, \infty)$ or $(-\infty, 0]$. But every function $f(x)$ satisfying $\int_{-\infty}^\infty (1 + x^2)|f(x)|^2 \, dx < \infty$ and with zero integral belongs to the usual H^1 space.

B. Maurey has proved the following remarkable theorem ([184]):

Theorem 5. *Dyadic H^1 is isomorphic to the usual H^1 space (as a Banach space).*

Maurey's proof depended on methods of Banach space geometry. The first proofs giving an explicit isomorphism were obtained by L. Carleson and P. Wojtaszczyk and, in some sense, prefigured the theory of wavelets ([46], [237]).

We shall give a new proof of Maurey's theorem. Using the Haar system, dyadic H^1 can be decomposed by repeating word for word the decomposition we have given of the usual H^1 space in terms of a wavelet basis.

Let $U : L^2(\mathbb{R}^n) \to L^2(\mathbb{R}^n)$ be the isometry given by $U(h_Q^\varepsilon) = \psi_Q^\varepsilon$, where h_Q^ε is the n-dimensional Haar system, obtained as a special case of the construction of n-dimensional wavelets based on the multiresolution approximation given by step functions.

A function f is in dyadic H^1 if and only if $f(x) = \sum_\varepsilon \sum_Q \alpha^\varepsilon(Q) h_Q^\varepsilon$ and $\left(\sum_{Q \ni x} |\alpha^\varepsilon(Q)|^2 |Q|^{-1} \right)^{1/2} \in L^1(dx)$ for $\varepsilon \in E$. This is exactly the same condition as that specifying whether the series $\sum_\varepsilon \sum_Q \alpha^\varepsilon(Q) \psi_Q^\varepsilon$ belongs to the usual H^1 space.

Maurey's theorem is thus proved.

When we study the theory of Calderón-Zygmund operators, we shall see that U extends to an isomorphism of $L^p(\mathbb{R}^n)$ with itself for $1 < p < \infty$. There is thus a break when $p = 1$. In the other direction ($p = \infty$), we similarly introduce dyadic BMO, which consists of the functions which are locally square integrable and satisfy the condition

$$(7.1) \qquad \sup_{Q \in Q} \left(\frac{1}{|Q|} \int_Q |f(x) - m_Q f|^2 \, dx \right)^{1/2} < \infty,$$

where the upper bound is taken over the set of all dyadic cubes.

Once again, the characterization of dyadic BMO using the Haar system coincides with the characterization of the usual BMO space using a wavelet basis. The operator U extends to an isomorphism between dyadic BMO and the usual BMO space.

8 Notes and complementary remarks

We have omitted some of the most important properties of the spaces H^1 and BMO. These properties are stated and studied in [75], which is an excellent reference.

One of the remarkable properties of $BMO(\mathbb{R}^n)$ is given by the theorem of John and Nirenberg. Let $b(x)$ be a function on $\mathbb{R}^n$ which is locally integrable. Suppose that there exists a constant C and that constants $\gamma(B)$ can be found for each ball $B \subset \mathbb{R}^n$ such that

$$\frac{1}{|B|} \int_B |b(x) - \gamma(B)| \, dx \le C.$$

Then John and Nirenberg proved that $b(x)$ lies in BMO. To be more precise, for every exponent $p > 1$, there exists a (finite) constant $C(p)$ such that

$$(8.1) \qquad \sup_B \left(\frac{1}{|B|} \int_B |b(x) - m_B b|^p \, dx \right)^{1/p} \le C(p).$$

The extreme case $p = \infty$ requires the replacement of the L^∞ norm by the expression

(8.2)
$$\sup_B \frac{1}{|B|} \int_B e^{\lambda|b(x)-m_B b|} \, dx,$$

which is finite as long as the product $\lambda\|b\|_{\text{BMO}}$ is less than a certain constant $c(n) > 0$ ([140]).

For properties of the space $H^1(\mathbb{R}^n)$, we may invoke duality to glean information about the definition of atoms. Instead of employing the L^2 normalization of atoms, we can use apparently more specialized atoms such that $\|a\|_\infty \le |B|^{-1}$, the support of a is contained in B, and $\int_B a(x) \, dx = 0$. The atomic H^1 space obtained from these special atoms is the same as the space we described earlier.

We shall give two complementary results relating the space $H^1(\mathbb{R}^n)$ to an r-regular multiresolution approximation V_j, $j \in \mathbb{Z}$, of $L^2(\mathbb{R}^n)$, with $r \ge 1$.

We take a function $f \in L^1(\mathbb{R}^n)$ and define the maximal operator $E_\star(f)$ by

(8.3)
$$E_\star f(x) = \sup_{-\infty < j < \infty} |E_j(f)(x)|.$$

The operators E_j are, as before, the orthogonal projections of $L^2(\mathbb{R}^n)$ onto V_j. We then have the following remarkable statement:

Theorem 6. *The condition $E_\star f \in L^1(\mathbb{R}^n)$ characterizes the space $H^1(\mathbb{R}^n)$.*

This characterization is called the maximal characterization.

Another characterization is given by $\left(\sum_{-\infty}^\infty |D_j f(x)|^2\right)^{1/2} \in L^1(\mathbb{R}^n)$, where $D_j = E_{j+1} - E_j$, as before.

For one real variable, the Hardy space $H^1(\mathbb{R})$ consists of the real parts of the complex Hardy space $\mathbb{H}^1(\mathbb{R})$, whose definition we now recall. A function $f(z)$ which is holomorphic in the upper half-plane $\Im z > 0$ belongs to $\mathbb{H}^1(\mathbb{R})$ if (and only if) $\sup_{y>0} \int_{-\infty}^\infty |f(x+iy)| \, dx < \infty$. Such a function always has a trace on $y = 0$, defined as the limit in $L^1(\mathbb{R})$ of the functions $f(\cdot + iy)$ as $y > 0$ tends to 0.

These traces form a closed subspace of $L^1(\mathbb{R})$, which, by abuse of language, we still call $\mathbb{H}^1(\mathbb{R})$. It is characterized by the Paley-Wiener theorem. This theorem states that f belongs to $\mathbb{H}^1(\mathbb{R})$ if and only if $f \in L^1(\mathbb{R})$ and $\hat{f}(\xi) = \int_{-\infty}^\infty e^{-ix\xi} f(x) \, dx$ is zero for $\xi \le 0$.

Now each function f in $H^1(\mathbb{R})$ (the real version given by Stein and Weiss) can be written uniquely as $f = u + \bar{v}$, where u and v belong to $\mathbb{H}^1(\mathbb{R})$. The decomposition is unique and amounts to writing $\hat{f} = \hat{f}\chi_+ + \hat{f}\chi_-$, where χ_+ and χ_- are the characteristic functions of $[0, \infty)$ and $(-\infty, 0)$, respectively. If, further, f is a real-valued function, then f

is the real part of a function in the complex Hardy space $\mathbb{H}^1(\mathbb{R})$. This is why we speak of the "real version of the complex Hardy space". If we want a Banach space over the complex field, it is enough to complexify the vector space of those real parts.

An equivalent way of saying the same thing is to use the Hilbert transform H. The Hilbert transform is defined by the condition that $H(u) = v$ if u and v are real-valued functions of a real variable and $u + iv$ is the trace on $y = 0$ of a holomorphic function $f(x + iy)$ on the upper half-plane.

The Hilbert transform can also be defined by using the Fourier transform: $H(f) = g$ if and only if the Fourier transforms $\hat{f}$ and $\hat{g}$ of f and g are related by $\hat{g}(\xi) = -i \operatorname{sgn} \xi \hat{f}(\xi)$.

We then come to the following definition of $H^1(\mathbb{R})$. We have $f \in H^1(\mathbb{R})$ if and only if f and $H(f)$ both belong to $L^1(\mathbb{R})$. The norm of f in $H^1(\mathbb{R})$ is defined to be the sum of the L^1 norms of f and $H(f)$.

We shall now indicate why the characterization of the complex Hardy space $\mathbb{H}^1(\mathbb{R})$ by the integrability of Lusin's area function (obtained by Calderón in 1965) and the characterization of the real Hardy space $H^1(\mathbb{R})$ by the condition

$$\left(\sum_{I \ni x} |(f, \psi_I)|^2 |I|^{-1} \right)^{1/2} \in L^1(dx)$$

look so similar. To do this, we shall describe a kind of dictionary enabling us to move from one "language" to the other. It will translate the function $(2\pi i)^{-1}(t + i)^{-2}$ into the wavelet $\psi(t)$, the points of the upper half-plane into dyadic intervals, points (u, v), with $v > 0$ in the upper half-plane such that $v \geq |u - x|$, into dyadic intervals containing x and, lastly, $(\iint_{\{v \geq |u-x|\}} |f'(z)|^2 \, du \, dv)^{1/2}$ into $(\sum_{I \ni x} |(f, \psi_I)|^2 |I|^{-1})^{1/2}$.

The heuristic justification for the translation is as follows. The function $(t + i)^{-2}$ is very similar to a wavelet, because it has the required properties of regularity, localization and cancellation: certainly $\int_{-\infty}^{\infty} (t + i)^{-2} \, dx = 0$.

As shown in Chapter 4, it is only necessary to know the values

$$f'(k2^{-j} + i2^{-j}), \quad j, k \in \mathbb{Z},$$

of a function f in complex Hardy space in order to construct the function. We then use the geometric correspondence between the dyadic intervals $I = [k2^{-j}, (k + 1)2^{-j})$ and the points $z = k2^{-j} + 2^{-j-1} + i2^{-j-1}$, $j, k \in \mathbb{Z}$, in the upper half-plane. According to this correspondence, $I \ni x$ is equivalent to $z \in \Gamma(x)$, where $\Gamma(x)$ is the cone with vertex x, defined by $v \geq |u - x|$.

These analogies lead to the following definitions and substitutions. We let S denote the set of points $z = k2^{-j} + 2^{-j-1} + i2^{-j-1}$, $j, k \in \mathbb{Z}$, in the upper half-plane; we put $\psi(t) = (2\pi i)^{-1}(t+i)^{-2}$ and, for $z = x + iy$, $x \in \mathbb{R}$, $y > 0$, we put $\psi_z(t) = y^{-1/2}\psi((t-x)/y)$. Then Cauchy's formula gives $f'(z) = y^{-3/2}(f, \psi_z)$, for $f \in \mathbb{H}^1(\mathbb{R})$. It turns out that the fluctuations of a function in $\mathbb{H}^1(\mathbb{R})$ are much more significant close to the boundary. This leads us to replace Lusin's area function $(\iint_{\Gamma(x_0)} |f'(z)|^2 \, dx \, dy)^{1/2}$ by the "Riemann sum"

$$\left(\sum_{z \in S \cap \Gamma(x_0)} 2^j |(f, \psi_z)|^2 \right)^{1/2}.$$

The second substitution consists of replacing $\psi(t)$ by the wavelet of compact support used in this chapter and, at the same time, replacing the points $z \in S$ by the corresponding dyadic intervals $I = [k2^{-j}, (k+1)2^{-j})$. Then, as claimed, $\sum_{z \in S \cap \Gamma(x_0)} 2^j |(f, \psi_z)|^2$ becomes $\sum_{I \ni x_0} |(f, \psi_I)|^2 |I|^{-1}$.

We return to examples of functions in $\mathbb{H}^1(\mathbb{R})$. Consider the branch of $\log z$ which is holomorphic in $\Im z > 0$ and such that $\log 1 = 0$. Then $f(z) = z^{-1}(\log z + i)^{-2}$ is in $\mathbb{H}^1(\mathbb{R})$. A related example is defined by

$$f(x) = \begin{cases} 0, & \text{if } |x| > \frac{1}{2}; \\ x^{-1}(\log|x|)^{-2}, & \text{if } |x| \le \frac{1}{2}. \end{cases}$$

It is easy to write down the atomic decomposition of this function, because the atoms $\lambda_j a_j(x)$ (written with their coefficients) are the restrictions of f to the dyadic "anuli" $U_j = [-2^{-j}, -2^{-j-1}] \cup [2^{-j-1}, 2^{-j}]$.

On the other hand, although the function $f(x) = x^{-1}(\log x + i)^{-2}$ for $x > 0$ and 0 for $x \le 0$ has zero integral, it does not belong to $\mathbb{H}^1(\mathbb{R})$. This can be seen by using the fact that $\log|x|$ belongs to BMO($\mathbb{R}$) and that the same is true for the function $g(x) = \log|x|$ when $|x| \le 1$ and 0 when $|x| \ge 1$, which is the same as $\inf(0, \log|x|)$. It is then easy to see that $\int f(x)g(x) \, dx = -\infty$.

The function $\log|x|$ is the best-known example of a function of BMO($\mathbb{R}^n$). This example has been generalized in two separate ways. Stein observed that, if $P(x_1, \ldots, x_n)$ is an arbitrary non-trivial polynomial, then $\log|P(x_1, \ldots, x_n)|$ belongs to BMO($\mathbb{R}^n$).

Another generalization takes a Borel measure $\mu \ge 0$ and lets $\mu^\star(x)$ denote the quantity $\sup_{B \ni x} |B|^{-1}\mu(B)$. If $\mu^\star(x)$ is finite almost everywhere, then $\log \mu^\star(x)$ belongs to BMO($\mathbb{R}^n$) The case where μ is Dirac measure at 0 gives the example $\log|x|$ ([75]).

We now pass to $H^1(\mathbb{R}^n)$. Initially, Stein and Weiss defined this space as the set of functions $f \in L^1(\mathbb{R}^n)$ such that $R_j f$ belongs to $L^1(\mathbb{R}^n)$, for

$1 \leq j \leq n$. Here, $R_j = (\partial/\partial x_j)(-\Delta)^{-1/2}$ denotes the Riesz transform. The vector $(f, R_1 f, \ldots, R_n f)$ can be written as

$$\left(-\frac{\partial}{\partial t} u(x,t), \frac{\partial}{\partial x_1} u(x,t), \ldots, \frac{\partial}{\partial x_n} u(x,t)\right)\Bigg|_{t=0},$$

where $u(x,t) = e^{-t(-\Delta)^{1/2}}((-\Delta)^{-1/2}f)$ is a harmonic function. We see here the basic idea of Stein and Weiss, which is to imitate holomorphic functions by the gradients of harmonic functions.

The decisive step was Fefferman's discovery of the duality between H^1 and BMO which led Coiffman and Weiss to change the way of looking at H^1 by taking the atomic definition as starting point. We refer the reader to [109] and [75].

In [231], A. Uchiyama resolved the fundamental problem of characterizing $H^1(\mathbb{R}^n)$ by operators other than the Riesz transforms. Let $m \in C^\infty(\mathbb{R}^n \backslash \{0\})$ be a homogeneous function of degree 0 and let M be the convolution operator corresponding to the multiplier $\mathcal{F}(M(f)) = m\hat{f}$, where $\mathcal{F}(f) = \hat{f}$ is the Fourier transform of f. Then H^1 is characterized by the property $M_0 f \in L^1, \ldots, M_k f \in L^1$ if and only if the corresponding multipliers $m_0(\xi), \ldots, m_k(\xi)$ (which are C^∞ except at 0 and homogeneous of degree 0) satisfy the following condition: for every $\xi \neq 0$, the rank of the matrix

$$\begin{pmatrix} m_0(\xi) & \cdots & m_k(\xi) \\ m_0(-\xi) & \cdots & m_k(-\xi) \end{pmatrix}$$

equals two.

Note that we do not require f to belong to L^1 (but this is necessarily satisfied in the event). The usual case thus corresponds to $k = n$, $m_0(\xi) = 1$ and $m_j(\xi) = i\xi_j/|\xi|$.

The connection with the problem of unconditionally convergent series in L^1 started with the pioneering work of Maurey, who showed that the space $H^1(\mathbb{R}^n)$ had an unconditional basis. It is perhaps worth noting that our approach to $H^1(\mathbb{R}^n)$ remains faithful to the historical one, because we have characterized H^1 by $f \in L^1$ and $\sup_{\omega \in \Omega} \|U_\omega(f)\|_1 < \infty$, where the U_ω denote the operators which change the signs of the wavelet coefficients. In other words, $U_\omega(\psi_\lambda) = \omega(\lambda)\psi_\lambda$, where $\omega(\lambda) = \pm 1$. The operators U_ω, $\omega \in \Omega = \{-1, 1\}^\Lambda$, are Calderón-Zygmund operators in the sense of Chapter 7. This means that the U_ω and the R_j belong to the same family, so our characterization of H^1 is faithful to that of the "founding fathers".

Before ending these remarks, let us indicate how the periodic BMO space is defined. We shall restrict ourselves to dimension 1, because the general case is similar. Although the functions in $L^2_{\text{loc}}(\mathbb{R})$ which are periodic, of period 1, are obtained by extending arbitrary functions of $L^2[0,1)$ by periodicity, the corresponding property for BMO is false. Periodic BMO is defined as the subspace of $BMO(\mathbb{R})$ which is composed of the periodic functions of period 1, and this subspace does not coincide with that obtained by extending the functions of $BMO[0,1]$ periodically. (We have denoted by $BMO[0,1]$ the restrictions of the functions in $BMO(\mathbb{R})$ to $[0,1)$.)

Denoting by $g_0(x) = 1$, $g_1(x), \ldots$ the periodic wavelets of Section 11 in Chapter 3, periodic BMO is characterized by the Carleson condition applied to the coefficients with respect to the basis composed of periodic wavelets. That is, $\sum_{I(k)\subset I} |\alpha(k)|^2 \leq C|I|$, where $I(k)$ denotes the interval $[r2^{-j}, (r+1)2^{-j})$, for $k = 2^j + r$, $0 \leq r < 2^j$, and where I denotes an arbitrary dyadic interval in $[0,1]$.

If, for example, $\sum_0^\infty |\alpha_k|^2 < \infty$ and the moduli $|\alpha_k|$ form a decreasing sequence, then $\sum_0^\infty \alpha_k g_k(x)$ belongs to periodic $BMO(\mathbb{R})$. If, however, $\sum_0^\infty |\alpha_k|^2 = \infty$ and the moduli $|\alpha_k|$ are decreasing, then $\sum_0^\infty \alpha_k g_k(x)$ is a distribution which cannot be a measure.

We finish by describing the space product-$H^1(\mathbb{R}^2)$ as studied by S.Y. Chang and R. Fefferman in their remarkable series of papers [51], [110] and [111].

The partial Hilbert transforms H_1 and H_2 are defined by the multipliers $-i\,\mathrm{sgn}\,\xi_1$ and $-i\,\mathrm{sgn}\,\xi_2$, where $\xi = (\xi_1, \xi_2) \in \mathbb{R}^2$. Then $f \in$ product-$H^1(\mathbb{R}^2)$ if the four functions f, $H_1(f)$, $H_2(f)$ and $H_1 H_2(f)$ belong to $L^1(\mathbb{R}^2)$.

Let $\mathcal{I}$ denote the set of all dyadic intervals I in $\mathbb{R}$ and let ψ_I, $I \in \mathcal{I}$ be a wavelet basis of $L^2(\mathbb{R})$ arising from an r-regular multiresolution approximation, with $r \geq 1$.

Then the functions $\psi_I \otimes \psi_J$, $(I,J) \in \mathcal{I} \times \mathcal{I}$, form an unconditional basis of product-$H^1(\mathbb{R}^2)$. More precisely, let $\mathcal{R}$ denote the set of all dyadic rectangles $R = I \times J$ and then put $\psi_R(x,y) = \psi_I(x)\psi_J(y)$. With this notation, a series $\sum_{R\in\mathcal{R}} \alpha(R)\psi_R$ belongs to product-$H^1(\mathbb{R}^2)$ if and only if $\left(\sum_{R\ni x} |\alpha(R)|^2 |R|^{-1}\right)^{1/2}$ belongs to $L^1(\mathbb{R}^2)$ and, similarly, the dual space product-BMO is characterized by the "Carleson condition" $\sum_{R\subset\Omega} |\alpha(R)|^2 \leq C|\Omega|$, which must now be satisfied for every open set Ω in $\mathbb{R}^2$.

These characterizations, which were rediscovered by Lemarié, follow easily from the work of Chang and Fefferman.

6

Wavelets and spaces of functions and distributions

1 Introduction

In the course of this chapter, we shall show that wavelet series which are orthonormal series in the standard space $L^2(\mathbb{R}^n)$ are just as effective in the analysis of other spaces, such as $L^p(\mathbb{R}^n)$, for $1 < p < \infty$, $L^{p,s}(\mathbb{R}^n)$, where s is an index of regularity, Hölder spaces, Hardy spaces, Besov spaces, etc.

The functions in these various spaces ("functions" which may be tempered distributions in some cases) can thus be written in the form

$$(1.1) \qquad f(x) = \sum_{\lambda \in \Lambda} (f, \psi_\lambda) \psi_\lambda(x),$$

in the case of homogeneous function spaces and

$$(1.2) \qquad f(x) = \sum_{\lambda \in \Gamma_0} (f, \phi_\lambda) \phi_\lambda(x) + \sum_{j \geq 0} \sum_{\lambda \in \Lambda_j} (f, \psi_\lambda) \psi_\lambda(x),$$

in the case of inhomogeneous spaces.

In both situations, the series converge unconditionally to $f(x)$ in the relevant norm.

As we have already indicated, to show that a family of vectors e_λ, $\lambda \in \Lambda$, in a Banach space B, is an unconditional basis of B amounts to verifying two things, one of which is immediate, while the other is often a delicate matter:

(a) the e_λ, $\lambda \in \Lambda$, form a complete (or total) subset of B, in the sense that the finite linear combinations of the e_λ, $\lambda \in \Lambda$, form a dense subspace of B;

(b) there exists a constant $C \geq 1$ such that, for every sequence m_λ, $\lambda \in \Lambda$ satisfying $\sup_{\lambda \in \Lambda} |m_\lambda| \leq 1$, we can find an operator T (unique, by (a)) which is continuous on B, has norm $\|T\| \leq C$ and is such that $T(e_\lambda) = m_\lambda e_\lambda$, for every $\lambda \in \Lambda$.

The operators which appear in the proofs in this chapter are Calderón-Zygmund operators. Their continuity on the appropriate function spaces will be a consequence of the real-variable methods of Chapter 7 or of Theorem T(1) of David and Journé which will not be given until Chapter 8.

We shall thus not follow a linear or logical path in our exposition, and we beg the reader's indulgence. We want the first six chapters to be entirely devoted to wavelets—with as few operators as possible—so that they will be accessible to a wider audience than that of the later chapters.

Chapter 6 begins with the criteria for wavelet series to belong to $L^p(\mathbb{R}^n)$, when $1 < p < \infty$. These criteria are a natural extension of those obtained for $H^1(\mathbb{R}^n)$, but clearly do not apply to the spaces L^1 or L^∞, which have no unconditional bases. In a certain sense, wavelets are stable under the operations of integration and differentiation and it is thus not surprising to get the expected criteria for the space $L^{p,s}$ of integrals or fractional derivatives of L^p functions, as long as s is not greater than the regularity r of the wavelets.

We shall then pass to the case of the Hardy spaces H^p for $0 < p < 1$.

The second part of this chapter deals with various, more or less well-known, examples of Besov spaces. These are also characterized by very simple conditions on the moduli of the wavelet coefficients.

The final part is devoted to two classical problems in the theory of Banach spaces (constructing a Schauder basis of the disc algebra and an unconditional basis of the complex Hardy space $\mathbb{H}^1$) to which the theory of wavelets brings a fresh and elegant solution.

2 Criteria for belonging to $L^p(\mathbb{R}^n)$ and to $L^{p,s}(\mathbb{R}^n)$

Throughout this section, we shall suppose that $1 < p < \infty$: it is essential to exclude the extreme cases $p = 1$ and $p = \infty$.

Recall that Λ is the set $\{\lambda = 2^{-j}k + \varepsilon 2^{-j-1} : j \in \mathbb{Z}, k \in \mathbb{Z}^n, \varepsilon \in E\}$, where E is the set $\{0,1\}^n$ excluding $(0,0,\ldots,0)$. Note that j, k and ε are then uniquely determined by λ. We let $Q(\lambda)$ denote the dyadic cube defined by $2^j x - k \in [0,1)^n$ and let $\chi_\lambda(x)$ denote the characteristic function of $Q(\lambda)$. The volume of $Q(\lambda)$ is 2^{-nj} and is denoted by

$|Q(\lambda)|$. Lastly, ψ_λ, $\lambda \in \Lambda$, is a wavelet basis arising from an r-regular multiresolution approximation of $L^2(\mathbb{R}^n)$.

Theorem 1. *The norms* $\|f\|_p$, $\|(\sum_{\lambda \in \Lambda} |\alpha(\lambda)|^2 |Q(\lambda)|^{-1} \chi_\lambda(x))^{1/2}\|_p$ *and* $\|(\sum_{\lambda \in \Lambda} |\alpha(\lambda)|^2 |\psi_\lambda(x)|^2)^{1/2}\|_p$ *are equivalent on the space* $L^p(\mathbb{R}^n)$, *when* $1 < p < \infty$.

Moreover, a wavelet series $\sum \alpha(\lambda) \psi_\lambda(x)$ belongs to $L^p(\mathbb{R}^n)$ if and only if one of the two latter norms is finite.

We shall first verify that the first and third norms are equivalent.

To do this, we let Ω denote the product set $\{-1, 1\}^\Lambda$ which we provide with the Bernoulli probability measure $d\mu(\omega)$, obtained by taking the product of the measures on each factor which give a mass of $1/2$ to each of the points -1 and 1. An element ω of Ω is a sequence $\omega(\lambda)$, $\lambda \in \Lambda$, consisting of ± 1s.

For each $\omega \in \Omega$ we define the operator $T_\omega : L^2(\mathbb{R}^n) \to L^2(\mathbb{R}^n)$ given by $T_\omega(\psi_\lambda) = \omega(\lambda) \psi_\lambda$.

By virtue of the results of Chapter 7, we then get

Lemma 1. *The set* $\{T_\omega : \omega \in \Omega\}$ *is a bounded set of Calderón-Zygmund operators.*

This lemma is obvious, because the T_ω are isometric on $L^2(\mathbb{R}^n)$ and the kernels $K_\omega(x, y) = \sum_{\lambda \in \Lambda} \omega(\lambda) \psi_\lambda(x) \bar{\psi}_\lambda(y)$ satisfy

$$|K_\omega(x, y)| \le C|x - y|^{-n}$$

and

$$\left| \frac{\partial K_\omega}{\partial x_j} \right| + \left| \frac{\partial K_\omega}{\partial y_j} \right| \le C|x - y|^{-n-1}$$

uniformly in ω.

We need a second lemma, namely, Khinchin's well-known inequality ([239]).

Lemma 2. *For* $1 < p < \infty$, *all the* $L^p(\Omega, d\mu(\omega))$ *norms are equivalent on the closed subspace of* $L^2(\Omega)$ *consisting of the functions* $S(\omega) = \sum_{\lambda \in \Lambda} \alpha(\lambda) \omega(\lambda)$. *Hence, for each* p, *there are constants* $C_p \ge C_p' > 0$ *such that*

$$C_p' \left(\sum |\alpha(\lambda)|^2 \right)^{1/2} \le \left(\int_\Omega |S(\omega)|^p \, d\mu(\omega) \right)^{1/p} \le C_p \left(\sum |\alpha(\lambda)|^2 \right)^{1/2}.$$

Let us proceed to the proof of the equivalence of the norms of Theorem 1. By Calderón-Zygmund theory, $\|T_\omega(f)\|_p \le C\|f\|_p$. We raise this inequality to the power p and take the mean over $\omega \in \Omega$ of the resulting inequality. We thus get a double integral over $\mathbb{R}^n \times \Omega$, with respect to

$dx\,d\mu(\omega)$. Applying Fubini's theorem and, for each $x \in \mathbb{R}^n$, Khinchin's inequality then gives

$$(2.1) \qquad \left\| \left(\sum_{\lambda \in \Lambda} |\alpha(\lambda)|^2 |\psi_\lambda(x)|^2 \right)^{1/2} \right\|_p \leq C' \|f\|_p .$$

To obtain the converse inequality, we raise the left-hand side of (2.1) to the power p. The resulting integral, which we denote by $N_p(f)$, can be bounded below by $C_p^{-p} \int_\omega \|T_\omega(f)\|_p^p \, d\mu(\omega)$. This time, we have used the second part of Khinchin's inequality. Finally, we observe that $T_\omega^2 = I$, so we can write $\|f\|_p \leq C\|T_\omega(f)\|_p$, for each ω, and this gives $\|f\|_p \leq C'(N_p(f))^{1/p}$.

To show that $\|f\|_p$ and $\|(\sum_{\lambda \in \Lambda} |\alpha(\lambda)|^2 |Q(\lambda)|^{-1} \chi_\lambda(x))^{1/2}\|_p$ are equivalent, we use the equivalence we have just proved, applied to the wavelets ψ_λ of compact support of Chapter 3, Sections 8 and 9.

Let m be an integer such that the support of each ψ_λ is contained in the cube $mQ(\lambda)$ (m times as big as $Q(\lambda)$ but with the same centre).

We write h_λ, $\lambda \in \Lambda$ for the n-dimensional Haar system. We have $|h_\lambda| = |Q(\lambda)|^{-1/2} \chi_\lambda$ and $h_\lambda(x)$ is constant on each of the 2^n dyadic subcubes obtained by subdividing $Q(\lambda)$.

We define the unitary operator $U : L^2(\mathbb{R}^n) \to L^2(\mathbb{R}^n)$ by $U(h_\lambda) = \psi_\lambda$, for each $\lambda \in \Lambda$.

Lemma 3. *If $1 < p < \infty$, then the operator U is an isomorphism on $L^p(\mathbb{R}^n)$.*

This lemma, together with the probabilistic interpretation of the norms appearing in Theorem 1, concludes its proof.

To prove Lemma 3, we can use the fact that U is an isomorphism between dyadic H^1 and its classical analogue, and that U is also an isometry on $L^2(\mathbb{R}^n)$. By interpolation, U and U^{-1} are bounded on $L^p(\mathbb{R}^n)$ when $1 < p < 2$. (We use the theorem which lets us interpolate between H^1 and L^2. The reader will find it in [109].)

But we can also show that U and U^{-1} are bounded on $L^p(\mathbb{R}^n)$, for $1 < p < 2$, by applying a different interpolation theorem—Marcinkiewicz's theorem ([217], p.21)—after showing that U and U^{-1} are bounded as operators from $L^1(\mathbb{R}^n)$ to weak-L^1. The reader is asked to consult Section 3 of the next chapter, where the method we use is described.

The continuity of $U^{-1} : L^1 \to$ weak-L^1 is a direct consequence of Chapter 7, Section 3. Indeed, $\sum_{\lambda \in \Lambda} h_\lambda(x) \bar{\psi}_\lambda(y)$ is the kernel of U^{-1} and its regularity with respect to y is guaranteed by that of the wavelets.

The continuity of U from L^1 to weak-L^1 seems problematic, since the kernel of U is $\sum_{\lambda \in \Lambda} \psi_\lambda(x) h_\lambda(y)$ which has no regularity with respect to *y*. But this difficulty is only apparent. To establish the weak continuity of U, we apply a Calderón-Zygmund decomposition to $f \in L^1(\mathbb{R}^n)$ at the level $t > 0$. This gives a bad function $b(x)$ which is the sum of components $b_j(x)$ supported by dyadic cubes Q_j; moreover, the integral of $b_j(x)$ over Q_j is zero and $\sum |Q_j| \le Ct^{-1}\|f\|_1$. For $x \notin mQ_j$, we then have

$$(2.2) \qquad U(b_j)(x) = \sum_{\lambda \in \Lambda} \psi_\lambda(x) \int_{Q_j} h_\lambda(y) b_j(y)\, dy = 0.$$

Indeed, in order that the integral be non-zero, the support $Q(\lambda)$ of h_λ should meet Q_j. Now if two dyadic cubes intersect, one must be contained in the other. If $Q(\lambda)$ is contained in Q_j, the support of ψ_λ is contained in mQ_j and thus $\psi_\lambda(x) = 0$, x lying outside mQ_j. If Q_j is contained in $Q(\lambda)$ and if $Q_j \neq Q(\lambda)$, Q_j is contained in one of the cubes subdividing $Q(\lambda)$. Hence, the function h_λ is constant on Q_j and thus $\int_{Q_j} h_\lambda(y) b_j(y) = 0$.

Still referring to the methods and the notation of Chapter 7, Section 3, the (L^1, weak-L^1)-continuity of the operator U follows from the identity (2.2). The proof is even easier than in the usual case, because there is no Marcinkiewicz function.

Since U is unitary on $L^2(\mathbb{R}^n)$, its adjoint coincides with its inverse, so the case $2 < p < \infty$ can be deduced immediately from the case $1 < p < 2$.

We now intend to study the generalized Sobolev spaces $L^{p,s}$ (also denoted by $W^{p,s}$), which we shall first define in the case where $1 < p < \infty$ and $0 \le s < r$: as usual, r is the index of regularity of the wavelets. We shall then move to the case where $-r < s < 0$ and we shall lastly examine the corresponding homogeneous spaces.

If $1 < p < \infty$ and $s \ge 0$ is a positive real number, then $L^{p,s}$ is the subspace of L^p consisting of functions $f \in L^p$ such that $(I - \Delta)^{s/2}f$, taken in the sense of distributions, still belong to L^p. In other words, there must be a function $g \in L^p$ such that $\hat{g}(\xi) = (1 + |\xi|^2)^{s/2}\hat{f}(\xi)$.

We let ψ_λ, $\lambda \in \Lambda$, be an orthonormal wavelet basis arising from an r-regular multiresolution approximation of $L^2(\mathbb{R}^n)$, with $r > s$. Then we have the following criterion for $L^{p,s}(\mathbb{R}^n)$.

Proposition 1. *Let $1 < p < \infty$ and $0 \le s < r$. A wavelet series belongs to $L^{p,s}(\mathbb{R}^n)$ if and only if*

$$(2.3) \qquad \left(\sum_{\lambda \in \Lambda} |\alpha(\lambda)|^2 (1 + 4^{js}) 2^{nj} \chi_\lambda(x) \right)^{1/2} \in L^p(\mathbb{R}^n).$$

We shall first prove this result for the case of the Littlewood-Paley multiresolution approximation (Chapter 3, Section 2). We shall then indicate what changes have to be made for the general case.

For the Littlewood-Paley multiresolution approximation, we get

$$\psi_\lambda(x) = 2^{nj/2}\psi^\varepsilon(2^j x - k), \quad j \in \mathbb{Z}, \; k \in \mathbb{Z}^n, \; \varepsilon \in E,$$

and the $2^n - 1$ functions ψ^ε belong to the Schwartz class $S(\mathbb{R}^n)$. Moreover, the Fourier transforms $\hat{\psi}^\varepsilon(\xi)$ of the ψ^ε vanish in a neighbourhood of 0 and have compact support. Thus, for each $\gamma \in \mathbb{R}$, we can construct functions $\psi^{\varepsilon,\gamma}$ in $S(\mathbb{R}^n)$ by putting $\psi^{\varepsilon,\gamma} = (-\Delta)^{\gamma/2}\psi^\varepsilon$. The Fourier transforms satisfy $\hat{\psi}^{\varepsilon,\gamma}(\xi) = |\xi|^\gamma \hat{\psi}^\varepsilon(\xi)$.

Once the $\psi^{\varepsilon,\gamma}$ have been constructed, we can define a corresponding family $\psi_\lambda^\gamma(x)$, $\lambda \in \Lambda$, by

$$(2.4) \qquad \psi_\lambda^\gamma(x) = 2^{nj/2}\psi^{\varepsilon,\gamma}(2^j x - k) \quad \text{where} \quad \lambda = 2^{-j}k + 2^{-j-1}\varepsilon.$$

Let $\mathcal{L}_\gamma$ be the operator defined by $\mathcal{L}_\gamma(\psi_\lambda) = \psi_\lambda^\gamma$. To prove Proposition 1, we start by establishing the following result:

Proposition 2. Let $1 < p < \infty$. For each $\gamma \in \mathbb{R}$, the operator $\mathcal{L}_\gamma$ is an isomorphism of $L^p(\mathbb{R}^n)$ onto itself.

In particular, to calculate the L^p norm of a series $\sum \alpha(\lambda)\psi_\lambda^\gamma(x)$, it is enough to estimate the L^p norm of $\alpha(\lambda)\psi_\lambda(x)$, which is obtained by applying Theorem 1.

Let us show how Proposition 2 implies Proposition 1. We first observe that if $f \in L^{p,s}$, $s \geq 0$, then $(-\Delta)^{s/2}f$ exists, belongs to L^p, and the norm of f in $L^{p,s}$ is equivalent to $\|f\|_p + \|(-\Delta)^{s/2}f\|_p$.

To see this, we go to the Fourier transforms, and the remarks we have just made follow from this observation: the function $1 - |\xi|^s(1 + |\xi|^2)^{-s/2}$ is the Fourier transform of a function $g_s \in L^1(\mathbb{R}^n)$.

To estimate the $L^{p,s}$ norm of $\sum \alpha(\lambda)\psi_\lambda(x)$, we must therefore calculate the L^p norm of $(-\Delta)^{s/2} \sum \alpha(\lambda)\psi_\lambda(x) = \sum \alpha(\lambda)2^{js}\psi_\lambda^s(x)$. This last identity follows from the definition of the functions ψ_λ^s. We now apply Proposition 2 and the result is given by Theorem 1.

In our proof of Proposition 2, we shall first consider the case $p = 2$ and calculate the inverse operator $\mathcal{M}_\gamma$ of $\mathcal{L}_\gamma$ explicitly. To deal with the case $p = 2$ we need the following lemma:

Lemma 4. For each $\gamma \in \mathbb{R}$, the functions $\psi_\lambda^\gamma(x)$ form a Riesz basis of $L^2(\mathbb{R}^n)$, and the dual basis is $\psi_\lambda^{-\gamma}$, $\lambda \in \Lambda$.

To prove Lemma 4, we shall use a result which will be proved in Chapter 8, Section 3 and which gives

$$(2.5) \qquad \left\|\sum \alpha(\lambda)\psi_\lambda^\gamma(x)\right\|_2 \leq C(\gamma)\left(\sum |\alpha(\lambda)|^2\right)^{1/2}.$$

This inequality means that the functions ψ_λ^γ, $\lambda \in \Lambda$, are almost orthogonal. The almost-orthogonality follows from the structure of the functions ψ_λ^γ, that is, from their localization, regularity and cancellation—see Definition 3 and Theorem 2 of Chapter 8.

Once (2.5) is established, we observe that $\psi_\lambda^\gamma = 2^{-j\gamma}(-\Delta)^{\gamma/2}\psi_\lambda$, which implies that $(\psi_\lambda^\gamma, \psi_{\lambda'}^{-\gamma}) = 0$, if $\lambda \neq \lambda'$, and $= 1$, if $\lambda = \lambda'$. Since (2.5) is also satisfied when γ is replaced by $-\gamma$, we immediately get the converse inequality

$$\left\| \sum \gamma(\lambda)\psi_\lambda^\gamma(x) \right\|_2 \geq \delta(\gamma) \left(\sum |\alpha(\lambda)|^2 \right)^{1/2}.$$

Finally, we must show that the functions ψ_λ^γ form a complete subset of $L^2(\mathbb{R}^n)$. Suppose that $f \in L^2(\mathbb{R}^n)$ is orthogonal to all the functions ψ_λ^γ. Then $(f, (-\Delta)^{\gamma/2}\psi_\lambda) = 0$, for every $\lambda \in \Lambda$. We rewrite this condition by introducing the distribution $S = (-\Delta)^{\gamma/2}f$, which belongs to the Sobolev space $H^{-\gamma}$. This distribution satisfies $\langle S, \psi_\lambda \rangle = 0$ for all $\lambda \in \Lambda$ ($\langle \cdot, \cdot \rangle$ is the bilinear form giving the duality between distributions and test functions; we have used the fact that the ψ_λ are real-valued). The characterization of Sobolev spaces by wavelet coefficients gives $S = 0$. Lemma 4 is proven.

The significance of Lemma 4 is that the operator $\mathcal{L}_\gamma$ is an isomorphism of $L^2(\mathbb{R}^n)$ with itself. The kernel (which is a distribution) of $\mathcal{L}_\gamma$ is $\sum_{\lambda \in \Lambda} \psi_\lambda^\gamma(x)\bar{\psi}_\lambda(y)$ and its restriction to the open set $\Omega \subset \mathbb{R}^n \times \mathbb{R}^n$, defined by $y \neq x$ satisfies the Calderón-Zygmund estimations of Chapter 7. Hence $\mathcal{L}_\gamma$ extends to a continuous operator on $L^p(\mathbb{R}^n)$ for $1 < p < \infty$.

We then consider the operator $\mathcal{M}_\gamma$, defined by $\mathcal{M}_\gamma(\psi_\lambda^\gamma) = \psi_\lambda$, $\lambda \in \Lambda$. Its kernel (also a distribution) is $\sum_{\lambda \in \Lambda} \psi_\lambda(x)\bar{\psi}_\lambda^{-\gamma}(y)$ and, just like $\mathcal{L}_\gamma$, $\mathcal{M}_\gamma$ extends to a continuous operator h on all the L^p spaces.

Finally, $\mathcal{L}_\gamma \mathcal{M}_\gamma = \mathcal{M}_\gamma \mathcal{L}_\gamma = I$ on $L^p(\mathbb{R}^n)$, as can be seen by a denseness argument in $L^p(\mathbb{R}^n)$, using continuous functions of compact support —which are in $L^2(\mathbb{R}^n)$. We have proved Proposition 2.

We now consider the case of a wavelet basis arising from a general r-regular multiresolution approximation with $r \geq 1$. To adapt the preceding proof we need to control the regularity, the rate of decrease at infinity and the cancellation of the functions $(-\Delta)^{s/2}\psi^\varepsilon$ and $(-\Delta)^{-s/2}\psi^\varepsilon$, where ψ^ε is one of the basic wavelets used to construct our orthonormal basis.

We use the notation and the result of Chapter 2, Lemma 12. In particular, we let S_r denote the set of functions which, together with their derivatives of order less than or equal to r, decrease rapidly at infinity. Since all the moments $\int x^\alpha \psi^\varepsilon(x)\,dx$ are zero for $|\alpha| \leq r$, we have

(2.6) $$\psi^\varepsilon(x) = \sum_{|\alpha|=r} \partial^\alpha \psi_\alpha^\varepsilon(x),$$

where

$$\psi_\alpha^\varepsilon(x) \in \mathcal{S}_r \qquad \text{and} \qquad \int \psi_\alpha^\varepsilon(x)\,dx = 0.$$

This identity allows us to cope with $(-\Delta)^{-s/2}\psi^\varepsilon$. Indeed, passing to the Fourier transform, we must look at $|\xi|^{-s}\hat\psi^\varepsilon(\xi)$. We write $1 = \phi_0(\xi) + \phi_1(\xi)$, where $\phi_0(\xi)$ is infinitely differentiable, has compact support and equals 1 in a neighbourhood of the origin. Then, for every $s > 0$, $|\xi|^{-s}\phi_1(\xi)$ is the Fourier transform of an integrable function $K(x)$ which, moreover, decreases rapidly at infinity. Convolution with $K(x)$ preserves $\mathcal{S}_r$. To deal with $|\xi|^{-s}\phi_0(\xi)\hat\psi^\varepsilon(\xi)$, we use (2.6) which suggests looking at the terms $\xi^\alpha|\xi|^{-s}\phi_0(\xi)\hat\psi_\alpha^\varepsilon(\xi)$. Since $|\alpha| = r > s$, the terms which appear are Fourier transforms of infinitely differentiable functions which, together with their derivatives, are of order $O(|x|^{-n-r+s})$ at infinity.

We conclude that $(-\Delta)^{-s/2}\psi^\varepsilon$ is a function of integral zero which, together with its derivatives of order $\leq r$, is $O(|x|^{-n-r+s})$ at infinity.

An analogous argument holds for $(-\Delta)^{s/2}\psi^\varepsilon$. This time, the regularity decreases as s increases, and we get a function in the Hölder class of order $r - s$ which is $O(|x|^{-n-r+s})$ at infinity and has integral zero.

We then repeat the argument of the Littlewood-Paley multiresolution approximation and come to the same conclusion for $0 \leq s < r$.

We now proceed to the case where $-r < s < 0$. Let us begin by defining the spaces $L^{p,s}(\mathbb{R}^n)$ for $s < 0$. We can take $L^{p,s}(\mathbb{R}^n) = (I - \Delta)^{|s|/2} L^p(\mathbb{R}^n)$, just as in the case of the ordinary Sobolev spaces. But we can also take $L^{p,s}$ to be the dual of $L^{q,-s}$, where q is the conjugate exponent of p $(1/p + 1/q = 1)$ and the first space is regarded as a space of distributions, the second as the space of corresponding test functions.

We intend to prove the following result.

Proposition 3. *A wavelet series belongs to $L^{p,s}(\mathbb{R}^n)$, $1 < p < \infty$, $-r < s \leq 0$, if and only if*

(2.7) $$\left(\sum_{\lambda \in \Lambda} |\alpha(\lambda)|^2 2^{nj}(4^{-js}+1)^{-1}\chi_\lambda(x) \right)^{1/2} \in L^p(\mathbb{R}^n).$$

The characterization we have just given is, quite clearly, the dual characterization to that of Proposition 1. Namely, if E is a reflexive Banach space and if e_λ, $\lambda \in \Lambda$, is an unconditional basis of E, then the dual basis $e_\lambda^\star = f_\lambda$, $\lambda \in \Lambda$, is an unconditional basis of the dual $E^\star$ of E, the dual basis being defined by $f_\lambda(e_{\lambda'}) = \delta_{\lambda,\lambda'}$.

Whether a series $\sum \alpha(\lambda)e_\lambda$ belongs to E, or not, can be read off from the sequence of moduli $|\alpha(\lambda)|$ of the coefficients. Hence whether a series $\sum \beta(\lambda)f_\lambda$ belongs to $E^\star$, or not, depends only on the moduli $|\beta(\lambda)|$. In fact, the sequences $\beta(\lambda)$ which occur are characterized by the convergence of the series $\sum_{\lambda\in\Lambda} |\alpha(\lambda)||\beta(\lambda)|$ for all $\sum \alpha(\lambda)e_\lambda \in E$.

In our present situation, $E = L^{q,-s}$, $E^\star = L^{p,s}$, and the duality between E and $E^\star$ is induced by the duality between distributions and test functions. As a consequence, the basis dual to ψ_λ, $\lambda \in \Lambda$, is the self-same basis. Proposition 3 thus results from the following lemma:

Lemma 5. *Let $p \in (1,\infty)$ and let q be the conjugate exponent of p. Let $\omega(\lambda)$, $\lambda \in \Lambda$, be a sequence of strictly positive numbers and let E be the Banach space of sequences $\alpha(\lambda)$, $\lambda \in \Lambda$, such that*

$$\left(\sum_{\lambda\in\Lambda} |\alpha(\lambda)|^2 \omega(\lambda) 2^{nj} \chi_\lambda(x)\right)^{1/2}$$

belongs to $L^p(\mathbb{R}^n)$: its L^p norm is the norm of $\alpha(\lambda)$ in E.

Then the dual $E^\star$ of E is the Banach space of all sequences $\beta(\lambda)$, $\lambda \in \Lambda$, such that $\left(\sum_{\lambda\in\Lambda} |\beta(\lambda)|^2 \omega^{-1}(\lambda) 2^{nj} \chi_\lambda(x)\right)^{1/2}$ is in $L^q(\mathbb{R}^n)$ and the duality between the sequence spaces is given by $\sum_{\lambda\in\Lambda} \alpha(\lambda)\bar{\beta}(\lambda)$.

To prove Lemma 5, we look at the auxiliary function

$$f(x) = \sum_{\lambda\in\Lambda} \alpha(\lambda)\omega^{1/2}(\lambda)\psi_\lambda(x)$$

for each sequence $\alpha(\lambda)$, $\lambda \in \Lambda$. Similarly, each sequence $\beta(\lambda)$ has a corresponding auxiliary fuction $g(x) = \sum_{\lambda\in\Lambda} \beta(\lambda)\omega^{-1/2}(\lambda)\psi_\lambda(x)$. Then $\int f\bar{g}\,dx = \sum_{\lambda\in\Lambda} \alpha(\lambda)\bar{\beta}(\lambda)$. The norm of $\alpha(\lambda)$, $\lambda \in \Lambda$, in E is equivalent to that of f in $L^p(\mathbb{R}^n)$, by Theorem 1. The lemma then follows, because the dual of L^p is L^q.

We shall take a break for a moment in order to paraphrase Propositions 1 and 3. The wavelets form an unconditional basis of each of the $L^{p,s}$ spaces, for $1 < p < \infty$, $|s| < r$. Further, if $0 < s < r$, if f is a function in $L^{p,s}$ and if S is a distribution in the dual space $L^{q,-s}$, then the duality of the distribution and the test function f is written in the form of an absolutely convergent series $\langle S, f \rangle = \sum_{\lambda\in\Lambda} \alpha(\lambda)\bar{\beta}(\lambda)$, where the $\alpha(\lambda)$, $\lambda \in \Lambda$, are the wavelet coefficients of S, and the $\beta(\lambda)$ are those of f.

Nothing like this can hold for Fourier series. If p and q are conjugate exponents, with $1 < p < 2 < q < \infty$, and if f belongs to $L^p(0, 2\pi)$ and g to $L^q(0, 2\pi)$, we do have $(2\pi)^{-1} \int_0^{2\pi} f(x)\bar{g}(x)\,dx = \sum_{-\infty}^{\infty} a_k\bar{b}_k$, but the series on the right-hand side is not in general absolutely convergent (however, the partial sums $\sum_{-m}^{m} a_k \bar{b}_k$ converge to $(2\pi)^{-1} \int_0^{2\pi} f\bar{g}\,dx$ as m tends to infinity).

Let us return to the representation of functions of $L^{p,s}$ by wavelet series. However, this time we shall use the orthonormal basis of $L^2(\mathbb{R}^n)$ formed by the functions $\phi(x - \lambda)$, $\lambda \in \mathbb{Z}^n$, and the functions $\psi_\lambda(x)$, $\lambda \in \Lambda_j$, $j \geq 0$.

That is, we want to give a characterization of the coefficients $\beta(\lambda)$, $\lambda \in \mathbb{Z}^n$, and $\alpha(\lambda)$, $\lambda \in \Lambda_j$, $j \geq 0$, of functions $f \in L^{p,s}$, using the representation

$$(2.8) \qquad f(x) = \sum_{\lambda \in \mathbb{Z}^n} \beta(\lambda)\phi(x - \lambda) + \sum_{j=0}^{\infty} \sum_{\lambda \in \Lambda_j} \alpha(\lambda)\psi_\lambda(x).$$

The characterization is given by the following theorem.

Theorem 2. *The series (2.8) belongs to $L^{p,s}$, $1 < p < \infty$, $|s| < r$, if and only if the sequence $\beta(\lambda)$ is in $l^p(\mathbb{Z}^n)$ and*

$$(2.9) \qquad \left(\sum_{0}^{\infty} \sum_{\lambda \in \Lambda_j} 2^{nj} 4^{js} |\alpha(\lambda)|^2 \chi_\lambda(x) \right)^{1/2} \in L^p(\mathbb{R}^n).$$

We can interpret condition (2.9) by distinguishing between between a local condition and the behaviour at infinity of local norms. In fact, $f(x)$ belongs to $L^{p,s}$ if and only if, when $f(x)$ is localized around the points k of the lattice $\mathbb{Z}^n$ by an appropriate function $\phi \in \mathcal{D}(\mathbb{R}^n)$, the localizations $f(x)\phi(x - k)$, $k \in \mathbb{Z}^n$, all belong to $L^{p,s}$ and the sequence of the $L^{p,s}$ norms of these functions lies in $l^p(\mathbb{Z}^n)$.

To see this, we let $\mathcal{Q}_0$ denote the collection of dyadic cubes of side 1. Then every dyadic cube $Q \in \mathcal{Q}$ of side $2^{-j} \leq 1$ is contained in a unique $R \in \mathcal{Q}_0$, and, equally, every $\lambda = k2^{-j} + \varepsilon 2^{-j-1}$, $k \in \mathbb{Z}^n$, $j \geq 0$, $\varepsilon \in E$, belongs to a unique R. The terms $2^{nj} 4^{js} |\alpha(\lambda)|^2 \chi_\lambda(x)$ of the series on the left-hand side of (2.9) can thus be rearranged to form $\sum_{R \in \mathcal{Q}_0} \omega_R^2(x)$, where $\omega_R(x) \geq 0$ and where the support of $\omega_R(x)$ is contained in the cube R.

Let $\sigma_R(x)$ denote the partial sum $\sum_{Q(\lambda) \subset R} \alpha(\lambda)\psi_\lambda(x)$ of the wavelet series under consideration. Then $\sigma_R(x)$ is an approximate localization to R of the whole series $f(x) = \sum_{\lambda \in \Lambda} \alpha(\lambda)\psi_\lambda(x)$. The norm of $\sigma_R(x)$ in $L^{p,s}$ is equivalent to the L^p-norm of ω_R and, finally, the $L^{p,s}$ norm of f is equivalent to the $l^p(\mathbb{Z}^n)$ norm of the sequence of local norms $\|\sigma_R\|_{L^{p,s}}$.

The proof of Theorem 2 is now obvious, because the series on the right-hand side of (2.8) is obtained from the decomposition of f into wavelet series by means of partial sums on the terms of index $j < 0$. It is enough to apply Proposition 1 or Proposition 3.

We shall now describe the corollary of Theorem 2 which is used in the calculations of *quantum field theory*. It is a question of evaluating $\|(-\Delta + I)^{-1/2} f\|_4$ when $f(x)$ is given by the series (2.8). But the L^4-norm of the fractional integral is precisely the norm of f in the space $L^{4,-1}$. We apply (2.9) and the norm in question is equivalent to the sum of the l^4 norm of $\beta(\lambda)$, $\lambda \in \mathbb{Z}^n$, and of

$$\left(\int_{\mathbb{R}^n} \left(\sum_0^\infty \sum_{\lambda \in \Lambda_j} 2^{(n-2)j} |\alpha(\lambda)|^2 \chi_\lambda(x) \right)^2 dx \right)^{1/4}.$$

Let us expand the squared term. We get "square" terms and "rectangular" terms. The "square" terms are $\sum_{j \geq 0} \sum_{\lambda \in \Lambda_j} 2^{(n-4)j} |\alpha(\lambda)|^4$ and the "rectangular" terms are

$$\sum \sum_{\lambda' < \lambda} 2^{(n-2)j} 2^{(n-2)j'} |\alpha(\lambda)|^2 |\alpha(\lambda')|^2,$$

where $\lambda' < \lambda$ means that $Q(\lambda') \subset Q(\lambda)$. $Q(\lambda)$ is the dyadic cube defined by $2^j x - k \in [0,1)^n$, where $\lambda = 2^{-j}k + 2^{-j-1}\varepsilon$, $\varepsilon \in E$.

Before leaving the $L^{p,s}$ spaces, it is appropriate to say something about the corresponding homogeneous spaces. Recall that the norm of f in $L^{p,s}$ is defined by $\|(I - \Delta)^{s/2} f\|_p$. For a Banach function space E which is invariant under translations and dilations and which satisfies $\mathcal{S} \subset E \subset \mathcal{S}'$ (the inclusions being continuous), the corresponding homogeneous space can be constructed by resolving the following problem: find an exponent $\alpha \in \mathbb{R}$ such that $\lim_{\lambda \to \infty} \lambda^{-\alpha} \|f(\lambda x)\|_E$ exists for a dense vector subspace V of E. By definition, this limit is the norm of f in the homogeneous space corresponding to E.

The space $L^p(\mathbb{R}^n)$ is homogeneous. The L^p norm is translation-invariant and the exponent α is $-n/p$. On the other hand, the space $L^{p,s}(\mathbb{R}^n)$, $s \in \mathbb{R}$, is not homogeneous. If f lies in the subspace $\mathcal{S}_0(\mathbb{R}^n)$ consisting of the functions in $\mathcal{S}(\mathbb{R}^n)$ all of whose moments are zero, then $\|f(\lambda x)\|_{L^{p,s}}$ is equivalent to $\lambda^{s-n/p} N_{p,s}(f)$ as λ tends to infinity, where $N_{p,s}f = \|(-\Delta)^{s/2} f\|_p$. We make sense of this last expression by taking Fourier transforms: $\mathcal{F}((-\Delta)^{s/2} f) = |\xi|^s \hat{f}(\xi) \in \mathcal{S}(\mathbb{R}^n)$ and all the derivatives of this function vanish at 0. Hence $(-\Delta)^{s/2} f \in \mathcal{S}_0(\mathbb{R}^n)$ and the L^p norm of this function is finite.

The homogeneous space corresponding to $L^{p,s}$ is denoted by $\dot{L}^{p,s}$ and is formally defined by keeping only the condition

$$N_{p,s}(f) = \|(-\Delta)^{s/2} f\|_p < \infty$$

in the definition of $L^{p,s}$ (which is $\|(I - \Delta)^{s/2} f\|_p < \infty$). But we don't expect the homogeneous space $\dot{L}^{p,s}$ to be the set of all possible functions f for which $N_{p,s}(f)$ is finite. For example, when $0 < s < n/p$, we don't want the function 1 to belong to $\dot{L}^{p,s}$, even though, if $s = 2$, it is clear that $\Delta(1) = 0$.

To define $\dot{L}^{p,s}$, we let q denote the conjugate exponent of p (given by $1/p + 1/q = 1$) and we look at three separate cases, namely, $-n/q < s < n/p$, $s \le -n/q$ and, lastly, $s \ge n/p$.

In the first case, we define $\dot{L}^{p,s}$ directly as the completion of $\mathcal{D}(\mathbb{R}^n)$ in $N_{p,s}$ norm. The problem is then to know whether this complete space is a space of functions or distributions, or equivalently, whether the convergence in $N_{p,s}$ norm of a sequence of test functions implies the convergence of this sequence in the sense of distributions. It is not hard to see that this is the case. However, if $s \ge n/p$, it is no longer true. For example, if $s > n/p$ and λ tends to $+\infty$, the $N_{p,s}$ norms of the functions $\phi(x/\lambda)$ tend to 0, for every test function ϕ, whereas the same functions tend to the constant $\phi(0)$ in the sense of distributions.

We come to the second case: $s \le -n/q$.

The difference is that $N_{p,s}(\phi) = \infty$ if the integral of the test function ϕ is non-zero. So we take the subspace $S_0 \subset S(\mathbb{R}^n)$ consisting of the functions ϕ all of whose moments are zero, and we take its completion under the norm $N_{p,s}$. We can get the same result by defining an integer $m \in \mathbb{N}$ by $-n/q - m - 1 < s \le -n/q - m$, then defining the space $\mathcal{D}_m$ as the subspace of $\mathcal{D}(\mathbb{R}^n)$ consisting of the functions ϕ satisfying $\int x^\alpha \phi(x)\, dx = 0$, for $|\alpha| \le m$, and lastly defining $\dot{L}^{p,s}$ as the completion of $\mathcal{D}_m$ under the norm $N_{p,s}$. Once again, $\dot{L}^{p,s}$ is a space of tempered distributions.

Finally, we must deal with the case where $s \ge n/p$. We use the duality of $L^{p,s}$ and $L^{p',s'}$, where $1/p + 1/p' = 1$, $s' = -s$, $-n/q < s < n/p$, and $-n/q' < s' < n/p'$ (the final two conditions are the same, because $p' = q$ and $q' = p$). We extend this duality to define the space $\dot{L}^{p,s}$, for $s \ge n/p$.

Fixing the integer $m \in \mathbb{N}$ by $n/p + m \le s < n/p + m + 1$, we define $\dot{L}^{p,s}$ as the dual of $\mathcal{D}_m$ with respect to the $L^{p',s'}$ norm (p' and s' defined as above). Then $\dot{L}^{p,s}$ is a space of distributions modulo the vector space of polynomials of degree $\le m$. It is thus no longer even a space of distributions.

We return to our programme of characterizing the classical function spaces of functions or distributions by the moduli of wavelet coefficients. Let $r \ge 1$ be an integer and let ψ_λ, $\lambda \in \Lambda$, be a wavelet basis arising out of an r-regular multiresolution approximation. To characterize the spaces $\dot{L}^{p,s}$ by the moduli of wavelet coefficients, the wavelets have to act as test functions analysing the distributions $S \in \dot{L}^{p,s}$. This condition will certainly be fulfilled if $r > |s|$, which we shall suppose to be the case. We must next avoid concluding that the function 1 belongs to every $\dot{L}^{p,s}$

simply because all the wavelet coefficients of 1 are zero. We avoid this trap by supposing that $S \in \mathcal{S}'(\mathbb{R}^n)$ is, in fact, a distribution of order less than r and that, if $s < n/p$, the wavelet series of S converges to S in the sense of distributions, whereas, if $n/p + m \le s < n/p + m + 1$, we shall need to suppose that the wavelet series of S converges to S in the quotient space of distributions modulo the polynomials of degree $\le m$.

After these precautions, we can repeat the proof of Theorem 2 word for word to get:

Theorem 3. *If $1 < p < \infty$ and $-r < s < r$, then S belongs to $\dot{L}^{p,s}(\mathbb{R}^n)$ if and only if its wavelet coefficients $\alpha(\lambda)$ are such that*

$$(2.10) \qquad \left(\sum_{\lambda \in \Lambda} |\alpha(\lambda)|^2 2^{nj} 4^{sj} \chi_\lambda(x) \right)^{1/2} \in L^p(\mathbb{R}^n),$$

where $\chi_\lambda(x)$ is the characteristic function of the cube $Q(\lambda)$.

We shall now deduce two Sobolev embeddings from this characterization.

$$(2.11) \qquad \text{If} \quad s > \frac{n}{p} \quad \text{and} \quad \gamma = s - \frac{n}{p} \quad \text{then} \quad \dot{L}^{p,s} \subset \dot{C}^\gamma,$$

where $\dot{C}^\gamma$ is the homogeneous Hölder space (which will be defined and studied systematically below).

$$(2.12) \qquad \text{If} \quad s = \frac{n}{p} \quad \text{then} \quad \dot{L}^{p,s} \subset \text{BMO}.$$

To prove (2.11), we find a lower bound for the L^p norm appearing in (2.10) by taking just one term of the series to get $|\alpha(\lambda)| \le C 2^{-nj/2} 2^{-\gamma j} \|f\|_{p,s}$. This characterizes the space $\dot{C}^\gamma$.

To verify (2.12), we distinguish between two cases, $2 \le p < \infty$ and $1 < p \le 2$.

In the first one, we let $g(x)$ denote the "left-hand side" of (2.10). Then, for every dyadic cube Q of side 2^{-j_0},

$$\int_Q g^2(x) \, dx \le \left(\int_Q g^p(x) \, dx \right)^{2/p} \left(\int_Q 1 \, dx \right)^{1/r},$$

where $2/p + 1/r = 1$.

Expanding $\int_Q g^2(x) \, dx$, we get precisely $\sum_{Q(\lambda) \subset Q} |\alpha(\lambda)|^2 4^{js}$. Moreover, $|Q|^{1/r} = |Q| 4^{j_0 s}$. It follows that

$$\sum_{Q(\lambda) \subset Q} |\alpha(\lambda)|^2 4^{js} \le C |Q| 4^{j_0 s}$$

which is much stronger than Carleson's condition.

Finally, we consider the case where $s = n/p$ and $1 < p \le 2$. Here we use the converse Minkowski inequality, that is, if $0 < \beta \le 1$ and

$f_\lambda(x) \geq 0$, then

$$\left\| \sum f_\lambda(x) \right\|_\beta \geq \sum_{\lambda \in \Lambda} \| f_\lambda \|_\beta \,,$$

where $\| f \|_\beta = \left(\int |f(x)|^\beta \, dx \right)^{1/\beta}$.

Taking $\beta = p/2$ and $f_\lambda(x) = |\alpha(\lambda)|^2 2^{nj} 4^{js} \chi_\lambda(x)$, we apply the above inequality to give

$$\left(\int_{\mathbf{R}^n} g^p(x) \, dx \right)^{2/p} = \left(\int_{\mathbf{R}^n} \left(\sum_{\lambda \in \Lambda} |\alpha(\lambda)|^2 2^{nj} 4^{js} \chi_\lambda(x) \right)^{p/2} dx \right)^{2/p}$$

$$\geq \sum_{\lambda \in \Lambda} |\alpha(\lambda)|^2 2^{nj}.$$

We have proved the Sobolev embedding $\dot{L}^{p,s} \subset \dot{L}^{2,n/2}$, and the latter space is contained in BMO by the first case.

3 Hardy spaces $H^p(\mathbf{R}^n)$ with $0 < p \leq 1$

We define the Hardy spaces H^p, $p \leq 1$, by the atomic decomposition method, following the point of view adopted in [75].

$H^p(\mathbf{R}^n)$, $0 < p \leq 1$, is defined as the vector space of tempered distributions S which can be written as

$$(3.1) \qquad\qquad S = \sum_0^\infty \lambda_k a_k(x) \,,$$

where $\sum_0^\infty |\lambda_k|^p < \infty$ and the a_k are p-atoms.

The p-atoms are defined by the following condition: for each atom $a_k(x)$, there exists a ball B_k, of volume $|B_k|$, such that the support of $a_k(x)$ is contained in B_k,

$$(3.2) \qquad \|a_k\|_\infty \leq \frac{1}{|B_k|^{1/p}} \qquad \text{and} \qquad \int a_k(x) x^\alpha \, dx = 0 \,,$$

for every multi-index $\alpha \in \mathbf{N}^n$ such that $|\alpha| \leq n(1/p - 1)$.

Under these conditions, the series (3.1) converges in the sense of distributions.

More precisely, if $f(x)$ is a Hölder function of exponent $\gamma = n(1/p-1)$, the series $\sum_0^\infty \lambda_k \langle f, a_k \rangle$ is absolutely convergent. The condition on f is that it belong to the homogeneous Hölder space $\dot{C}^\gamma$, which is the usual space, if $0 < \gamma < 1$, the Zygmund class, if $\gamma = 1$, and is defined by $f \in \dot{C}^\gamma \iff \partial f/\partial x_j \in \dot{C}^{\gamma-1}$ (for $1 \leq j \leq n$), if $\gamma > 1$.

The convergence of our series is a consequence of the following basic lemma.

Lemma 6. *If $f \in \dot{C}^\gamma$, if the function $a(x)$ is supported by the ball B of radius R and satisfies*

$$(3.3) \qquad \|a\|_\infty \leq 1 \qquad \text{and} \qquad \int a(x) x^\alpha \, dx = 0,$$

for $|\alpha| \leq \gamma$, then $|\langle f, a \rangle| \leq C \|f\|_{\dot{C}^\gamma} R^{\gamma + n}$.

This lemma is more or less obvious when $0 < \gamma < 1$ and the case $\gamma \geq 1$ will be proved in the next section.

By the lemma, $|\langle f, a_k \rangle| \leq C \|f\|_{\dot{C}^\gamma}$, and the series converges because $\sum_0^\infty |\lambda_k| \leq (\sum_0^\infty |\lambda_k|^p)^{1/p}$ when $0 < p \leq 1$. We can define a distribution S by $\langle S, f \rangle = \sum_0^\infty \lambda_k \langle f, a_k \rangle$ for all $f \in \dot{C}^\gamma$, and H^p can be regarded as a subspace of the dual of $\dot{C}^\gamma$.

We now want to characterize the space H^p by its decomposition into wavelet series. To do this, we must use wavelets lying in H^p, that is, those having enough of their moments zero. This means that we must restrict ourselves to wavelets of regularity $r > n(1/p - 1)$. If this condition is fulfilled, the wavelets will belong to $\dot{C}^\gamma$, $\gamma = n(1/p - 1)$, in other words, the scalar products $\langle S, \psi_\lambda \rangle$ will make sense for every distribution $S \in H^p(\mathbb{R}^n)$.

Having taken these precautions, we get

Theorem 4. *The distributions $S \in H^p(\mathbb{R}^n)$ are the sums of the wavelet series $\sum_{\lambda \in \Lambda} \alpha(\lambda) \psi_\lambda(x)$ such that*

$$(3.4) \qquad \left(\sum_{\lambda \in \Lambda} |\alpha(\lambda)|^2 |\psi_\lambda(x)|^2 \right)^{1/2} \in L^p(\mathbb{R}^n).$$

The proof of this result is a straightforward rewriting of that of Theorem 1. It depends on the fact that, for every sequence $\omega(\lambda)$, $\lambda \in \Lambda$, of ± 1s, the operator $U_\omega : L^2(\mathbb{R}^n) \to L^2(\mathbb{R}^n)$, defined by $U_\omega(\psi_\lambda) = \omega(\lambda) \psi_\lambda$, $\omega = (\omega(\lambda))_{\lambda \in \Lambda}$, extends to a continuous linear operator on H^p for each $p \in (0, 1]$ satisfying $n(1/p - 1) < r$.

This remarkable property follows from the cancellation condition satisfied by the wavelets and from the following general result of Chapter 7:

Proposition 4. *Let $T : L^2(\mathbb{R}^n) \to L^2(\mathbb{R}^n)$ be a continuous linear operator. Let $r \geq 1$ be an integer and suppose that the restriction to the complement of the diagonal of the kernel-distribution of T satisfies $|\partial_y^\alpha K(x, y)| \leq C |x - y|^{-n - |\alpha|}$, for every multi-index $\alpha \in \mathbb{N}^n$ such that $|\alpha| \leq r$. Suppose further that $T^\star(x^\beta) = 0$, modulo polynomials of degree $\leq \beta$ for every multi-index $\beta \in \mathbb{N}^n$ satisfying $|\beta| \leq r$.*

Then T extends to a continuous linear operator on $H^p(\mathbb{R}^n)$ for every $p \in (0, 1]$ such that $n(1/p - 1) < r$.

Let us explain what the condition $T^\star(x^\beta) = 0$ means. If the kernel of T were given by a function $K(x, y)$ which was continuous on $\mathbb{R}^n \times \mathbb{R}^n$ and $O(|x - y|^{-n - |\beta| - 1})$ at infinity, then the condition $T^\star(x^\beta) = 0$ would just be $\int x^\beta K(x, y) \, dx = 0$ identically in $y \in \mathbb{R}^n$. In our case, the integral does not converge but there is a convergence procedure which consists of multiplying by a test function $f(y) \in \mathcal{D}(\mathbb{R}^n)$, all of whose moments of order $\leq |\beta|$ are zero, and then integrating with respect to y. On integrating with respect to x, we divide the domain into the (overlapping) sets $|x| \leq 3R$ and $|x| \geq 2R$, where $R > 0$ is defined by the condition that the support of f is included in $|y| \leq R$. We use a partition of unity $1 = \phi_0(x) + \phi_1(x)$ adapted to these sets. Then the double integral $\iint x^\beta \phi_0(x) K(x, y) f(y) \, dx \, dy$ makes sense because it is written as the value of the distribution K (the kernel of T) on the test function $x^\beta \phi_0(x) f(y) \in \mathcal{D}(\mathbb{R}^n \times \mathbb{R}^n)$. The second integral is dealt with by initially integrating by parts $|\beta|$ times, with respect to y. This can be done, because all the moments of f of order $\leq |\beta|$ are 0. This improves the behaviour of the kernel $K(x, y)$ at infinity.

Having made these remarks, we get Theorem 4 by repeating the proof of Theorem 1.

The proof of Proposition 4 depends on the idea of a molecule, introduced by Coifman and Weiss. A function $g(x)$ is a p-molecule, with centre x_0 and width d and of type $s > n(1/p - 1/2)$ if

$$(3.5) \qquad \left(\int_{\mathbb{R}^n} |g(x)|^2 \left(1 + \frac{|x - x_0|}{d} \right)^{2s} dx \right)^{1/2} \leq d^{-n(1/p - 1/2)}$$

and $\int x^\alpha g(x) \, dx = 0$, for every multi-index α satisfying $|\alpha| \leq n(1/p - 1)$.

Up to normalization, p-atoms are p-molecules. One shows, following Coifman and Weiss, that p-molecules of type $s > n(1/p - 1/2)$ belong to a bounded subset of $H^p(\mathbb{R}^n)$. Finally, a linear operator which takes p-atoms to p-molecules extends by continuity to $H^p(\mathbb{R}^n)$. Up to a multiplicative renormalization factor, this is the case for the operator T of Proposition 4. The method of proof will be given in detail in Section 3 of Chapter 7.

4 Hölder spaces

We start by describing the inhomogeneous Hölder spaces. They are simpler than the homogeneous spaces, because they are function spaces. In fact $\mathcal{S}(\mathbb{R}^n) \subset E \subset \mathcal{S}'(\mathbb{R}^n)$ when E is a non-homogeneous Hölder space, and these inclusions are continuous. This property is no longer true for homogeneous spaces which are quotients of function spaces, so we shall, perforce, construct lifting operators which can be studied by means of either traditional Littlewood-Paley decompositions or by wavelet series.

If $0 < s < 1$, the space $C^s(\mathbb{R}^n)$ is defined to be the Banach space of bounded, continuous functions on $\mathbb{R}^n$ whose modulus of continuity $\omega(h)$ satisfies the inequality $\omega(h) \le Ch^s$, for some constant C. The modulus of continuity is defined by $\omega(h) = \sup\{|f(x) - f(y)| : |x - y| \le h\}$. The norm of $f \in C^s(\mathbb{R}^n)$ is $\|f\|_\infty + \sup_{0 < h \le 1} \omega(h)h^s$.

If $s = 1$, we replace $C^1(\mathbb{R}^n)$ by the Zygmund class, defined by the following conditions: f is bounded and continuous and there exists a constant C such that , for all $x, y \in \mathbb{R}^n$,

(4.1) $|f(x + y) + f(x - y) - 2f(x)| \le C|y|$.

Finally, if $m < s \le m + 1$, we write $f \in C^s(\mathbb{R}^n)$ when f is a function of class C^m (in the usual sense), all of whose derivatives $\partial^\alpha f$, $|\alpha| \le m$, belong to C^{s-m}.

It is impossible to characterize the inhomogeneous spaces $C^s(\mathbb{R}^n)$ by the moduli of the coefficients of wavelet series of type (1.1). We shall see that this is because $L^\infty(\mathbb{R}^n)$ is a "component" of $C^s(\mathbb{R}^n)$ and that $L^\infty(\mathbb{R}^n)$ cannot be characterized by the moduli of wavelet coefficients.

The characterization, via series of type (1.2), of inhomogeneous Hölder spaces C^s—adjusted using the Zygmund class when s is an integer—is given by the following theorem.

Theorem 5. *A function $f \in L^1_{\text{loc}}(\mathbb{R}^n)$ belongs to $C^s(\mathbb{R}^n)$ if and only if, in a multiresolution approximation of regularity $r > s$, the wavelet coefficients*

$$\beta(\lambda) = \int f(x)\bar{\phi}(x - \lambda)\,dx, \qquad \lambda \in \mathbb{Z}^n$$

and

$$\alpha(\lambda) = \int f(x)\bar{\psi}_\lambda(x)\,dx, \qquad \lambda \in \Lambda_j, \qquad j \ge 0$$

satisfy

$$|\beta(\lambda)| \le C_0, \qquad \lambda \in \mathbb{Z}^n$$

and

$$|\alpha(\lambda)| \le C_1 2^{-nj/2} 2^{-js}, \qquad \lambda \in \Lambda_j \qquad j \in \mathbb{N}.$$

The condition $|\beta(\lambda)| \le C_0$ is an immediate consequence of $f \in L^\infty(\mathbb{R}^n)$ and $\phi \in L^1(\mathbb{R}^n)$. On the other hand, the conditions on the $\alpha(\lambda)$, $\lambda \in \Lambda_j$, $j \ge 0$ are connected to the regularity of f and need no control on $\|f\|_\infty$. That is, the conditions on the $\alpha(\lambda)$ come from the fact that f belongs to the corresponding homogeneous space, which we are about to define.

We shall therefore come back to the proof of Theorem 5 after exploring the homogeneous Hölder spaces and proving the basic lemma which gives an upper bound for the scalar product of a (class of) function(s) f belonging to the homogeneous Hölder space $\dot{C}^s$ and an integrable function g certain of whose moments are 0.

We recall the definition of the homogeneous Hölder spaces $\dot{C}^s(\mathbb{R}^n)$ when $0 < s < 1$. The modulus of continuity $\omega_f(h)$ of a continuous function $f : \mathbb{R}^n \to \mathbb{C}$ is defined by $\omega_f(h) = \sup\{|f(x) - f(y)| : |x - y| \le h\}$. Then f belongs to $\dot{C}^s$ if and only if there exists a constant C such that, for every $h > 0$, we have $\omega_f(h) \le Ch^s$. Here, the large values of h play as much of a part as the small ones. The norm of f in $\dot{C}^s$ is, by definition, the lower bound of the constants C. We see straightaway that the norm in question is not a norm, because the constant functions have norm 0. We rescue the situation by defining $\dot{C}^s$, $0 < s < 1$, to be the quotient space of the above function space, modulo the constant functions.

To show that $\dot{C}^s$ is a Banach space, we cannot do without the construction of a lifting operator which, for each class $f \in \dot{C}^s$, gives a corresponding function f_0 which is a representative of the class f and which satisfies the inequalities $\sup_K |f_0(x)| \le C(K)\|f_0\|_{\dot{C}^s}$, where K is an arbitrary compact subset of $\mathbb{R}^n$ and $C(K)$ depends only on K. We need these inequalities when we want to show that $\dot{C}^s$ is complete. We take a series $\sum_0^\infty f_j$, with $\|f_j\|_{\dot{C}^s} \le C/2^j$ and we want to show that the series converges to a function $f \in \dot{C}^s$. To do this, we lift each f_j to the special representative $f_{j,0}$ which satisfies $\sup_K |f_{j,0}| \le C(K)C/2^j$. The series $\sum_0^\infty f_{j,0}$ is thus uniformly convergent on every compact set and defines a continuous function f_0.

It is easily verified that f_0 is a representative of an equivalence class of $\dot{C}^s$, and thus $\dot{C}^s$ is a Banach space.

The canonical lifting of a class $f \in \dot{C}^s$ is obtained by requiring that $f(0) = 0$. We then have $|f(x)| \le C|x|^s$, for all $x \in \mathbb{R}^n$, and this is the best growth estimate possible, because $|x|^s \in \dot{C}^s$.

Our last remark about the homogeneous spaces $\dot{C}^s$, for $0 < s < 1$, is that the homogeneous space is derived from the inhomogeneous space C^s by the general rule which consists of trying to find an expression equivalent to $\|f(\lambda x)\|_{C^s}$ as λ tends to $+\infty$. We have $\lim_{\lambda \to \infty} \lambda^{-s}\|f(\lambda x)\|_{C^s} = \|f\|_{\dot{C}^s}$.

We now consider the subtler case of the homogeneous version of the Zygmund class.

We say that a function f, which is continuous on $\mathbb{R}^n$, belongs to Λ_* if there exists a constant C such that, for all $x, y \in \mathbb{R}^n$, $|f(x + y) +$

$f(x-y) - 2f(x)| \leq C|y|$. We do not require f to belong to $L^\infty(\mathbf{R}^n)$. In dimension 1, the function $f(x) = x \log|x|$ belongs to the Zygmund class, whereas $f(x) = x|\log|x||^\alpha$ does not belong to Λ_* when $\alpha > 1$.

Once again, Λ_* is not a Banach space, because the affine functions have norm 0 in Λ_*. In fact, we have to define Λ_* as the quotient of the space of functions satisfying (4.1) modulo the affine functions.

The lifting problem is not as easy as for the spaces $\dot{C}^s$ with $0 < s < 1$. Indeed, the difference $f(x) - f(0)$ cannot be what we are looking for, because it is not defined modulo the affine functions. This constraint suggests lifting f to $f(x) - f(0) - \sum_1^n x_j \partial f/\partial x_j(0)$, but this lifting is not defined when $f \in \Lambda_*$. Indeed, the partial derivatives of a function in Λ_* are not necessarily Borel measures. For example, in dimension 1, the lacunary Fourier series $\sum_0^\infty 2^{-k} \sin 2^k x$ belongs to Λ_*, but its derivative $\sum_0^\infty \cos 2^k x$ is a distribution.

To resolve the problem of realizing functions in the Zygmund class, we use the Littlewood-Paley decomposition, which here shows itself to be a suppler and more powerful instrument than Taylor's formula.

In order to be sure of the convergence of certain integrals, we must be precise about the behaviour at infinity of functions of the Zygmund class.

Lemma 7. *Let $f(x)$ be a continuous function on $\mathbf{R}^n$ satisfying (4.1). Then $f(x) = O(|x| \log|x|)$ as $|x|$ tends to infinity, and this estimate is best possible.*

The optimality of the estimate comes from the fact that $f(x) = x_n \log|x_n|$ (where x_n is the nth coordinate of $x \in \mathbf{R}^n$) belongs to Λ_*, as can easily be verified.

To get Lemma 7, we use the special case of (4.1) when $x = 0$ and $x = y$ to give $|f(2x) - 2f(x) + f(0)| \leq C|x|$. This implies $|f(2x)| \leq 2|f(x)| + |f(0)| + C|x|$. We then put

$$\eta_j = 2^{-j} \sup\{|f(x)| : 2^j \leq |x| < 2^{j+1}\}$$

which gives $\eta_{j+1} \leq \eta_j + 2C + 2^{j-1}|f(0)|$. Hence $\eta_j \leq \eta_0 + |f(0)| + 2Cj$, which gives the result.

Now we come to the definition of the Littlewood-Paley decomposition which we use here. Let ϕ be a *radial* function in the Schwartz class $\mathcal{S}(\mathbf{R}^n)$, whose Fourier transform $\hat{\phi}(\xi)$ equals 1, if $|\xi| \leq 1/2$, and equals 0, if $|\xi| \geq 1$. We let S_j, $j \in \mathbf{Z}$, denote the operator defined by convolution with the function $2^{nj}\phi(2^j x)$ and put $\Delta_j = S_{j+1} - S_j$. Then Δ_j is the operation of convolution with the radial function $2^{nj}\psi(2^j x)$, where $\psi(x) = 2^n \phi(2x) - \phi(x)$.

The following lemma gives a remarkable characterization of the Zygmund class.

Lemma 8. *Let* $f \in S'(\mathbb{R}^n)$ *be a tempered distribution such that* $f = \sum_{-\infty}^{\infty} \Delta_j(f)$, *in the sense of convergence in* $S'(\mathbb{R}^n)$. *Then* f *belongs to* Λ_* *if and only if* $\|\Delta_j f\|_\infty \leq C' 2^{-j}$, *for some constant* C' *and for all* $j \in \mathbb{Z}$.

Let us first suppose that $f \in \Lambda_*$. Lemma 7 tells us that $f(x)$ grows slowly at infinity (in fact, it describes the behaviour at infinity much more precisely). So a function on $\mathbb{R}^n$ which is continuous and belongs to Λ_* is a tempered distribution. We have

$$\Delta_j f(x) = \int f(x - y)\psi_j(y)\,dy$$

$$= \frac{1}{2} \int (f(x + y) + f(x - y) - 2f(x))\psi_j(y)\,dy\,,$$

because ψ_j is radial and its integral is zero. Thus

$$\|\Delta_j f\|_\infty \leq \frac{C}{2} \int |y||\psi_j(y)|\,dy = C' 2^{-j}\,.$$

The operator Δ_j is, in effect, defined on the quotient space of the Zygmund class modulo the affine functions. Indeed, we have $\int \psi_j(x)\,dx = 0$, and, equally, $\int x_1 \psi_j(x)\,dx = \cdots = \int x_n \psi_j(x)\,dx = 0$.

We shall now prove the converse implication of Lemma 8. In fact, we intend to do better, by taking a sequence of functions f_j, $j \in \mathbb{Z}$, belonging to $C^2(\mathbb{R}^n)$ and satisfying $\|f_j\|_\infty \leq C_0 2^{-j}$ and $\|\partial^\alpha f_j\|_\infty \leq C_1 2^j$, for $|\alpha| = 2$. With these conditions, we shall define the sum $g(x)$ of the series $\sum_{-\infty}^{\infty} f_j$ and show that the sum satisfies

(4.2) $$|g(x)| \leq C(1 + |x|)\log(2 + |x|)\,,$$

where C depends only on C_0, C_1, and the dimension n.

We shall then have resolved the lifting problem for an equivalence class, modulo the affine functions, of functions of Λ_*.

Let us begin by verifying that $f_j = \Delta_j f$ satisfies the conditions $\|\partial^\alpha f_j\|_\infty \leq C_1 2^j$, when $|\alpha| = 2$. This is a consequence of the support of the Fourier transform of f_j being contained in the ball $|\xi| \leq 2^{j+1}$, so that we can apply Bernstein's inequality (Chapter 1).

For $j \leq 0$ we set

$$r_j(x) = f_j(x) - f_j(0) - \sum_1^n x_k \frac{\partial f_j}{\partial x_k}(0)$$

and Taylor's theorem gives the estimate $|r_j(x)| \leq C|x|^2 2^j$. We put

(4.3) $$g(x) = \sum_1^\infty f_j(x) + \sum_{-\infty}^0 r_j(x)\,.$$

This series is uniformly convergent on compacta. Let us show that the behaviour of $g(x)$ at infinity is given by (4.2). To do this, we use the inequality $\|\partial^\alpha f_j\|_\infty \leq C'$, when $|\alpha| = 1$, which is another of Bernstein's inequalities when $f_j = \Delta_j f$, but results from the logarithmic convexity of the L^∞ norms of the successive derivatives of a function in the general case. So we have $|r_j(x)| \leq C|x|$. If $|x| \geq 1$, we define the integer $m \in \mathbf{N}$ by $2^m \leq |x| < 2^{m+1}$ and divide $\sum_{-\infty}^{0} r_j$ into $\sum_{-\infty}^{-m} + \sum_{-m+1}^{0}$. For the terms of the first series, we use the upper bound $|r_j(x)| \leq c|x|^2 2^j$, and for the second we apply the upper bound $|r_j(x)| \leq C|x|$. Inequality (4.2) follows immediately.

The above correction of the series $\sum_{-\infty}^{\infty} f_j(x)$ is called a *renormalization*. This renormalization has turned a divergent series into a convergent one. Further, the (infinitely many) correction terms are affine functions, compatible with the equivalence relation on $\Lambda_\star$.

We shall verify that $g(x)$ belongs to $\Lambda_\star$. We have

$$g(x+y) + g(x-y) - 2g(x) = \sum_{1}^{\infty} (f_j(x+y) + f_j(x-y) - 2f_j(x))$$

$$+ \sum_{-\infty}^{0} (r_j(x+y) + r_j(x-y) - 2r_j(x)).$$

When $j \leq 0$, we can replace $r_j(x+y) + r_j(x-y) - 2r_j(x)$ by $f_j(x+y) + f_j(x-y) - 2f_j(x)$ and this gives

$$g(x+y) + g(x-y) - 2g(x) = \sum_{-\infty}^{\infty} (f_j(x+y) + f_j(x-y) - 2f_j(x)).$$

We define $m \in \mathbf{N}$ by $2^{-m} \leq |y| < 2^{-m+1}$ and divide the series into $\sum_{j \leq m}$ and $\sum_{j > m}$.

For the first sum, Taylor's theorem gives

$$|f_j(x+y) + f_j(x-y) - 2f_j(x)| \leq C|y|^2 \sum_{|\alpha|=2} \|\partial^\alpha f_j\|_\infty \leq C'|y|^2 2^j.$$

The contribution of the corresponding terms is $2C'|y|^2 2^m \leq 4C'|y|$.

For the second sum $\sum_{j>m}$, we replace f_j by the simple upper bound $\|f_j\|_\infty$ and once more get $O(2^{-m}) = O(|y|)$. The function g thus satisfies (4.1).

We have finished the proof of Lemma 8.

Here is an important consequence.

Proposition 5. Let $g \in L^1(\mathbf{R}^n)$ satisfy $\int_{|x| \geq 2} |g(x)| \, |x| \log |x| \, dx < \infty$ and

$$(4.4) \qquad \int g(x) \, dx = \int x_1 g(x) \, dx = \cdots = \int x_n g(x) \, dx = 0.$$

Then there is a constant $C(g)$ such that, for every function $f \in \Lambda_*$, every $\lambda > 0$, and every $x_0 \in \mathbb{R}^n$,

(4.5) $$\left| \int f(x)g(\lambda x + x_0)\,dx \right| \leq C(g)\lambda^{-n-1}\|f\|_{\Lambda_*}.$$

To prove (4.5), we use the homogeneity of Λ_*. If $f(x)$ belongs to Λ_*, then $\lambda^{-1}f(\lambda(x - x_0))$ also belongs to Λ_*, with the same norm, when $\lambda > 0$ and $x_0 \in \mathbb{R}^n$. It is therefore enough to prove Proposition 5 when $x_0 = 0$ and $\lambda = 1$. In this case, we replace $f(x)$ by its realization $F(x)$ satisfying $|F(x)| \leq C(1 + |x|)\log(2 + |x|)\|f\|_{\Lambda_*}$. Doing this does not change the integral $\int f(x)g(x)\,dx$, and the conclusion follows.

We now continue the presentation of the homogeneous spaces $\dot{C}^s$, $s > 0$, and the description of the algorithms for realizing these homogeneous spaces.

Suppose $m < s < m + 1$, $m \in \mathbb{N}$. Then a function f of class C^m belongs to $\dot{C}^s$ if and only if all its derivatives $\partial^\alpha f$, $|\alpha| = m$, belong to the homogeneous Hölder space $\dot{C}^r$ defined above, with $r = s - m$.

The algorithm for realizing a function $f \in \dot{C}^s$ is Taylor's expansion to order m about x_0, say. This gives

$$f(x) = \sum_{|\alpha| \leq m} \frac{x^\alpha}{\alpha!} \partial^\alpha f(x_0) + R(x_0; x)$$

where

$$|R(x_0; x)| \leq C|x - x_0|^s.$$

Conversely, suppose that $f(x)$ is a continuous function such that, for a certain constant C and for all $x_0 \in \mathbb{R}^n$, we can find a polynomial $P(x_0; x)$ whose degree in x is at most m and a function $R(x_0; x)$ such that $f(x) = P(x_0; x) + R(x_0; x)$ and $|R(x_0; x)| \leq C|x_0 - x|^s$. Then $f(x)$ belongs to $\dot{C}^s$, which is thus characterized by the manner of approximation by polynomials of degree $\leq m$.

If $s = m + 1$, we define $\dot{C}^s$ by means of the Zygmund class (and it must be done this way so that we do not get "discontinuous" results when s is an integer).

So, when $s = m + 1$, we define the homogeneous space $\dot{C}^s$ as the set of functions of class C^m, all of whose derivatives $\partial^\alpha f$ of order $|\alpha| = m$ belong to the Zygmund class Λ_*. In fact this space should be defined modulo the polynomials of degree at most $m + 1$.

Adapting the arguments used for the Zygmund class itself, we can show that there is a constant C such that, for each $x_0 \in \mathbb{R}^n$ and every function $f \in \dot{C}^s$, $s \in \mathbb{N}$, we can construct a polynomial $P(x_0; x)$ and a function $g(x_0; x)$, with the following properties:

(4.6)   the degree of $P(x_0; x)$ is less than or equal to s;

(4.7) $|g(x_0; x)| \le C|x - x_0|^s \omega(|x - x_0|)\|f\|_{\dot{C}^s}$,   where

$$\omega(u) = \begin{cases} |\log u|, & \text{if } 0 < u \le 1/2, \\ \log 2, & \text{if } 1/2 \le u \le 2, \\ \log u, & \text{if } u \ge 2; \end{cases}$$

(4.8) $f(x) = P(x_0; x) + g(x_0; x)$.

These algorithms for the realizations of functions in the homogeneous spaces $\dot{C}^s$ give the following property, which we shall use in Section 4 of Chapter 10.

Theorem 6. *Let s be a positive real number and let $\gamma > 0$ be a real exponent. Let S denote a tempered distribution with the following three properties (describing its behaviour at infinity, its order and its oscillations):*

(4.9)   *the restriction of S to $|x| > 1$ is a function $b(x)$ belonging to $L^\infty\{|x| > 1\}$ and satisfying $|b(x)| \le C_0|x|^{-n-s-\gamma}$, for $|x| > 1$;*

(4.10)   *$|\langle S, u \rangle| \le C_1$, whenever u is a C^s-function of norm ≤ 1 with support contained in the ball $|x| \le 2$;*

(4.11)   *$\langle S, x^\alpha \rangle = 0$, for all $\alpha \in \mathbf{N}^n$ such that $|\alpha| \le s$.*

Then, for every function f in the homogeneous Hölder space $\dot{C}^s$, $\langle S, f \rangle$ is well defined and $|\langle S, f \rangle| \le C\|f\|_{\dot{C}^s}$.

To see this, we split f using (4.8), with $x_0 = 0$. We get $\langle S, f \rangle = \langle S, g \rangle$. Then we let $\phi_0 \in \mathcal{D}(\mathbf{R}^n)$ denote a function which is 1 in a neighbourhood of $|x| \le 1$ and whose support is contained in $|x| \le 2$. We set $\phi_1 = 1 - \phi_0$ and write $g = \phi_0 g + \phi_1 g$. The estimate of $|\langle S, \phi_0 g \rangle|$ comes from (4.10) and that of $|\langle S, \phi_1 g \rangle|$ is a result of (4.7) and (4.9).

The next theorem characterizes the homogeneous Hölder space $\dot{C}^s$ in terms of the order of magnitude of wavelet coefficients. We take a multiresolution approximation V_j, $j \in \mathbf{Z}$, of regularity $r \ge 1$. We suppose that $0 < s < r$, so that, if $f \in \dot{C}^s$ and ψ_λ, $\lambda \in \Lambda$, is a wavelet arising from the V_j, then, for each $j \in \mathbf{Z}$ and each $\lambda \in \Lambda_j$,

(4.12) $$|(f, \psi_\lambda)| \le C2^{-(nj/2 + js)}\|f\|_{\dot{C}^s}.$$

To get this, we apply Theorem 6. The oscillatory properties of the wavelets give (4.11).

The converse of (4.12) is described in the following statement:

Theorem 7. *A wavelet series $\sum_{\lambda \in \Lambda} \alpha(\lambda)\psi_\lambda(x)$ defines a function in the homogeneous space $\dot{C}^s$, $0 < s < r$, if and only if there exists a constant C such that, for all $j \in \mathbf{Z}$ and all $\lambda \in \Lambda_j$,*

(4.13) $$|\alpha(\lambda)| \le C2^{-nj/2}2^{-sj}.$$

To see this, we let $f_j(x)$ denote the series $\sum_{\lambda \in \Lambda_j} \alpha(\lambda)\psi_\lambda(x)$, which is convergent because the wavelets are localized. Again, by the localization and regularity of the wavelets, $\|\partial^\alpha f_j\|_\infty \leq C2^{|\alpha|j}2^{-sj}$. The series $\sum_{-\infty}^{\infty} f_j(x)$ cannot then converge in the usual sense, and must be renormalized. The renormalization is carried out by a variant of the method we used in the case of the Zygmund class. The details are left to the reader.

A very simple application of the preceding characterization is the "Sobolev embedding" $\dot{L}^{p,s} \subset \dot{C}^\sigma$, for $s - n/p = \sigma > 0$.

5   The Beurling algebra

The Beurling algebra was invented by A. Beurling as a substitute for the Wiener algebra $A(\mathbb{R}^n) = \mathcal{F}L^1(\mathbb{R}^n)$. Beurling told the author that he knew that spectral synthesis was impossible in the Wiener algebra well before P. Malliavin's justly celebrated discovery of a counter-example. As we shall see, the Beurling algebra is very close to the Wiener algebra and spectral synthesis is satisfied for the Beurling algebra. Our intention is to characterize the Beurling algebra by the moduli of wavelet coefficients. We shall later see that this is just a special case of a more general statement about Besov spaces.

We first define a family Ω of weights: we write $\omega \in \Omega$ if $\omega(x)$ is a strictly positive, real-valued function defined on $\mathbb{R}^n$, which is radial (that is, depends only on $|x|$), is an increasing function of $|x|$ and, lastly, is such that

$$(5.1) \qquad \int_{\mathbb{R}^n} \frac{dx}{\omega(x)} < \infty.$$

An example is $\omega(x) = (1 + |x|^2)^s$, with $s > n/2$. We then let $L^2(\mathbb{R}^n, \omega\,dx)$ denote the Hilbert space of (equivalence classes of) measureable functions f such that $\int_{\mathbb{R}^n} |f(x)|^2\omega(x)\,dx < \infty$. By the Cauchy-Schwarz inequality, $L^2(\mathbb{R}^n, \omega\,dx) \subset L^1(\mathbb{R}^n)$.

Definition 1. *By $\mathcal{A}$, we denote the collection of all continuous functions f on $\mathbb{R}^n$ of the form $f = \hat{F}$, where $F \in L^2(\mathbb{R}^n, \omega\,dx)$ for some $\omega \in \Omega$.*

In other words, $\mathcal{A} = \bigcup_{\omega \in \Omega} \mathcal{F}(L^2(\mathbb{R}^n, \omega\,dx))$, where $\mathcal{F}$ denotes the Fourier transform.

Theorem 8. *Let ψ_λ, $\lambda \in \Lambda$, be a wavelet basis of regularity $r > n/2$. Then the wavelet coefficients of the functions $f \in \mathcal{A}$ are characterized* *by the condition*

$$(5.2) \qquad \sum_{-\infty}^{\infty} 2^{nj/2} \left(\sum_{\lambda \in \Lambda_j} |\alpha(\lambda)|^2 \right)^{1/2} < \infty.$$

To establish this result, we first prove an intermediate result, given by the following lemma:

Lemma 9. *For every $j \in \mathbf{Z}$, we let Γ_j denote the dyadic anulus $2^j \le |\xi| \le 2^{j+1}$ and we consider the rudimentary version of the Littlewood-Paley decomposition, namely, $f = \sum_{-\infty}^{\infty} f_j$, where $\hat{f}_j$ is the product of $\hat{f}$ by the characteristic function of Γ_j. Then*

$$(5.3) \qquad f \in \mathcal{A} \iff \sum_{-\infty}^{\infty} 2^{nj/2} \|f_j\|_2 < \infty.$$

We first suppose $f \in \mathcal{A}$ and let g_j denote the function $\hat{f}_j$. By definition,

$$(5.4) \qquad \sum_{-\infty}^{\infty} \int_{\Gamma_j} |g_j(\xi)|^2 \omega(\xi) \, dx < \infty.$$

Let ω_j be the greatest lower bound of $\omega(\xi)$ on Γ_j: $\omega_j = \omega(\xi)$, for $|\xi| = 2^j$, and ω_j is the least upper bound of $\omega(\xi)$ on Γ_{j-1}. We then have

$$\int_{\mathbf{R}^n} \frac{dx}{\omega(x)} = \sum_{-\infty}^{\infty} \int_{\Gamma_j} \frac{dx}{\omega(x)} \simeq \sum_{-\infty}^{\infty} \frac{2^{nj}}{\omega_{j+1}}.$$

So $\omega \in \Omega$ if and only if $\sum_{-\infty}^{\infty} 2^{nj}/\omega_j < \infty$. The convergence of the series (5.4) is equivalent to that of $\sum_{-\infty}^{\infty} \|g_j\|_2^2 \omega_j$. By the Cauchy-Schwarz inequality, the convergence of both the series $\sum_{-\infty}^{\infty} \|g_j\|_2^2 \omega_j$ and $\sum_{-\infty}^{\infty} 2^{nj}/\omega_j$ implies that $\sum_{-\infty}^{\infty} 2^{nj/2} \|g_j\|_2 < \infty$.

Conversely, define $\omega(\xi)$ for $2^j \le |\xi| < 2^{j+1}$ by $\omega(\xi) = \omega_j$, where $\omega_j^{-1} = \sum_{k \ge j} \|g_k\|_2 2^{-kn/2}$. The function $\omega(x)$ is certainly radial and increasing. Further, $\int (\omega(\xi))^{-1} \, d\xi < \infty$, since

$$\sum_{-\infty}^{\infty} \frac{2^{nj}}{\omega_j} = \sum_{k \ge j} \sum 2^{nj} 2^{-kn/2} \|g_k\|_2 = (1 - 2^{-n})^{-1} \sum_{-\infty}^{\infty} 2^{kn/2} \|g_k\|_2 < \infty.$$

Finally,

$$\int |\hat{f}(\xi)|^2 \omega(\xi) \, d\xi = \sum_{-\infty}^{\infty} \int_{\Gamma_j} |g_j(\xi)|^2 \omega_j \, d\xi = \sum_{-\infty}^{\infty} \omega_j \|g_j\|_2^2$$

$$\le \sum_{-\infty}^{\infty} \frac{2^{nj}}{\omega_j} < \infty.$$

We have used the inequality $\|g_j\|_2 \le 2^{nj/2} \omega_j^{-1}$, which follows from the definition of ω_j by discarding all but the first term of the defining series.

Naturally, (5.3) remains true if, instead of the rudimentary Little-wood-Paley decomposition of Lemma 9, we use the usual smoothed versions, where $\hat{f}_j$ is the product of $\hat{f}$ by $\theta(2^{-j}\xi)$, θ being a function in $\mathcal{D}(\mathbb{R}^n)$ which vanishes in a neighbourhood of 0, but is non-vanishing on $1 \le |\xi| \le 2$.

In particular, the equivalence (5.3) is preserved when, starting with the Littlewood-Paley multiresolution approximation, we define f_j by $D_j(f)$. Then we get $\|D_j(f)\|_2 = (\sum_{\lambda \in \Lambda_j} |\alpha(\lambda)|^2)^{1/2}$, which completes the proof of the characterization of $\mathcal{A}$ by (5.2).

We still have to check that (5.2) does not depend on the particular choice of wavelet basis, as long as the regularity of the multiresolution approximation used satisfies $r > n/2$.

Let $\tilde{\psi}_\lambda$, $\lambda \in \Lambda$, be another wavelet basis, and let $U : L^2(\mathbb{R}^n) \to L^2(\mathbb{R}^n)$ be the unitary operator defined by $U(\psi_\lambda) = \tilde{\psi}_\lambda$. The kernel of U is then the distribution $\sum_{\lambda \in \Lambda} \tilde{\psi}_\lambda(x)\bar{\psi}_\lambda(y)$. Let $K(x, y)$ denote its restriction to the open set $O \subset \mathbb{R}^n \times \mathbb{R}^n$ defined by $y \ne x$ and, by a straightforward calculation, we get

(5.5) $|\partial_x^\alpha K(x,y)| \le C|x - y|^{-n-|\alpha|}$ if $0 \le |\alpha| \le r$.

Moreover, if $0 \le |\alpha| \le r - 1$, we get $U(x^\alpha) = 0$, modulo polynomials of degree $\le |\alpha|$, because the moments of the wavelets ψ_λ are zero. That we can define $U(x^\alpha)$ for $|\alpha| \le r - 1$ is a consequence of (5.5), as will be seen in Chapter 10. Relying on results which will not be proved until Chapter 10, the properties of the operator U which we have described, together with the continuity of U on $L^2(\mathbb{R}^n)$, give the continuity of U on all the homogeneous Besov spaces $\dot{B}_p^{s,q}$, where $0 < s < r$ and $1 \le p, q \le \infty$. Now, the Beurling algebra coincides with the Besov space $\dot{B}_2^{n/2,2}$. By the invariance specified by the operator U, the characterization of the Beurling algebra does not depend upon the choice of wavelet basis.

To finish, we need to verify that $\mathcal{A}$ is an algebra when the product is the usual multiplication of functions. We return to the characterization of the Fourier transforms of the functions of $\mathcal{A}$ and show that the convolution product of $f \in \mathcal{F}\mathcal{A}$ and $g \in \mathcal{F}\mathcal{A}$ lies in $\mathcal{F}\mathcal{A}$.

To make the notation simpler, we put $B = \mathcal{F}\mathcal{A}$ and

$$\|u\|_B = \sum_{-\infty}^{\infty} 2^{nj/2} \|u_j\|_2,$$

where u_j is the product of u with the characteristic function of the anulus Γ_j.

We observe that, if $f \in L^2(\mathbb{R}^n)$ vanishes outside the ball $|x| \le R$, then

(5.6) $\|f\|_B \le CR^{n/2}\|f\|_2$.

Here's why. We can obviously restrict ourselves to $R = 2^m$, $m \in \mathbb{Z}$. We decompose f as $\sum_{-\infty}^{m-1} f_j$. Then

$$\|f\|_B = \sum_{-\infty}^{m-1} 2^{nj/2} \|f_j\|_2 \le \sum_{-\infty}^{m-1} 2^{nj/2} \|f\|_2$$

$$= (1 - 2^{-n/2})^{-1} 2^{nm/2} \|f\|_2 = CR^{n/2} \|f\|_2 .$$

We return to the convolution product $f * g = h$. We write $f = \sum_{-\infty}^{\infty} f_j$, $g = \sum_{-\infty}^{\infty} g_j$, with f_j and g_j supported by Γ_j. This leads to $h = \sum_{-\infty}^{\infty} h_j$, where $h_j = f_j * g_j + \sum_{k<j} f_k * g_j + \sum_{k<j} g_k * f_j$. The support of a convolution product is contained in the algebraic sum of the supports of the multiplicands. The support of h_j is thus contained in $|x| \le 2^{j+2}$, and to show that h belongs to B, it is enough to prove that $\sum_{-\infty}^{\infty} 2^{nj/2} \|h_j\|_2 < \infty$ and to apply (5.6).

We get a bound for $\|h_j\|_2$ by applying the obvious inequality $\|u * v\|_2 \le \|u\|_1 \|v\|_2$, noting that $\|f_j\|_1 \le |\Gamma_j|^{1/2} \|f_j\|_2 = C 2^{nj/2} \|f_j\|_2$. We finally get

$$\|\sum_{k<j} g_k * f_j\|_2 \le \sum_{k<j} \|g_k\|_1 \|f_j\|_2 \le C \sum_{k<j} 2^{nk/2} \|g_k\|_2 \|f_j\|_2 \le C \|g\|_B \|f_j\|_2$$

and deal with the other terms making up h_j in a similar fashion. This gives

$$\sum_{-\infty}^{\infty} 2^{nj/2} \|h_j\|_2 \le 3C \|f\|_B \|g\|_B .$$

6 The hump algebra

We intend to describe another approximation to the Wiener algebra $\mathcal{F}L^1(\mathbb{R}^n)$, namely the "hump algebra", whose representation in terms of a wavelet basis is particularly simple.

We consider, for each $x_0 \in \mathbb{R}^n$ and each $\delta > 0$, the Gaussian function $e^{-|x-x_0|^2/2\delta}$, whose "width" is $\delta > 0$, whose centre is x_0 and whose height is 1 (this is not the usual normalization). We shall denote this Gaussian function by $g_{(x_0,\delta)}(x)$. An immediate observation is the identity

(6.1) $$g_{(x_0,\delta_0)}(x) g_{(x_1,\delta_1)}(x) = \gamma g_{(x_2,\delta_2)}(x) ,$$

where $0 < \gamma \le 1$, $1/\delta_2 = 1/\delta_0 + 1/\delta_1$, and where x_2 is the barycentre of the points x_0 and x_1 with coefficients δ_2/δ_0 and δ_2/δ_1.

Identity (6.1) suggests forming the set S of convex linear combinations $\sum p_j g_{t_j}(x)$, where $t_j = (x_j, \delta_j)$ lies in $\mathbb{R}^n \times (0, \infty)$, and where $\sum p_j \ge 1$ and $p_j \ge 0$. This convex set is stable under multiplication.

This very simple observation is the starting point for the construction of the Banach algebra of Gaussian humps. This is the Banach algebra B consisting of the continuous functions on $\mathbb{R}^n$ which tend to zero at infinity and can be written in the form

$$(6.2) \qquad f(x) = \sum_0^\infty \lambda_j g_{(x_j, \delta_j)}(x) \,,$$

where

$$(6.3) \qquad \sum_0^\infty |\lambda_j| < \infty \,.$$

The norm of f in B is defined to be the greatest lower bound of all the series (6.3), taken over the set of all representations (6.2).

If, in (6.2), we had decided to restrict ourselves to finite sums, the representation (6.2) would have been unique, but the resulting space would have been pathological. For one thing, it would not have contained the space $\mathcal{S}(\mathbb{R}^n)$ of functions in the Schwartz class. For another, the sequence

$$f_k(x) = e^{-|x|^2} - e^{-|x-x_k|^2} \,, \qquad x_k \to 0,$$

which ought to tend to 0, would have norm 2.

The introduction of the "atomic representations" (6.2) makes this pathology disappear, but the uniqueness is lost. But this is of no importance, because we shall, in the end, be able to write down unique canonical representations in which the initial Gaussians—whose amusing simplicity is, in fact, deceptive—are replaced by wavelets. The canonical representation is given by the following statement.

Let V_j, $j \in \mathbb{Z}$ be a multiresolution approximation of regularity $r > n$ and let ψ_λ, $\lambda \in \Lambda$ be the corresponding wavelet basis.

Theorem 9. *A continuous function f which vanishes at infinity belongs to the Banach algebra B if and only if its wavelet coefficients $\alpha(\lambda)$, $\lambda \in \Lambda$, satisfy the condition*

$$(6.4) \qquad \sum_{-\infty}^\infty \sum_{\lambda \in \Lambda_j} 2^{nj/2} |\alpha(\lambda)| < \infty \,,$$

that is, if the wavelet series of f converges absolutely in the L^∞ norm.

To prove this, we let $N(f)$ denote the sum of the series in (6.4). We must show that $N(f)$ is finite when f is given by (6.2). By linearity and convexity, it is enough to examine the case of a Gaussian $e^{-\delta |x-x_0|^2}$, $\delta > 0$, $x_0 \in \mathbb{R}^n$. We reduce to the case $1 \le \delta \le 4$, by replacing x by $2^m x$, $m \in \mathbb{Z}$, if necessary. In this case ($1 \le \delta \le 4$ and $x_0 \in \mathbb{R}^n$), the verification of the convergence in (6.4) is straightforward and is left to the reader.

In the opposite direction, we have to show that every absolutely convergent series $\sum \alpha(\lambda)\psi_\lambda$, where $\sum |\alpha(\lambda)| 2^{nj/2} < \infty$, converges to a function of B. For simplicity, we shall restrict ourselves to the Littlewood-Paley multiresolution approximation. We use the basic observation that, if $f(x) \in B$, then, for every $\delta > 0$ and each $x_0 \in \mathbb{R}^n$, $f(\delta x + x_0) \in B$, and that the B norms of the functions are the same.

To prove the converse implication in Theorem 9, it is thus enough to show that $\mathcal{S}(\mathbb{R}^n)$ is contained in B. To this end, we use the following lemmas.

Lemma 10. *There is a constant C such that every function $f \in L^1(\mathbb{R}^n)$ can be written $f(x) = \sum_0^\infty \lambda_k g_k(x)$, where $\sum_0^\infty |\lambda_k| \le C\|f\|_1$ and where the $g_k(x)$ are Gaussians, centred on x_k, of width δ_k and normalized by $\int g_k(x) = 1$.*

In other words,

$$g_k(x) = (2\pi\delta_k)^{-n/2} e^{-|x-x_k|^2/2\delta_k} .$$

This lemma is elementary and is left to the reader.

A corollary of Lemma 10 is the following observation.

Lemma 11. *If $f \in L^1(\mathbb{R}^n)$ and if $g(x) = e^{-|x|^2/2}$, then $f * g$ belongs to B and $\|f * g\|_B \le C\|f\|_1$.*

We use the atomic decomposition of $L^1(\mathbb{R}^n)$ given by Lemma 10, reducing to the case where f is a Gaussian normalized in L^1 norm. Then $h = f * g$ is also a Gaussian, normalized in L^∞ norm of width (or standard deviation) ≥ 1. Hence h is in B as well, and $\|h\|_B \le C$.

Lemma 12. *If f belongs to $L^1(\mathbb{R}^n)$ and the Fourier transform $\hat{f}$ of f has compact support, then f lies in B.*

Indeed, we define $h \in L^1(\mathbb{R}^n)$ by $\hat{h}(\xi) = e^{|\xi|^2/2}\hat{f}(\xi)$, so that $f = g * h$, where $g(x) = (2\pi)^{-n/2} e^{-|x|^2/2}$.

Lemma 13 $\mathcal{S}(\mathbb{R}^n) \subset B$.

Indeed, we apply the Littlewood-Paley decomposition to $\mathcal{S}(\mathbb{R}^n)$. We thus have, for $f \in \mathcal{S}(\mathbb{R}^n)$,

$$f(x) = \sum_0^\infty f_j(2^j x) ,$$

where the Fourier transforms of the f_j are supported by the ball $|x| \le 1$ and $\|f_j\|_1$ is a rapidly decreasing sequence. We then apply Lemma 12, using the dilation-invariance of the B norm.

Theorem 9 is thus proved for the case of the Littlewood-Paley multiresolution approximation. The invariance of (6.4) under change of wavelet basis follows from the operator-theoretical remarks made in the case of the Beurling algebra.

The hump algebra may be interpreted in terms of a minimality condition. Consider the Banach algebras B satisfying $\mathcal{D}(\mathbb{R}^n) \subset B \subset C_0(\mathbb{R}^n)$, where $C_0(\mathbb{R}^n)$ is the algebra of continuous functions vanishing at infinity, and where B is a subalgebra of $C_0(\mathbb{R}^n)$. We require the norm on B to satisfy the same conditions of translation- and dilation-invariance as the norm on $C_0(\mathbb{R}^n)$, that is, $\|f(\delta x + x_0)\|_B = \|f(x)\|_B$, for all $\delta > 0$ and each $x_0 \in \mathbb{R}^n$. Then, among all these Banach algebras B, there is a minimal algebra, which is precisely the hump algebra, and which is contained in every other Banach algebra satisfying the above conditions.

In particular, the hump algebra is contained in the Beurling algebra, which is, anyway, obvious from Theorems 8 and 9.

7 The space generated by special atoms

The Banach space which we intend to study is an approximation to the Hardy space $H^1(\mathbb{R}^n)$ of Stein and Weiss. This approximation is obtained by replacing the atoms used in constructing $H^1(\mathbb{R}^n)$ by "special atoms", which we now describe.

We start with the situation in dimension 1, following the account of R. O'Neill, G. Sampson and G. Soares de Souza ([198]). For each interval $I = [a, b] \subset \mathbb{R}$, we define the function $\mathcal{I}_I(x)$ equal to $|I|^{-1}$ on the left half of I, to $-|I|^{-1}$ on the right half, and to 0 outside I. Then $\mathcal{I}_I$ is an atom in the Stein and Weiss space $H^1(\mathbb{R})$, but it is a special atom because the cancellation condition $\int_I \mathcal{I}_I(x)\, dx = 0$ is guaranteed by its very special geometry.

Let f be a function in $L^1(\mathbb{R})$. We say $f \in B_1^{0,1}$ if and only if there is a (finite or infinite) sequence of intervals I_k, $k \in \mathbb{N}$, and a sequence λ_k of scalar coefficients such that

$$(7.1) \qquad \sum_0^\infty |\lambda_k| < \infty$$

and

$$(7.2) \qquad f(x) = \sum_0^\infty \lambda_k \mathcal{I}_{I_k}(x)\,.$$

Then the series (7.2) converges in $L^1(\mathbb{R})$ as well as in $H^1(\mathbb{R})$.

The result which we shall prove for an arbitrary dimension gives the following characterization, in dimension 1, of $B_1^{0,1}$ by the moduli $|\alpha(\lambda)|$

of the wavelet coefficients of f (as long as the wavelets arise from a multiresolution approximation of regularity $r > 1$):

$$(7.3) \qquad \sum \alpha(\lambda)\psi_\lambda(x) \in B_1^{0,1} \iff \sum_{-\infty}^{\infty} \sum_{\lambda \in \Lambda_j} 2^{-j/2}|\alpha(\lambda)| < \infty.$$

The meaning of (7.3) is that the wavelet series of f gives a special atomic representation of f when we slightly widen the definition of the atoms used to construct $B_1^{0,1}$.

Moreover, the primitives (vanishing at infinity) of the functions of $B_1^{0,1}$ are the functions of the hump algebra of Section 6. This last remark suggests studying the vector space of primitives, vanishing at infinity, of the functions of the Hardy space $H^1(\mathbb{R})$. This space is, in fact, a Banach algebra, whose symbolic calculus has been worked out by S. Janson: every Lipschitz function operates on the Banach algebra of primitives of functions of H^1 ([135]).

We turn to special atoms in an arbitrary dimension. Let E denote the closed, convex subset of $L^1(\mathbb{R}^n)$ consisting of functions satisfying

$$(7.4) \qquad \int_{\mathbb{R}^n} |a(x)| \, dx \le 1, \quad \int_{|x| \ge 2} |a(x)| \log|x| \, dx \le 1, \quad \int_{\mathbb{R}^n} a(x) \, dx = 0,$$

and

$$(7.5) \qquad \int_{|y| \le 1} \|a(x+y) - a(x)\|_{L^1(dx)} |y|^{-n} \, dy \le 1.$$

We then let $A \subset L^1(\mathbb{R}^n)$ be the set of functions $b(x) = t^{-n} a((x - x_0)/t)$, where $t > 0$, $x_0 \in \mathbb{R}^n$ and $a \in E$. The functions $b \in A$ are called *special atoms*.

We now must define *very special atoms*. The construction is the same as for special atoms, but E is replaced by $F \subset E$, where F is the set of C^1-functions supported by the unit ball of $\mathbb{R}^n$, whose integrals vanish and which satisfy $\|a\|_\infty \le 1$ and $\|\partial a/\partial x_j\|_\infty \le 1$, for $1 \le j \le n$.

We take a wavelet basis ψ_λ, $\lambda \in \Lambda$, arising from a multiresolution approximation of regularity $r > n$. We then have

Theorem 10. *Let* $f \in L^1(\mathbb{R}^n)$. *The following three properties of* f *are equivalent:*

(7.6) *there exists a sequence of special atoms* $a_j(x)$, $j \in \mathbb{N}$, *and a sequence* λ_j *of scalars such that*

$$\sum_{-\infty}^{\infty} |\lambda_j| < \infty \qquad \text{and} \qquad f(x) = \sum_0^{\infty} \lambda_j a_j(x);$$

(7.7) *we can suppose that the* $a_j(x)$ *in the above representation are very special atoms;*

(7.8) $f(x) = \sum_{\lambda \in \Lambda} \alpha(\lambda)\psi_\lambda(x)$ and $\sum_{\lambda \in \Lambda} |\alpha(\lambda)|2^{-nj/2} < \infty.$

We should make a few remarks before proving this result. The series of (7.8) are just the wavelet series which converge absolutely in L^1. Moreover, the wavelets $2^{nj/2}\psi_\lambda(x)$ (normalized in L^1 norm) are special atoms.

We come now to the proof of the theorem. To simplify the notation, we shall let $\mathcal{B}$ denote the Banach space of functions $f(x) = \sum_{\lambda \in \Lambda} \alpha(\lambda)\psi_\lambda(x)$ such that $\|f\|_\mathcal{B} = \sum_{\lambda \in \Lambda} |\alpha(\lambda)|2^{-nj/2} < \infty.$

The equivalence of this condition with (7.6) will show that the definition is independent of the particular wavelet basis used.

The first remark to make is that the C^1-functions of compact support, whose integrals vanish, belong to $\mathcal{B}$. This is straightforward and left to the reader. In particular, $F \subset \mathcal{B}$, and this embedding is continuous. We immediately deduce that, if $a(x) \in F$, then the collection of functions $\delta^{-n}a((x - x_0)/\delta)$, $\delta > 0$, $x_0 \in \mathbb{R}^n$, is a bounded subset of $\mathcal{B}$. Finally, the series $\sum_0^\infty \lambda_j a_j(x)$, where the $a_j(x)$ are functions constructed by the preceding rule and where $\sum_0^\infty |\lambda_j| < \infty$, define functions of $\mathcal{B}$. We have proved the implication (7.7)$\Rightarrow$(7.8).

The fact that (7.8)$\Rightarrow$(7.6) is trivial, since the wavelets $2^{nj/2}\psi_\lambda(x)$, once they are normalized in L^1 norm, are special atoms.

We come now to the delicate part of Theorem 10: showing that every special atom is an absolutely convergent series of very special atoms. To see this, we may suppose that the special atom in question satisfies (7.4) and (7.5).

We let ψ denote a real-valued, radial, C^1-function supported by the unit ball and not identically zero. Then we set $\psi_t(x) = t^{-n}\psi(x/t)$, $t > 0$, and define Q_t to be the operation of convolution with ψ_t. We can then use Calderón's identity

(7.9) $f = \int_0^\infty f * \psi_t * \psi_t \, \frac{dt}{t}$ for $f \in L^2(\mathbb{R}^n)$

(up to a normalization of ψ given by multiplication by a positive constant). This identity of Calderón's follows from $\int_0^\infty (\hat{\psi}(t\xi))^2 t^{-1} dt = 1$ for $\xi \neq 0$.

The first thing to do is to observe that, if f satisfies (7.4) and (7.5), then $N_\psi(f) = \int_0^\infty \int_{\mathbb{R}^n} |f * \psi_t(x)| \, dx \, t^{-1} dt$ is finite. Then we shall put $w(x,t) = f * \psi_t(x)$ and (7.9) will give

$$f(x) = \int_0^\infty \int_{\mathbb{R}^n} \psi_t(x - y)w(y,t) \, dy \, t^{-1} dt.$$

So the function $f(x)$ is represented as an average of translations and dilations of the function $\psi(x)$. All that remains is to transform the double integral into a series giving the representation in terms of very special atoms of the form $\psi_{t_j}(x - y_j)$, $t_j > 0$, $y_j \in \mathbb{R}^n$.

Now for the details of the proof.

The first result to be used is the following lemma.

Lemma 14. *Suppose that a function ψ satisfies*

$$|\psi(x)| \leq C_0(1 + |x|)^{-n-1}, \qquad \int_{\mathbb{R}^n} \psi(x)\, dx = 0$$

and

$$\|\psi(x - y) - \psi(x)\|_{L^1(dx)} \leq C_0|y| \qquad \text{for} \qquad |y| \leq 1.$$

Then, for $a(x)$ satisfying (7.4) and (7.5), we have

$$(7.10) \qquad N_\psi(a) = \int_0^\infty \|Q_t(a)\|_1\, \frac{dt}{t} \leq C_1,$$

where C_1 depends only on C_0 and the dimension n.

We start by finding an upper bound for $\int_0^1 \|Q_t a\|_1\, t^{-1} dt$, using the regularity of the special atom and the oscillatory character of the function ψ. We set $\omega(y) = \|a(x) - a(x - y)\|_{L^1(dx)}$, so that $\omega(y) \leq 2\|a\|_1$, for $|y| \geq 1$, while $\int_{|y| \leq 1} \omega(y)|y|^{-n}\, dy \leq 1$.

Then $Q_t a(x) = \int (a(x - y) - a(x))\psi_t(y)\, dy$ gives

$$\|Q_t a\|_1 \leq \int \omega(y)|\psi_t(y)|\, dy\,.$$

Finally,

$$\int_0^1 |\psi_t(y)|\, \frac{dt}{t} \leq \frac{C_2}{|y|^n} \qquad \text{if } |y| \leq 1$$

and

$$\int_0^1 |\psi_t(y)|\, \frac{dt}{t} \leq \frac{C_2}{|y|^{n+1}} \qquad \text{if } |y| \geq 1.$$

This leads to

$$\int_0^1 \|Q_t a\|_1\, \frac{dt}{t} \leq C\|a\|_1 + C \int_{|y| \leq 1} \omega(y)|y|^{-n}\, dy \leq 2C\,.$$

We come now to $\int_1^\infty \|Q_t a\|_1\, t^{-1} dt$, where the roles of ψ and a are, as it were, reversed. We write

$$Q_t a(x) = \int \psi_t(x - y)a(y)\, dy = t^{-n} \int \psi((x - y)/t)a(y)\, dy\,.$$

Then $\|Q_t(a)\|_1 = t^n \int |Q_t a(tx)|\, dx$, and this change of variable, followed by changing t to $1/t$, reduces the calculation of $\int_1^\infty \|Q_t a\|_1\, t^{-1} dt$ to that of

$$\int_0^1 \left\| \int (\psi(x - ty) - \psi(x))a(y)\, dy \right\|_{L^1(dx)} \frac{dt}{t}\,.$$

We put $\theta(y) = \|\psi(x-y)-\psi(x)\|_1$, so that $\int_0^1 \theta(ty)\,t^{-1}dt \leq C\log(2+|y|)$ which enables us to bound

$$\int_0^1 \int \|\psi(x-ty)-\psi(x)\|_{L^1(dx)}|a(y)|dy\,\frac{dt}{t}$$

by

$$C\int |a(y)|\log(2+|y|)\,dy\,.$$

The second result needed is the following, fairly obvious, observation:

Lemma 15. *Let E be a Banach space (with norm $\|\cdot\|$) and let $A \subset E$ be a bounded subset. Suppose that there exists a locally compact space Ω such that each $x \in E$ can be written*

$$(7.11) \qquad\qquad x = \int_\Omega f(\omega)\,d\mu(\Omega)\,,$$

where $f : \Omega \to A$ is a continuous function and μ is a real, or complex bounded Borel measure, of total mass $\|\mu\| \leq C_0\|x\|$, where C_0 is a constant. Then every $x \in E$ can be written as

$$(7.12) \qquad\qquad x = \sum_0^\infty \lambda_j a_j\,, \qquad a_j \in A\,,$$

where

$$(7.13) \qquad\qquad \sum_0^\infty |\lambda_j| \leq 2C_0\,.$$

Indeed, we can approximate the integral $\int_\Omega f(\omega)\,d\mu(\omega)$ by Riemann sums, because the bounded function $f : \Omega \to A$ is continuous. This gives $x = \sum_0^\infty \lambda_j a_j + y$, where $\sum_0^\infty |\lambda_j| \leq C_0$ and $\|y\| \leq \|x\|/2$. To get (7.12), it is enough to iterate this decomposition.

We come back to the representation of special atoms by very special atoms. Suppose that $f(x)$ satisfies (7.4) and (7.5). Then Lemma 14 gives $\int\int_0^\infty |w(x,t)|\,dx\,t^{-1}dt \leq C$ and this brings us to the following interpretation of Calderón's identity. Let Ω denote the upper half-space $\mathbb{R}^n \times (0,\infty)$ and let $d\mu(\omega)$ denote the measure $w(x,t)\,dx\,t^{-1}dt$, for whose total mass we have just found an upper bound. The function $f(\omega)$ appearing in (7.11) is $t^{-n}\psi((x-y)/t)$, regarded as a mapping from Ω to $\mathcal{B}$; here, $\omega = (t,y) \in \Omega$. The function f is continuous, because the C^1-functions of compact support and zero integral belong to $\mathcal{B}$, and this remark leads to the expected estimate of the norm on $\mathcal{B}$. Finally, A is precisely the complete set of functions $t^{-n}\psi((x-y)/t)$, $(t,y) = \omega \in \Omega$.

We have proved Theorem 10. It is easy to modify the choice of ψ. For example, in dimension 1, we can take $\psi(x) = 1$, for $0 \leq x < 1/2$, $= -1$, for $1/2 \leq x < 1$ and $= 0$ outside the interval $[0,1)$. Then the arguments we have used remain in force and we get the representation of functions of $B_1^{0,1}$ by the "special atoms" of Soares de Souza, Sampson and 0'Neill ([198]).

The space $B_1^{0,1}$ is like $L^1(\mathbb{R}^n)$ as far as the translation- and dilation-invariance of the norm is concerned. Among the various equivalent norms defining $B_1^{0,1}$, let us pick that norm which is obtained from the atomic representation (7.7). Then, if $f \in B_1^{0,1}$ and if $g(x) = t^n f(tx+x_0)$, $t > 0$, $x_0 \in \mathbb{R}^n$, we have $\|g\| = \|f\|$. Another remark about $B_1^{0,1}$ is the inclusion $\mathcal{D}_0 \subset B_1^{0,1}$, where $\mathcal{D}_0$ denotes the topological vector space of test functions with zero integral. Then the minimality property of $B_1^{0,1}$ can be stated as: $B_1^{0,1}$ is the smallest Banach space containing $\mathcal{D}_0$ whose norm satisfies the homogeneity properties of the L^1-norm. If we were to replace $\mathcal{D}_0$ by $\mathcal{D}$, the minimal Banach space would be $L^1(\mathbb{R}^n)$.

The minimality property results from Theorem 10. Indeed, it is enough to choose ψ in the class $\mathcal{D}_0$.

To finish, we will give an indication, without proof, of the connection between $B_1^{0,1}$ and the hump algebra. A function f belongs to the algebra of Gaussian humps if and only if all the derivatives $\partial^\alpha f$, with $|\alpha| = n$, belong to $B_1^{0,1}$. This property is an easy consequence of the character-ization of the Besov spaces by the moduli of wavelet coefficients. This will be given in Section 10, to which we refer the reader.

8 The Bloch space $B_\infty^{0,\infty}$

Initially, the *Bloch space* was defined as the space of functions $f(z)$ which are holomorphic in the half-plane $y = \Im z > 0$ and satisfy

$$\sup_{y>0} y|f'(z)| < \infty.$$

A function $g(z)$ which is holomorphic on the upper half-plane and satis-fies the condition $\iint_{y>0} |g'(z)| \, dx \, dy < \infty$ belongs to the *Bergman space*. The Bloch space is the dual of the Bergman space, with the duality realized by

$$\int_{\mathbb{R}^n} f\bar{g} \, dx = 2 \iint f'(x + iy)\bar{g}'(x + iy)y \, dx \, dy .$$

These spaces are not very far removed from their real versions. Indeed, we let H denote the Hilbert transform and $P = (I+iH)/2$ the orthogonal projection operator of $L^2(\mathbb{R})$ onto the complex Hardy space $\mathbb{H}^2(\mathbb{R})$. It is easy to show that P can be extended to a continuous, surjective, linear mapping of $B_1^{0,1}(\mathbb{R})$ onto the Bergman space. More precisely, every function $f \in B_1^{0,1}(\mathbb{R})$ can be written uniquely as $f = g + \bar{h}$, where g and h belong to the Bergman space on the upper half-plane. The Fourier transforms $\hat{f}$ and $\hat{g}$ of f and g are related by $\hat{g}(\xi) = \hat{f}(\xi)$, if $\xi \geq 0$, and $\hat{g}(\xi) = 0$, otherwise.

Let us return to the "real" space $B_1^{0,1}$ and to the study of its dual. If $\alpha(\lambda)$ are the wavelet coefficients of a distribution S and if $\beta(\lambda)$ are those of f, then $\langle S, \bar{f} \rangle = \sum_{\lambda \in \Lambda} \alpha(\lambda) \bar{\beta}(\lambda)$, which lets us study the duality using wavelet series. If we decide to let $B_\infty^{0,\infty}$ denote the dual of $B_1^{0,1}$ (when we study Besov spaces, we shall show that this notation is well founded), then the distributions $S \in B_\infty^{0,\infty}$ are characterized by the condition $|\alpha(\lambda)| \leq C 2^{-nj/2}$, $\lambda \in \Lambda_j$, applied to the wavelet coefficients of S. The wavelet series of a distribution $S \in B_\infty^{0,\infty}$ is thus of the form

$$ S = \sum_{\varepsilon \in E} \sum_j \sum_k c(\varepsilon, j, k) \psi^\varepsilon (2^j x - k) \qquad \text{where } |c(\varepsilon, j, k)| \leq C $$

and where the ψ^ε are the $2^n - 1$ analysing wavelets which are used in the construction of the wavelet basis.

In dimension 1, the distributions $S \in B_\infty^{0,\infty}$ coincide with the derivatives (in the sense of distributions) of the functions of the Zygmund class. Here is an amusing way of seeing this.

The functions $F(x)$ in the Zygmund class are characterized by $|F(x + h) + F(x - h) - 2F(x)| \leq Ch$, $x \in \mathbb{R}$, $h > 0$. We introduce the special atom $a(t)$ which is $1/2h$ on $[x - h, x)$ and $-1/2h$ on $[x, x + h)$ and we integrate by parts. This gives $|\langle F', a \rangle| \leq C/2$. Since the special atoms generate the space $B_1^{0,1}$ (by convex combinations and multiplication by coefficients of modulus 1), we have shown that the functionals in the dual of $B_1^{0,1}$ coincide with the derivatives of the functions in the Zygmund class.

9 Characterization of continuous linear operators
$$ T : B_1^{0,1} \to B_1^{0,1} $$

Let ψ_λ, $\lambda \in \Lambda$, be an orthonormal wavelet basis arising from a multiresolution approximation of $L^2(\mathbb{R}^n)$ of regularity $r \geq 1$. It is remarkable that it is possible to give a very simple characterization of the matrices $M = (\tau(\lambda, \lambda'))$, $(\lambda, \lambda') \in \Lambda \times \Lambda$, corresponding to the continuous linear operators $T : B_1^{0,1} \to B_1^{0,1}$.

We have $\tau(\lambda, \lambda') = (T\psi_{\lambda'}, \bar{\psi}_\lambda) = \langle T\psi_{\lambda'}, \psi_\lambda \rangle$ and the characterization is given by the following statement:

Proposition 6. The operator T is continuous on $B_1^{0,1}$ if and only if there is a constant C such that, for all $\lambda' \in \Lambda$,

$$ (9.1) \qquad \sum_{\lambda \in \Lambda} |\tau(\lambda, \lambda')| 2^{-nj/2} \leq C 2^{-nj'/2} . $$

Of course, λ and j are related by $\lambda \in \Lambda_j$, and the same is true for λ' and j'.

The proof of the proposition is immediate, because (9.1) means that, for each $\lambda' \in \Lambda$, $T(2^{nj'/2}\psi_{\lambda'})$ belongs to $B_1^{0,1}$ and that its norm is not greater than C. But the functions $2^{nj'/2}\psi_{\lambda'}$ generate $B_1^{0,1}$ by convex combinations and multiplication by complex numbers of modulus 1.

In Chapter 8, Section 3, we shall use the following corollary of the proposition:-

Corollary. *Suppose that an operator T and its adjoint $T^\star$ are bounded on $B_1^{0,1}$. Then T is continuous on $L^2(\mathbb{R}^n)$.*

Indeed,

$$\sum_{\lambda \in \Lambda} |\tau(\lambda, \lambda')| 2^{-nj/2} \leq C 2^{-nj'/2}$$

and

$$\sum_{\lambda' \in \Lambda} |\tau(\lambda, \lambda')| 2^{-nj'/2} \leq C 2^{-nj/2},$$

which enables us to apply Schur's lemma and conclude that the matrix M of T is continuous on $l^2(\Lambda)$. We are anticipating Chapter 8, to which we refer the reader.

10 Wavelets and Besov spaces

The introduction of homogeneous and inhomogeneous Besov spaces enables us to regroup all the statements about special cases from Section 4 onwards into a coherent whole. The Besov spaces $B_p^{s,q}$ are generalizations of the Sobolev spaces $H^s = B_1^{s,2}$ and of the Hölder spaces $C^s = B_\infty^{s,\infty}$.

To begin with, we shall characterize the inhomogeneous Besov spaces $B_p^{s,q}$ (which were defined in Section 9 of Chapter 2) by the moduli of the wavelet coefficients.

We take a multiresolution approximation of regularity $r \geq 1$ and let ψ_λ denote the corresponding wavelets, indexed by the set Λ. We set $\Lambda_j = 2^{-j-1}\mathbb{Z}^n \setminus 2^{-j}\mathbb{Z}^n$, so that the Λ_j, $j \in \mathbb{Z}$, form a partition of Λ. The wavelets ψ_λ, $\lambda \in \Lambda_j$, form an orthonormal basis of the orthogonal complement W_j of V_j in V_{j+1}.

If $|s| < r$, which we shall suppose be the case in all that follows, the space $B_p^{s,q}$ is characterized by the conditions $E_0(f) \in L^p(\mathbb{R}^n)$ and $\|D_j f\|_p \leq \varepsilon_j 2^{-js}$, where $\varepsilon_j \in l^q(\mathbb{N})$. We shall use the following remark to reformulate these conditions.

Proposition 7. *There exist two constants $C' \geq C > 0$ such that, for every exponent $p \in [1, \infty]$, for each $j \in \mathbb{Z}$ and for every finite sum $f(x) = \sum_{\lambda \in \Lambda_j} \alpha(\lambda) \psi_\lambda(x)$ in W_j,*

$$C\|f\|_p \leq 2^{nj/2} 2^{-nj/p} \left(\sum_{\lambda \in \Lambda_j} |\alpha(\lambda)|^p \right)^{1/p} \leq C'\|f\|_p .$$

This result is proved in exactly the same way as Lemma 8 of Chapter 2, Section 5.

Armed with this result, we immediately get the following characterization of $B_p^{s,q}$: the functions $f \in B_p^{s,q}$ are the sums of those series

$$f(x) = \sum_k \beta(k) \phi(x - k) + \sum_{j \geq 0} \sum_{\lambda \in \Lambda_j} \alpha(\lambda) \psi_\lambda(x)$$

which satisfy

(10.1)
$$\left(\sum_k |\beta(k)|^p \right)^{1/p} < \infty$$

and

(10.2)
$$\left(\sum_{\lambda \in \Lambda_j} |\alpha(\lambda)|^p \right)^{1/p} = 2^{-nj(1/2 - 1/p)} 2^{-js} \varepsilon_j ,$$

with $\varepsilon_j \in l^q(\mathbb{N})$.

The case of the homogeneous Besov spaces is similar, except that the index j now belongs to $\mathbb{Z}$ and the function ϕ (the "father wavelet") no longer appears in the wavelet series.

We shall define the homogeneous Besov spaces by means of the Littlewood–Paley decomposition $I = \sum_{-\infty}^{\infty} \Delta_j$, where Δ_j is the operation of convolution with ψ_j (Lemma 8 of Section 4 of this chapter).

If $s < n/p$ or if $s = n/p$ and $q = 1$, the homogeneous space $\dot{B}_p^{s,q}$ is a space of distributions. This means that $\dot{B}_p^{s,q}$ is a vector-subspace of $\mathcal{S}'(\mathbb{R}^n)$ and that the inclusion $\dot{B}_p^{s,q} \subset \mathcal{S}'(\mathbb{R}^n)$ is continuous. A distribution $f \in \mathcal{S}'(\mathbb{R}^n)$ belongs to $\dot{B}_p^{s,q}$ if and only if the following two conditions are satisfied:

(10.3) the partial sums $\sum_{-m}^{m} \Delta_j(f)$ converge to f for the $\sigma(\mathcal{S}', \mathcal{S})$-topology;

(10.4) the sequence $\varepsilon_j = 2^{sj} \|\Delta_j(f)\|_p$ belongs to $l^q(\mathbb{Z})$.

For example, the function which is identically equal to 1 does not belong to $\dot{B}_p^{s,q}$ when $s < n/p$, since $\Delta_j(1) = 0$ for every $j \in \mathbb{Z}$.

If $s = n/p$ and $q > 1$, or if $s > n/p$, $\dot{B}_p^{s,q}$ is no longer even a space of distributions: the function which is identically equal to 1 belongs to $\dot{B}_p^{s,q}$ but is identified with the function which is identically zero.

We put $\sigma = s - n/p$ and define $\dot{B}_p^{s,q}$ as a subspace of the homogeneous Hölder space $\dot{C}^\sigma$. This inclusion is justified by observing that $\|\Delta_j(f)\|_\infty \leq C2^{nj/p}\|\Delta_j(f)\|_p = C\varepsilon_j 2^{-j\sigma}$. If $m \in \mathbb{N}$ is the integer part of σ, then $\dot{B}_p^{s,q}$ is the space of tempered distributions f, modulo the polynomials of degree $\leq m$, such that $f = \sum_{-\infty}^{\infty} \Delta_j(f)$ (the convergence takes place in the quotient space) and $\|\Delta_j(f)\|_p = 2^{-js}\varepsilon_j$, with $\varepsilon_j \in l^q(\mathbb{Z})$.

If $n/p \leq s < r$, the wavelet coefficients (f, ψ_λ) of an element $f \in \dot{B}_p^{s,q}$ are well defined, because $\int x^\alpha \psi_\lambda(x)\,dx = 0$, for $|\alpha| \leq r$, and because $m < r$.

The elements $f \in \dot{B}_p^{s,q}$ are the sums of wavelet series $\sum \alpha(\lambda)\psi_\lambda(x)$ such that

$$(10.5) \qquad 2^{js}2^{nj(1/2-1/p)}\left(\sum_{\lambda \in \Lambda_j} |\alpha(\lambda)|^p\right)^{1/p} = \varepsilon_j \in l^q(\mathbb{Z}).$$

A renormalization is necessary for the wavelet series $\sum \alpha(\lambda)\psi_\lambda(x)$ of an element $f \in \dot{B}_p^{s,q}$ to converge. Indeed, wavelets ψ_λ such that $\lambda \in \Lambda_j$, with $j \leq 0$, are very flat and, if $s > n/p$, have coefficients which tend to infinity exponentially. The renormalization is the same as that used in the case of the homogeneous Hölder spaces; the details are left to the reader.

We finish this section by considering the special case of the periodic wavelets of Chapter 3, Section 11.

The series $\sum_0^\infty \alpha_m g_m(x)$ belongs to $B_p^{s,q}$ if and only if

$$(10.6) \qquad 2^{js}2^{j(1/2-1/p)}\left(\sum_{2^j}^{2^{j+1}} |\alpha_m|^p\right)^{1/p} \in l^q(\mathbb{N}).$$

There is clearly no homogeneous version of this result, in which we suppose that the condition $|s| < r$ holds.

For example, suppose that the sequence $|\alpha_m|$ varies slowly, that is, that there exists a constant C such that $|\alpha_m| \leq C|\alpha_{m'}|$ if $m \leq 2m'$ or if $m' \leq 2m$. Then $|\alpha_m| \sim \omega_j$, if $2^j \leq m < 2^{j+1}$, and condition (10.6) becomes $\varepsilon_j = 2^{j(1/2+s)}\omega_j \in l^q(\mathbb{N})$. That is, whether these special wavelet series belong to $B_p^{s,q}$, or not, does not depend on p.

To illustrate this, we choose $\omega_j = 2^{-j/2}j^{-1/2}$, $j \geq 1$, and get a whole family of distributions which are in $B_p^{0,q}$, for $1 \leq p \leq \infty$ and $q > 2$, none of which are Borel measures.

11 Holomorphic wavelets and Bochkariev's theorem

We do not know whether there is a fuction ψ which belongs both to the complex Hardy space $\mathbb{H}^2(\mathbb{R})$ and to the Schwartz class $\mathcal{S}(\mathbb{R})$ and is such that $2^{j/2}\psi(2^j x - k)$, $j, k \in \mathbb{Z}$, is an orthonormal basis of $\mathbb{H}^2(\mathbb{R})$. If we did not require ψ to be rapidly decreasing, the answer to this question would be obvious—it would be enough to take $\psi(x) = (e^{4\pi ix} - e^{2\pi ix})/2\pi ix$. This choice would not work at all for what we have in mind.

We shall therefore abandon the search for orthonormal bases of $\mathbb{H}^2$ with the extremely simple structure of wavelets. We shall instead follow Bochkariev's line and construct "wavelets with two humps".

These "wavelets with two humps" will form an orthonormal basis of $\mathbb{H}^2(\mathbb{R})$ and an unconditional basis of $\mathbb{H}^p(\mathbb{R})$, for $0 < p < \infty$.

Now for the details of the construction. We take the Littlewood-Paley wavelet basis $2^{j/2}\psi(2^j x - k)$. We recall that these functions form an orthonormal basis of $L^2(\mathbb{R})$ and that

(11.1) $\psi(x) \in \mathcal{S}(\mathbb{R})$ and ψ is real-valued;

(11.2) $\hat{\psi}(\xi)$ has support in $\dfrac{2\pi}{3} \le |\xi| \le \dfrac{8\pi}{3}$;

(11.3) $\psi(1 - x) = \psi(x)$.

Let $S : L^2(\mathbb{R}) \to L^2(\mathbb{R})$ denote the isometry defined by $(Sf)(x) = f(-x)$. Then the collection $\psi_{j,k}$ of wavelets is globally invariant under S and

$$S(\psi_{j,k}) = \psi_{j,k^\star} \qquad \text{where} \qquad k^\star = -k - 1.$$

This global symmetry of the wavelet basis follows from (11.3).

Let E be the subspace of $L^2(\mathbb{R})$ consisting of even functions. Then the "wavelets with two humps" $(\psi_{j,k} + \psi_{j,k^\star})/\sqrt{2}$ form a remarkable orthonormal basis of E.

We let $P : L^2(\mathbb{R}) \to \mathbb{H}^2(\mathbb{R})$ be the orthogonal projection. Using the Fourier transform, this operator can be defined by

$$(Pf)\hat{\ }(\xi) = \hat{f}(\xi)\chi_+(\xi) \qquad \text{where} \qquad \chi_+(\xi) = \begin{cases} 1, & \text{if } \xi \ge 0; \\ 0, & \text{if } \xi < 0. \end{cases}$$

Then $\sqrt{2}P : E \to \mathbb{H}^2(\mathbb{R})$ is an isometric isomorphism. Indeed, if $f \in E$, its Fourier transform $\hat{f}(\xi)$ is an even function which is completely defined by its restriction to $[0, \infty)$.

We put $P(\psi) = \mathcal{I}$. To calculate $\mathcal{I}$, it is convenient to write $\hat{\psi}(\xi) = e^{-i\xi/2}\omega(\xi)$, where $\omega(\xi)$ is in $\mathcal{D}(\mathbb{R})$ and is real-valued—in fact, $\omega(\xi) \ge 0$ in the special case of Littlewood-Paley wavelets. We then get $\mathcal{I}(x) = (1/2\pi) \int_0^\infty e^{i(x-1/2)\xi}\omega(\xi)\,d\xi$. When we restrict to $\xi \ge 0$, ω vanishes outside the interval $[2\pi/3, 8\pi/3]$ and is infinitely differentiable. As a result, $\mathcal{I}(x)$ belongs to $\mathcal{S}(\mathbb{R})$ and $\bar{\mathcal{I}}(x) = \mathcal{I}(1 - x)$.

We define $\mathcal{I}_{j,k}$, for $j, k \in \mathbf{Z}$, by

(11.4) $$\mathcal{I}_{j,k} = 2^{j/2}\mathcal{I}(2^j x - k).$$

Then, for $k^* = -k - 1$, we have $\mathcal{I}_{j,k^*}(x) = \bar{\mathcal{I}}_{j,k}(-x)$ (note that, if $f \in \mathbf{H}^2(\mathbf{R})$, then so is $\bar{f}(-x)$, whose Fourier transform is $\hat{f}(\xi)$).

We shall prove the following theorem:

Theorem 11. *The collection of functions* $\mathcal{I}_{j,k} + \mathcal{I}_{j,k^*}$, $j, k \in \mathbf{Z}$, *is an orthonormal basis of* $H^2(\mathbf{R})$ *and is also an unconditional basis of each space* $\mathbf{H}^p(\mathbf{R})$, $0 < p < \infty$.

The first assertion of Theorem 11 is an obvious consequence of the construction of "wavelets with two humps".

To prove the second, we take an arbitrary sequence $\lambda(j, k)$, where $j, k \in \mathbf{Z}$, subject only to the condition $|\lambda(j,k)| \leq C_0$. If we define $T : \mathbf{H}^2 \to \mathbf{H}^2$ by the condition

$$T(\mathcal{I}_{j,k} + \mathcal{I}_{j,k^*}) = \lambda(j,k)(\mathcal{I}_{j,k} + \mathcal{I}_{j,k^*}),$$

then the second assertion will be established if we can show that T is bounded on each $\mathbf{H}^p$-space, for $0 < p < \infty$.

To prove the continuity of T on $\mathbf{H}^p$, we use, once again, the Calderón-Zygmund theory of Chapter 7. More precisely, we use Proposition 4 of Section 3 of that chapter.

Because of the two humps of the function $\mathcal{I}_{j,k} + \mathcal{I}_{j,k^*}$, the situation is slightly more complicated. The function $\mathcal{I}(x)$ satisfies the identity $\bar{\mathcal{I}}(x) = \mathcal{I}(1 - x)$, which gives $\mathcal{I}_{j,k^*} = \bar{\mathcal{I}}_{j,k}(-x)$.

We now calculate the kernel $T(x, y)$ of the operator T. It is given by

(11.5) $$T(x,y) = \sum_{-\infty}^{\infty} \sum_{0}^{\infty} \lambda(j,k)(\mathcal{I}_{j,k}(x) + \bar{\mathcal{I}}_{j,k}(-x))(\bar{\mathcal{I}}_{j,k}(y) + \mathcal{I}_{j,k}(-y)).$$

On expanding the products, two distributions appear, which we call $T_1(x, y)$ and $T_2(x, -y)$. These distributions define the operators, T_1 and $T_2 S$. It is not difficult to see that T_1 and T_2 are Calderón-Zygmund operators satisfying the conditions of Proposition 4.

Theorem 11 is proved. We shall now give the periodified version of the preceding results, which will enable us to prove the theorems of Wojtaszczyk and Bochkariev ([237],[20]).

We return to the basic Littlewood-Paley wavelet ψ and form the periodic wavelets of period 1 defined, for $j \in \mathbf{N}$ and $0 \leq k < 2^j$, by

(11.6) $$g_{j,k}(x) = 2^{j/2} \sum_{-\infty}^{\infty} \psi(2^j x + 2^j l - k).$$

Then the sequence $1, g_{0,0}, g_{1,0}, g_{1,1}, g_{2,0}, g_{2,1}, g_{2,2}, g_{2,3}, \ldots$ is an orthonormal basis of $L^2[0,1]$.

Further, $g_{j,k}(x)$ is real, and its graph is symmetric about $x = k2^{-j} + 2^{-j-1}$.

Let $J : L^2[0,1] \to L^2[0,1]$ be the unitary symmetry operator defined by $Jf(x) = f(1-x)$. Then $Jg_{j,k} = g_{j,k^*}$, where $k^* = 2^j - k - 1$. In other words, $J(g_{0,0}) = g_{0,0}$, $J(g_{1,0}) = g_{1,1}$, etc.

An elementary calculation shows that $g_{j,k}(x) = g_j(x - k2^{-j})$, where

$$g_j(x) = (2\pi)^{-1}2^{-j/2} \sum_{-\infty}^{\infty} \hat{\psi}(2k\pi 2^{-j})e^{2k\pi ix}$$

$$= (2\pi)^{-1}2^{-j/2} \sum_{-\infty}^{\infty} \omega(2k\pi 2^{-j})e^{2k\pi i(x-2^{-j-1})}.$$

(This sum has only a finite number of terms, since $2^{j+1}/3 \le |k| \le 2^{j+2}/3$.)

We form the holomorphic part h_j of g_j defined by

$$(11.7) \qquad h_j(x) = (2\pi)^{-1}2^{-j/2} \sum_{0}^{\infty} \omega(2k\pi 2^{-j})e^{2k\pi i(x-2^{-j-1})}$$

and get $h_j(2^{-j} - x) = \bar{h}_j(x)$.

We let $\mathbb{H}^2[0,1]$ denote the subspace of $L^2[0,1]$ consisting of the functions whose negative Fourier coefficients are zero. In other words, $f \in \mathbb{H}^2[0,1]$ means that there is a Taylor series $F(z) = \sum_{0}^{\infty} c_k z^k$ such that $\sum_{0}^{\infty} |c_k|^2 < \infty$ and such that $f(t) = F(e^{2\pi it})$, when $z = e^{2\pi it}$ and $0 \le t \le 1$. We also let $\mathbb{H}^2$ denote the Hilbert space of functions $F(z)$ holomorphic in $z < 1$ and satisfying

$$(11.8) \qquad \sup_{0 \le r < 1} \int_0^1 |F(re^{2\pi it})|^2 \, dt < \infty.$$

We define the sequence of functions $G_m \in \mathbb{H}^2$, $m \in \mathbb{N}$, by $G_0 = 1$, $G_1(x) = e^{2\pi ix}$ and

$$(11.9) \quad G_m(x) = \frac{1}{\pi}2^{-j/2} \sum_{0}^{\infty} \omega(2k\pi 2^{-j}) \cos\left(2k\pi\left(l+\frac{1}{2}\right)2^{-j}\right) e^{2k\pi ix},$$

for $m = 2^{-j-1} + l + 1$ and $0 \le l < 2^{j-1}$.

In other words, $G_m = P(g_{j,l} + g_{j,l^*})$, where $P : L^2 \to \mathbb{H}^2$ is the orthogonal projection.

Theorem 12. *The sequence G_m, $m \in \mathbb{N}$, is an unconditional basis of every $\mathbb{H}^p$-space, for $0 < p < \infty$, and is a Schauder basis for the algebra $A(D)$.*

The algebra $A(D)$ consists of the functions which are holomorphic in the open unit disc and continuous on the closed unit disc. If $F(z)$ belongs to $A(D)$, then $f(t) = F(e^{2\pi it})$, $t \in \mathbb{R}$, is a continuous function on the real line, all of whose negative Fourier coefficients are zero (and conversely).

The proof of the first statement is a straightforward adaptation of the proof given in the non-periodic case (and is left to the reader).

The proof of the second part of the theorem contains an interesting novelty.

For $q \in \mathbf{N}$, let $E_q : L^2[0,1] \to L^2[0,1]$ denote the partial sum operator associated with the wavelet series of functions f, given by

$$E_q(f) = (f,1)1 + (f,g_{0,0})g_{0,0} + \sum_{j=1}^{q}\sum_{k=0}^{2^j}(f,g_{j,k})g_{j,k}.$$

We know (Chapter 3, Theorem 5) that, if $f(x)$ is continuous on the real line and periodic, of period 1, then $E_q(f)$ converges uniformly to f.

In the light of that result, let us examine the operator $\sigma_q : \mathbf{H}^2 \to \mathbf{H}^2$, defined by

$$\sigma_q(f) = (f,1)1 + \cdots + (f,G_{2^q})G_{2^q}.$$

Using the identities $G_m = P(g_{j,k} + g_{j,k^*})$ and $g_{j,k^*} = J(g_{j,k})$, we get

(11.10) $$\sigma_q(f) = PE_q(f) + PE_q(Jf) - (f,1)1.$$

To show that $\sigma_q(f)$ converges to f uniformly, as q tends to infinity, it is enough to do so in the case when f is a trigonometric polynomial (a finite linear combination of functions $e^{2li\pi x}$, $l \in \mathbf{N}$) and to verify that the norms of the operators $\sigma_q : A(D) \to A(D)$ are uniformly bounded.

The first verification is very easy. Indeed, by (11.9), the frequencies of G_m satisfy $2^j/3 \le |k| \le 2^{j+2}/3$ and $(f,G_m) = 0$, for large enough m, when f is a trigonometric polynomial.

The second verification depends on a property which is not in the least obvious. This property is given by the next proposition.

Proposition 8. *The commutators $[E_q, P]$ are uniformly bounded on the Banach space of functions which are continuous on the real line and periodic, of period 1.*

Let us assume this result and finish the proof of Theorem 12. For $f \in A(D)$, we have

$$PE_q(f) = E_q P(f) - [E_q, P]f = E_q(f) - [E_q, P]f,$$

so there is a constant C such that, for $f \in A(D)$,

$$\|PE_q(f)\|_\infty \le C\|f\|_\infty.$$

Now we consider the second term on the right-hand side of (11.10). We have $Jf(x) = \sum_0^\infty c_k e^{-2k\pi ix}$, so $PJf = c_0$. Then $PE_q(Jf) = E_q PJ(f) - [E_q, P]Jf = c_0 - [E_q, P]Jf$, and again we get

$$\|PE_q(Jf)\|_\infty \le C\|f\|_\infty,$$

for $f \in A(D)$.

The last term in (11.10) requires no proof.

To conclude the proof of Theorem 12, we need to consider the arbitrary partial sums

$$(11.11) \qquad S_N(f) = \sum_0^N (f, G_m) G_m .$$

We must show that the norms of the operators $S_N : A(D) \to A(D)$ are uniformly bounded. We know this already when $N = 2^q$, since then $S_N = \sigma_q$, so we have to give a bound for the error terms appearing in $S_N - \sigma_q$, when $2^q < N < 2^{q+1}$. We use the excellent localization properties of the functions G_m ("wavelets with two humps"), which give

$$(11.12) \qquad \left\| \sum_{2^q < m < 2^{q+1}} |G_m(x)| \right\|_\infty \leq C 2^{q/2},$$

where C is a constant. Hence, for $2^q < N < 2^{q+1}$,

$$\left\| \sum_{2^q}^N (f, G_m) G_m \right\|_\infty \leq C \sup_{2^q \leq m \leq 2^{q+1}} \|G_m\|_1 \left\| \sum_{2^q}^{2^{q+1}} |G_m| \right\|_\infty \|f\|_\infty$$

$$\leq C' \|f\|_\infty.$$

We have concluded the proof of Theorem 12, modulo the proof of Proposition 8.

We go back to identity (12.1) of Chapter 2. After periodification, this becomes

$$E_q = S_q S_q^\star + M_q \Delta_q + \Delta_q M_q^{-1} ,$$

where S_q is a convolution operator whose explicit form does not concern us, where M_q is the operator given by pointwise multiplication by $\exp(2\pi i 2^q x)$, and where Δ_q is a convolution operator such that

$$\Delta_q(e^{i\xi x}) = \eta(2^{-q}\xi)e^{i\xi x},$$

where $\xi \in 2\pi \mathbf{Z}$ and $\eta(\xi) \in \mathcal{D}$ has support in $[-2/3, -1/3]$.

So $\Delta_q P = P\Delta_q = 0$, but $PM_q\Delta_q = M_q\Delta_q$. Similarly, we have $P\Delta_q M_q^{-1} = 0$, while $\Delta_q M_q^{-1} P = \Delta_q M_q^{-1}$.

Finally, $E_q P - PE_q = \Delta_q M_q^{-1} - M_q \Delta_q$, and each of these operators is bounded on the space $L^\infty(\mathbf{R})$.

The proof we have given applies, without alteration, to the algebra $A_r(D)$ of functions which are holomorphic on the open disc $|z| < 1$ and extend to a C^r-function on the closed disc $|z| \leq 1$. The sequence G_m, $m \in \mathbf{N}$, is a Schauder basis for $A_r(D)$.

12 Conclusion

We can summarize the results of this chapter by noting that everything takes place as if the wavelets ψ_λ, $\lambda \in \Lambda_j$, were eigenvectors of the differential operators ∂^α, $|\alpha| \leq r$, with corresponding eigenvalues $2^{j|\alpha|}$. Similarly, everything takes place as if the wavelets enjoyed the strong orthogonality condition of the Haar system, namely, that ψ_λ, $\lambda \in \Lambda_j$, is "almost orthogonal" to $F(f)$, for $f \in V_j$ and F infinitely differentiable.

Let us illustrate these remarks by an example. To calculate the norm of $\sum \alpha_\lambda \psi_\lambda(x)$ in $\dot{L}^{4,s}$, $s \in \mathbb{R}$, we apply a fractional differentiation operator of order s to the series. This gives $\sum \alpha_\lambda 2^{js} \psi_\lambda(x)$, taking account of the preceding heuristic remarks. Then we evaluate the L^4 norm of $g(x) = \sum \alpha_\lambda 2^{js} \psi_\lambda(x)$ by expanding $\int g^2(x) \bar{g}^2(x) \, dx$. This leads to integrals $I(\lambda_1, \lambda_2, \lambda_3, \lambda_4) = \int \psi_{\lambda_1}(x) \psi_{\lambda_2}(x) \bar{\psi}_{\lambda_3}(x) \bar{\psi}_{\lambda_4}(x) \, dx$. The "strong orthogonality" property shows that $I(\lambda_1, \lambda_2, \lambda_3, \lambda_4)$ can be replaced by 0, except when $\lambda_1 = \lambda_3$ and $\lambda_2 = \lambda_4$, or when $\lambda_1 = \lambda_4$ and $\lambda_2 = \lambda_3$. Lastly, to calculate the integrals $\int |\psi_\lambda(x)|^2 |\psi_{\lambda'}(x)|^2 \, dx$, we use the wavelets' localization properties to replace $|\psi_\lambda(x)|^2$ by $2^{nj} \chi_\lambda(x)$, where $\chi_\lambda(x)$ is the characteristic function of the cube $Q(\lambda)$ corresponding to ψ_λ. Then $\int |\psi_\lambda(x)|^2 |\psi_{\lambda'}(x)|^2 \, dx$ is replaced by 0, except when $Q(\lambda) \subset Q(\lambda')$ or $Q(\lambda') \subset Q(\lambda)$, and, in this case, the integral becomes $\inf(|Q(\lambda)|, |Q(\lambda')|) 2^{n(j+j')}$.

This heuristic approach gives exactly the same results as those described by Theorem 3.

References

[1] AHLFORS, L. Zur Theorie der Überlagerungsflächen. *Acta Mathematica*, **65**, 1935, 157-194.

[2] ARSAC, J. *Transformation de Fourier et théorie des distributions*. Dunod, Paris, 1961.

[3] AUSCHER, P. Wavelets on chord-arc curves. *Ondelettes, Méthodes Temps-Fréquences et Espaces des Phases*, Combes, J.M., Grossman, A. and Tchamitchian, P. eds., C.P.T., C.N.R.S.-Luminy, Case 907, 13288-Marseille-Cedex 9.

[4] AUBIN, J.P. *Approximation of elliptic boundary-value problems*. Pure and Applied Mathematics, Krieger Publishing Co., Huntington, N.Y., 1980.

[5] BAERNSTEIN II, A. and SAWYER, E.T. *Embedding and multiplier theorems for $H^p(\mathbf{R})$*, Memoirs of the A.M.S., **53**, 1985.

[6] BALIAN, R. Un principe d'incertitude fort en théorie du signal ou en mecanique. *C.R. Acad. Sci. Paris*, **292** *Série II*, 1981, 1357-1361.

[7] BALSLEV, E., GROSSMANN, A. and PAUL, T. A characterization of dilation analytic operators. *Annales I.H.P., Physique Théorique*, **45**, 1986.

[8] BATTLE, G. A block spin construction of ondelettes, Part I: Lemarié functions. *Comm. Math. Phys.* **110**, 1987, 601-615.

[9] BATTLE, G. A block spin construction of ondelettes, Part II: the QFT connection. *Comm. Math. Phys.* **114**, 1988, 93-102.

[10] BATTLE, G. and FEDERBUSH, P. Ondelettes and phase cluster expansion, a vindication. *Comm. Math. Phys.* **109**, 1987, 417-419.

[11] BEALS, R. Characterization of pseudo-differential operators and applications. *Duke Math. J.* **44**, 1977, 45-57.

[12] BENEDEK, A. CALDERÓN, A. and PANZONE, R. Convolution operators on Banach space valued functions. *Proc. Nat. Acad. Sci. U.S.A.*, **48**, 1962, 356-365.

[13] BERKSON, E. On the structure of the graph of the Franklin analyzing wavelet. In *Analysis at Urbana*, (Berkson et al, eds.), Cambridge University Press, Cambridge, 1989.

[14] BESOV, O.V. Théorèmes de plongement des espaces fonctionnels. *International Math. Congress, Nice* 1970, II, 467-463.

[15] BEURLING, A. Construction and analysis of some convolution algebras. *Ann. Inst. Fourier (Grenoble)*, **14**, 1962, 1-32.

[16] BONY, J.M. Calcul symbolique et propagation des singularités pour les équations aux dérivés partielles non-linéaires. *Ann. Scient. E.N.S.*, **14**, 1981, 209-246.

[17] BONY, J.M. Propagation et interaction des singularités pour les solutions des équations aux dérivés partielles non-linéaires. *Proc. Int. Congress Math.*, *Warzawa*, 1983, 1133-1147.

[18] BONY, J.M. Interaction des singularités pour les équations aux dérivés partielles non-linéaires. *Séminaire E.D.P.*, 1979-80, no. 22, 1981-82, no. 2 and 1983-84, no. 10, Centre de Mathématique, Ecole Polytechnique, 91128-Palaiseau, France.

[19] BOOLE, G. On the comparison of transcendents with certain applications to the theory of definite integrals., *Phil. Trans. Roy. Soc.*, **47**, 1857, 745-803.

[20] BOCHKARIEV, S.V. Existence of bases in the space of analytic functions and some properties of the Franklin system. *Mat. Sbornik*, **98**, 1974, 3-18.

[21] BOURDAUD, G. *Sur les opérateurs pseudo-différentiels à coéfficients peu réguliers.* Université de Paris VII, Mathématique, tour 45-55, 5ième étage, 2 Place Jussieu, 75251 Paris Cedex 05.

[22] BOURDAUD, G. Réalisations des espaces de Besov homogènes. *Arkiv för Mat.*, **26**, 1988, 41-54.

[23] BOURDAUD, G. Localisation et multiplicateurs des espaces de Sobolev homogènes. *Manuscripta. Math.* **60**, 1988, 93-103.

[24] BOURGAIN, J. On the L^p-bounds for maximal functions associated to convex bodies. *Israel J. Math.*, **54**, 1986, 307-316.

[25] BOURGAIN, J. Geometry of Banach spaces and Harmonic Analysis. *Proc. Int. Congress Math., Berkeley, Ca.*, 1986, 871-878.

[26] BOURGAIN, J. Extension of a result of Benedek, Calderon and Panzone. *Arkiv för Math.*, **22**, 1984, 91-95.

[27] BOURGAIN, J. *Some remarks on Banach spaces in which martingale difference sequences are unconditional.* Preprint.

[28] BOURGAIN, J. *On square functions on the trigonometric system.* Preprint.

[29] BOURGAIN, J. Vector-valued singular integrals and the H^1-BMO duality. *Probability theory and harmonic analysis*, Chao, J.-A. and Woyczinski, W.A., eds., Marcel Dekker, New York, 1986, 1-19.

[30] BURKHOLDER, D.L. A geometric condition that implies the existence of certain singular integrals of Banach space valued functions. *Conference in Harmonic Analysis in Honor of Antoni Zygmund*, 270-286, Beckner, W., Calderón, A.P., Fefferman, R. and Jones, P.W. eds., Wadsworth, Belmont Ca., 1983.

[31] BURKHOLDER, D.L. Martingale theory and harmonic analysis in Euclidean spaces. *Harmonic analysis in Euclidean spaces. Proc. Symp. Pure Math.*, **35**, 1979, 2, 283-301.

[32] BURKHOLDER, D.L. Martingales and Fourier analysis in Banach spaces. *Probability and analysis, Varenna (Como)*, 1985, Lecture Notes in Mathematics, **1206**, 61-108, Springer-Verlag, Berlin, 1986,

[33] BURKHOLDER, D. A geometric characterization of Banach spaces in

which martingale difference sequences are unconditional. *Annals of Probability, Vol.* **9**, **6**, 1981, 997-1011.

[34] BURKHOLDER, D. GUNDY, R. and SILVERSTEIN, M. A maximal function characterization of the class H^p. *Trans. Amer. Math. Soc.*, **157**, 1971, 137-153.

[35] CALDERÓN, A.P. Intermediate spaces and interpolation, the complex method. *Studia Math.* **24**, 1964, 113-190.

[36] CALDERÓN, A.P. Uniqueness in the Cauchy problem for partial differential equations. *Amer. J. Math.* **80**, 1958, 15-36.

[37] CALDERÓN, A.P. Algebra of singular integral operators. *Singular Integrals. Proc. Symp. Pure Math.*, **10**, 1967, 18-55.

[38] CALDERÓN, A.P. Cauchy integrals on Lipschitz curves and related operators. *Proc. Nat. Acad. Sci. U.S.A.*, **74**, 1977, 1324-1327.

[39] CALDERÓN, A.P. Commutators, singular integrals on Lipschitz curves and applications. *Proc. Int. Congress Math., Helsinki*, 1978, 85-96.

[40] CALDERÓN, A.P. Boundary value problems for the Laplace equation in Lipschitzian domains. *Recent progress in Fourier analysis*, North Holland Mathematical Studies, **111**, Peral, I. and Rubio de Francia, J.L., eds., North-Holland, Amsterdam 1983, 33-49.

[41] CALDERÓN, A.P. and TORCHINSKY, A. Parabolic maximal functions associated with a distribution. *Advances in Math.*, **16**, 1975, 1-63 and **24**, 1977, 101-171.

[42] CALDERÓN, A.P. and ZYGMUND.A. On the existence of certain singular integrals. *Acta Mathematica*, **88**, 85-139.

[43] CALDERÓN, A.P. and ZYGMUND.A. Singular integrals and periodic functions. *Studia Math.*, **14**, 1954, 249-271.

[44] CALDERÓN, A.P. and ZYGMUND.A. On singular integrals. *Amer. J. Math.*, **78**, 1956, 289-309 and 310-320.

[45] CALDERÓN, A.P. and ZYGMUND.A. Singular integral operators and differential equations. *Amer. J. Math.*, **79**, 1957, 901-921.

[46] CARLESON, L. An explicit unconditional basis in H^1. *Bull. des Sciences Math.*, **104**, 1980, 405-416.

[47] CARLESON, L. Interpolation of bounded analytic functions and the corona problem. *Annals of Math.*, **76**, 1962, 547-559.

[48] CARLESON, L. Two remarks on H^1 and BMO. *Analyse Harmonique*, Orsay, 1975, preprint no. 164.

[49] CHANG, A. Two remarks about H^1 and BMO on the bidisc. *Conference in Harmonic Analysis in Honor of Antoni Zygmund*, 373-393, Beckner, W., Calderón, A.P., Fefferman, R. and Jones, P.W. eds., Wadsworth, Belmont Ca., 1983.

[50] CHANG, A. and CIESIELSKI, Z. Spline characterizations of H^1. *Studia Math.*, **75**, 1983, 183-192.

[51] CHANG, S.Y. and FEFFERMAN, R. A continuous version of the duality of H^1 with BMO. *Annals of Math.*, **112**, 1980, 179-201.

[52] CHRIST, M. Weighted norm inequalities and Schur's lemma. *Studia Math.*, **78**, 1984, 309-319.

[53] CHRIST, M. and JOURNÉ, J.L. Polynomial growth estimates for multilinear singular integral operators. *Acta Mathematica*, **159**, 1987, 51-80.

[54] CHRIST, M. and RUBIO DE FRANCIA, J.L. Weak type (1,1)-bounds for rough operators. *Inv. Math.*, **93**, 1988, 225-237.

[55] CIESIELSKI, Z. Properties of the orthonormal Franklin system. *Studia Math.*, **23**, 1963, 141-157 and **27**, 1966, 289-323.

[56] CIESIELSKI, Z. Bases and approximations by splines. *Proc. Int. Congress of Math., Vancouver*, 1974, Vol. II, 47-51.

[57] CIESIELSKI, Z. Haar orthogonal functions in analysis and probability. *Alfred Haar Memorial Conference*, 25-26, Colloquia Mathematica Societatis Janos Bolyai, Budapest, 1985.

[58] CIESIELSKI, Z. and FIGIEL,T. Spline approximations and Besov spaces on compact manifolds. *Studia Math.*, **75**, 1982, 13-16.

[59] CIESIELSKI, Z. and FIGIEL,T. Spline bases in classical function spaces on compact manifolds. *Studia Math.*, **76**, 1983, 95-136.

[60] COIFMAN, R.R. A real variable characerization of H^p. *Studia Math.*, **51**, 1974, 269-274

[61] COIFMAN, R.R., DAVID, G. and MEYER, Y. La solution des conjectures de Calderón. *Advances in Math.*, **48**, 1983, 144-148.

[62] COIFMAN, R.R., DENG, D.G. and MEYER, Y. Domaine de la racine carée de certains opérateurs différentiels accrétifs. *Ann. Inst. Fourier (Grenoble)*, **33**, 2, 1983, 123-134.

[63] COIFMAN, R.R. and FEFFERMAN, C. Weighted norm inequalities for maximal functions and singular integrals. *Studia Math.*, **54**, 1974, 241-250.

[64] COIFMAN, R.R., JONES, P.W. and SEMMES, S. *Two elementary proofs of the L^2-boundedness of the Cauchy integral on Lipschitz curves.* Preprint, Dept. of Math., Yale University, New Haven, CT 06520, U.S.A.

[65] COIFMAN, R.R., MCINTOSH, A. and MEYER, Y. L'intégrale de Cauchy définit un opérateur borné sur les courbes Lipschitziennes. *Annals of Math.*, **116**, 1982, 361-387.

[66] COIFMAN, R.R., MCINTOSH, A. and MEYER, Y. The Hilbert transform on Lipschitz curves. *Miniconference on partial differential equations, Canberra*, July 9-10, 1981, Price, P.F., Simon, L.M. and Trudinger, N.S., eds.

[67] COIFMAN, R.R. and MEYER, Y. Lavrentiev curves and conformal mappings. *Institut Mittag-Leffler, Report no.* **5**, 1983.

[68] COIFMAN, R.R. and MEYER, Y. Au delà des opérateurs pseudo-différentiels. *Astérisque*, **57**, Soc. Math. France, 1978.

[69] COIFMAN, R.R. and MEYER, Y. Fourier analysis of multilinear convolutions., Calderon's theorem and analysis on Lipschitz curves. *Euclidean harmonic analysis, proceedings*, 109-122, Lecture Notes in Mathematics, **779**, Springer-Verlag, Berlin, 1979.

[70] COIFMAN, R.R. and MEYER, Y. Non-linear harmonic analysis, operator theory and P.D.E., *Beijing Lectures in Harmonic Analysis*, Stein, E.M., ed., *Annals of Math. Studies*, **112**, 1986.

[71] COIFMAN, R.R., MEYER, Y. and STEIN, E.M. Some new function spaces and their applications to harmonic analysis. *J. Funct. Anal.*, **62**, 1985, 304-335.

[72] COIFMAN, R.R., MEYER, Y. and STEIN, E.M. Un nouvel espace fonctionnel adapté à l'étude des opérateurs définis par des intégrales singlières. *Harmonic Analysis, Cortona 1982*, Lecture Notes in Math., **992**, Springer-Verlag, Berlin, 1983.

[73] COIFMAN, R.R., ROCHBERG, R. et al., The molecular characterization of certain Hardy spaces. *Astérisque* **77**, Soc. Math. France, 1980.

[74] COIFMAN, R.R., ROCHBERG, R. and WEISS, G. Factorization theorems for Hardy spaces in several complex variables. *Annals of Math.*, **103**, 1976, 611-635.

[75] COIFMAN, R.R. and WEISS, G. Extensions of Hardy spaces and their use in analysis. *Bull. Amer. Math. Soc.*, **83**, 1977, 569-645.

[76] COIFMAN, R.R. and WEISS, G. *Analyse harmonique non commutative sur certains espaces homogènes.* Lecture Notes in Math., **242**, Springer-Verlag, Berlin, 1971.

[77] COIFMAN, R.R. and WEISS, G. Transference methods in analysis. *Regional conference series in mathematics, no.* **31**, Amer. Math. Soc., Providence RI.

[78] CORDOBA, A. Maximal functions, covering lemmas and Fourier multipliers. *Harmonic analysis in Euclidean spaces, Proc. Symp. Pure Math.*, **35**, 1979, I, 29-50.

[79] CORDOBA, A. and FEFFERMAN, R. A geometric proof of the strong maximal theorem. *Annals of Math.*, **102**, 1975.

[80] COWLING, M. Harmonic analysis on semigroups. *Annals of Math.*, **117**, 1983, 267-283.

[81] DAHLBERG, B. Estimates of harmonic measures. *Arch. for Rat. Mech. and Anal.*, **65**, 1977, 275-288.

[82] DAHLBERG, B. Real analysis and potential theory. *Proc. Int. Congress Math., Warzawa*, 1983, 2, 953-959.

[83] DAHLBERG, B. Poisson semigroups and singular integrals. *Proc. Amer. Math. Soc.*, **97**, 1986, 41-48.

[84] DAHLBERG, B., JERISON, D. and KENIG, C. Area integral estimates for elliptic differential operators with non-smooth coefficients. *Arkiv för Mat.*, **22**, 1984, 97-108.

[85] DAHLBERG, B. and KENIG, C. The L^p Neumann problem for Laplace's equation on Lipschitz domains. *Annals of Math.*, **125**, 1987, 435-465.

[86] DAHLBERG, B., KENIG, C. and VERCHOTA, G. The Dirichlet problem for the bi-Laplacian on Lipschitz domains. *Ann. Inst. Fourier (Grenoble)*, **36**, 1986, 109-135.

[87] DAUBECHIES, I. The wavelet transform, time-frequency localization and signal analysis. *IEEE Trans. Inf. Th.*, **36**, 1990, 961-105.

[88] DAUBECHIES, I. Orthonormal basis of compactly supported wavelets. *Comm. Pure Appl. Math.*, **46**, 1988, 909-996.

[89] DAUBECHIES, I., GROSSMANN, A. and MEYER, Y. Painless nonorthonormal expansions. *J. Math. Phys.*, **27**, 5, May 1986, 1271-1283.

[90] DAUBECHIES, I. and PAUL, T. Wavelets and applications. *Proc. VIIIth Int. Congress on Math. Phys.*, Mebkhout, M. and Seneor, R., eds., World Scientific Publishers, Teaneck NJ, 1987.

[91] DAUBECHIES, I., KLAUDER, J.R. and PAUL, T. Wiener measures for path integrals with affine kinematic variables. *J. Math. Phys.*, **28**, 1987, 85-102.

[92] DAVID, G. Opérateurs de Calderon-Zygmund. *Proc. Int. Congress Math., Berkeley, Ca.*, 1986, 890-899.

[93] DAVID, G. Opérateurs intégraux singuliers sur certaines courbes du plan complexe. *Ann. Sci. E.N.S.*, **17**, 1984, 157-189.

[94] DAVID, G. *A lower estimate for the norm of the Cauchy integral operator on Lipschitz curves.* Preprint.

[95] DAVID, G. Opérateurs d'intégrale singulierère sur les surfaces régulières. *Ann. Sci. E.N.S.*, **21**, 1988, 225-258.

[96] DAVID, G. and JOURNÉ, J.L. A boundedness criterion for generalized Calderon-Zygmund operators. *Annals of Math.*, **120**, 1984, 371-397.

[97] DAVID, G., JOURNÉ, J.L. and SEMMES, S. Opérateurs de Calderon-Zygmund, fonctions para-accrétives at interpolation. *Revista Matematica Ibero-Americana*, **1**, 1985, 4, 1-56.

[98] DAVID, G., JOURNÉ, J.L. and SEMMES, S. *Calderon-Zygmund operators, para-accretive functions, and interpolation.* Preprint, Thèse, Centre de Mathé-matique, Ecole Polytechnique and Dept. Math., Yale, U.S.A.

[99] DAVID, G. and SEMMES, S. L'opérateur défini par PV $\int |A(x) - A(y)|/ |x - y|(x - y)^{-1} f(y) \, dy$ est borné sur $L^2(\mathbb{R})$ lorsque A est Lipschitzienne. *C.R. Acad. Sci., Paris*, **303, Série I**, 1986, 499-502.

[100] DESLAURIERS, G. and DUBUC, S. Interpolation dyadique. *Fractals, dimensions non entières et applications*, G.Cherbit, ed., Masson, Paris, 1987.

[101] DUBUC, S. Interpolation through an iterative scheme. *J. Math. Anal. and Appl.*, **114**, 1, 1986, 185-204.

[102] DUOANDIKOETXEA, J. and RUBIO DE FRANCIA, J.L. Maximal and singular integral operators via Fourier transform estimates. *Inv. Math.*, **84**, 1986, 541-561.

[103] DUREN, P. *Theory of H^p spaces*, Academic Press, New York, 1970.

[104] FABES, E., JODEIT, M. and RIVIÈRE, N. Potential techniques for boundary value problems on C^1 domains. *Acta Mathematica*, **141**, 1978, 165-186.

[105] FABES, E., JERISON, D. and KENIG, C. Multilinear Littlewood-Paley estimates with applications to partial differential equations. *Proc. Nat. Acad. Sci., U.S.A.*, **79**, 1982, 5746-5750.

[106] FABES, E., JERISON, D. and KENIG, C. Multilinear square functions and partial differential equations. *Amer. J. Math.*, **107**, 1985, 1325-1367.

[107] FEDERBUSH, P. Quantum theory in ninety minutes. *Bull. Amer. Math. Soc.*, **17**, 1987,93-103.

[108] FEFFERMAN, C. Recent progress in classical Fourier analysis. *Proc. Int. Congress Math., Vancouver*, 1974, vol. I, 95-118.

[109] FEFFERMAN, C. and STEIN, E.M. H^p spaces of several variables. *Acta Mathematica*, **129**, 1972, 137-193.

[110] FEFFERMAN, R. Multiparameter Fourier analysis. *Beijing Lectures in Harmonic Analysis*, Stein, E.M., ed., *Annals of Math. Studies*, **112**, 1986, 47-130.

[111] FEFFERMAN, R. Functions of bounded mean oscillation on the bi-disc. *Annals of Math.*, **10**, 1979.

[112] FEFFERMAN, R. Calderon-Zygmund theory for product domains H^p spaces. *Proc. Nat. Acad. Sci., U.S.A.* **83**, 1986, 840-843.

[113] FRAZIER, M. and JAWERTH, B. Decomposition of Besov spaces. *Indiana University Math. J.*, **34**, 1985, 777-799.

[114] FRAZIER, M. and JAWERTH, B. The transform and decomposition of distributions. *Function spaces and applications, Lund, 1986,* Lecture Notes in Math., **1302** Springer-Verlag, Berlin, 1988.

[115] GARCIA-CURVA, J. and RUBIO DE FRANCIA, J.L. *Weighted norm inequalities and related topics.* North Holland Math. Studies, North-Holland, Amsterdam, 1985.

[116] GARNETT, J. *Bounded analytic functions* Academic Press, New York, 1981.

[117] GARNETT, J. Corona problems, interpolation problems and inhomogeneous Cauchy-Riemann equations. *Proc. Int. Congress Math., Berkeley, Ca.,* 1986.

[118] GILBERT, J.E. Nikishin-Stein theory and factorization with applications. *Harmonic analysis in Euclidean spaces, Proc. Symp. Pure Math.,* **35**, 2, 1979, 233-267.

[119] GIRAUD, G. Equations à intégrales principales. *Ann. Sci. E.N.S.,* 1934, 251-372.

[120] GLIMM, J. and JAFFE, A. *Quantum physics, a functional integral point of view.* Springer-Verlag, New York, 1981.

[121] GOUPILLAUD, P., GROSSMANN, A. and MORLET, J. Cycle-octave and related transforms in seismic signal analysis. *Geoexploration,* **23**, 1984-85, 85-102, Elsevier, Amsterdam.

[122] GRÖCHENIG, K. Analyse multiéchelle et bases d'ondelettes. *C.R. Acad. Sci. Paris,* **305** *Série I,* 1987, 13-17.

[123] GROSSMAN, A., HOLSCHNEIDER, M., KRONLAND-MARTINET, R. and MORLET, J. *Detection of abrupt changes in sound signals with the help of the wavelet transform.* Preprint, Centre de Physique Théorique, C.N.R.S. Luminy Case 907, 13288 Marseilles Cedex 9, France.

[124] GROSSMAN, A. and MORLET, J. Decomposition of Hardy functions into square integrable wavelets of constant shape. *SIAM J. Math. Anal.,* **15**, 1984, 723-736.

[125] GROSSMAN, A., MORLET, J. and PAUL, T. Integral transforms associated to square integrable representations I *J. Math. Phys,* **26**, 1985, 2473-2479, and II *Ann. Inst. Henri Poincaré, Phys. Théorique,* **45**, 1986, 293-309.

[126] DE GUZMAN, M. *Differentiation of integrals in* $\mathbf{R}^n$. Lecture Notes in Mathematics, **481**, Springer-Verlag, Berlin, 1975.

[127] DE GUZMAN, M. *Real variable methods in Fourier analysis.* Notas de Matematica, North-Holland Math. Studies, **46**, North-Holland, Amsterdam.

[128] HAAR, A. Zur Theorie der orthogonalen Funktionensysteme. *Math. Ann.,* **69**, 1910, 331-371.

[129] HARDY, G.H. and LITTLEWOOD, J.E. A maximal theorem with function theoretic applications. *Acta Mathematica,* **54**, 1930, 81-116.

[130] HELSON, H. *Harmonic Analysis.* Addison-Wesley, Reading, Mass., 1983.

[131] HOLSCHNEIDER, M. On the wavelet transformation of fractal objects. *J. Stat. Phys.,* **50**, 1988, 963-993.

[132] HOLSCHNEIDER, M., KRONLAND-MARTINET, R., MORLET, J. and TCHAMITCHIAN, P. L'algorithme à trous. *Ondelettes, Méthodes Temps-Fréquences et Espaces des Phases,* Combes, J.M., Grossmann, A. and Tchamitchian, P. eds., C.P.T., C.N.R.S.-Luminy, Case 907, 13288-Marseille-Cedex 9.

[133] HÖRMANDER, L. *The analysis of linear partial differential equations*, Vols I, II and III. Springer-Verlag, New York, 1983-85.

[134] JAFFARD, S. and MEYER, Y. Bases d'ondelettes dans des ouverts de $\mathbf{R}^n$. *J. Math. Pures et Appliqués*, **68**, 1989, 95-108.

[135] JANSON, S. Verbal communication. *Congrès d'Analyse de Fourier, El Escorial, 1987*

[136] JERISON, D. and KENIG, C. An identity with applications to harmonic measures. *Bull. Amer. Math. Soc.*, **2**, 1980, 447-451.

[137] JERISON, D. and KENIG, C. The Dirichlet problem in non-smooth domains. *Annals of Math.*, **113**, 1981, 367-382.

[138] JERISON, D. and KENIG, C. Boundary value problems on Lipschitz domains. *Studies in partial differential equations*, Littman, W., ed., *MAA Studies in Math.*, **23**,1982, 1-68.

[139] JERISON, D. and KENIG, C. The Neumann problem on Lipschitz domains. *Bull. Amer. Math. Soc.*, **4**, 1981, 203-207.

[140] JOHN, F. and NIRENBERG, L. On functions of bounded mean oscillation. *Comm. Pure Appl. Math.*, **18**, 1965, 415-426.

[141] JOHNSON, R. Application of Carleson measures to partial differential equations and Fourier multiplier problems. *Harmonic Analysis, Cortona 1982*, Lecture Notes in Math., **992**, Springer-Verlag, Berlin, 1983.

[142] JONES, P.W. Some topics in the theory of Hardy spaces. *Topics in modern harmonic analysis*, Proceedings of a seminar held in Torino, May-June 1982, Istituto Nazionale di alta matematica Francesco Severi, Vol II, 551-569.

[143] JONES, P.W. Recent advances in the theory of Hardy spaces. *Proc. Int. Congress Math., Warzawa*, 1983, 829-838.

[144] JONES, P.W. *Congrès d'Analyse de Fourier, El Escorial, 1987*, Lecture Notes in Mathematics, Springer-Verlag, Berlin.

[145] JONES, P.W. and ZINSMEISTER, W. Sur la transformation conforme des domaines de Lavrentiev. *C.R. Acad. Sci. Paris, Série I*, **295**, 1982, 563-566.

[146] JOURNÉ, J.L. *Calderon-Zygmund operators, pseudo-differential operators and the Cauchy integral of Calderon.* Lecture Notes in Mathematics, **994**, Springer Verlag, Berlin, 1983.

[147] JOURNÉ, J.L. Calderon-Zygmund operators on product spaces. *Revista Math. Ibero-Americana*, **1**, 1985, 55-91.

[148] KAHANE, J.P. *Séries de Fourier absolument convergentes.* Ergebnisse der Math., **50**, Springer-Verlag, Berlin, 1970.

[149] KAHANE, J.P. *Some random series of functions*, Cambridge studies in advanced mathematics, **5**, Cambridge University Press, Cambridge, 1968.

[150] KAHANE, J.P., KATZNELSON, Y. and DE LEEUW, K. Sur les coefficients de Fourier des fonctions continues. *C.R. Acad. Sci. Paris*, **285**, Série I, 1977, 1001-1003.

[151] KATO, T. *Perturbation theory for linear operators.*, 1966, Springer-Verlag, New York.

[152] KATO, T. Scattering Theory. *MAA studies in Math.*, **7**, 1971, TAUB, A.H., ed., 90-115.

[153] KATO, T. and PONCE, G. *On the Euler and Navier-Stokes equations in Lebesgue spaces $L^{p,s}(\mathbf{R}^n)$.* Preprint, Dept. of Math., University of California, Berkeley, Ca., 94720, USA.

[154] KATZNELSON, Y. *An introduction to Harmonic Analysis.* John Wiley, New York, 1968.

[155] KELDYSH, M.V. and LAVRENTIEV, M.A. Sur la représentation conforme des domaines limités par des courbes rectifiables. *Ann. Sci. E.N.S.*, **54**, 1937, 1-38.

[156] KENIG, C. Weighted Hardy spaces on Lipschitz domains. *Harmonic analysis in Euclidean spaces, Proc. Symp. Pure Math.*, **35**, 1979, I, 263-274.

[157] KENIG, C. Weighted H^p spaces on Lipschitz domains. *Amer. J. Math.*, **102**, 1980, 129-163.

[158] KENIG, C. Recent progress on boundary value problems on Lipschitz domains. *Proc. Symp. Pure Math.*, **43**, 1985, 175-205.

[159] KENIG, C. Elliptic boundary value problems on Lipschitz domains. *Beijing Lectures in Harmonic Analysis*, STEIN, E.M., ed., *Annals of Math. Studies*, **112**, 1986.

[160] KENIG, C. and MEYER, Y. The Cauchy integral on Lipschitz curves and the square root of second order accretive operators are the same. *Recent progress in Fourier analysis*, North Holland Math. Studies, **111**, North-Holland, Amsterdam, 1985, 123-145.

[161] KENIG, C. and TOMAS, P. Maximal operators defined by Fourier multipliers. *Studia Math.*, **68**, 1980, 79-83.

[162] KOOSIS, P. *Introduction to H^p-spaces.* London Math. Soc. Lecture Notes, **40**, Cambridge University Press, Cambridge, 1980.

[163] LATTER, R.H. A characterization of $H^p(\mathbb{R})$ in terms of atoms. *Studia Math.* **62**, 1978, 93-101.

[164] LEMARIÉ, P.G. Continuité sur les espaces de Besov des opérateurs définis par des intégrales singulières. *Ann. Inst. Fourier (Grenoble)*, **35**, 1985, 4, 175-187.

[165] LEMARIÉ, P.G. Bases d'ondelettes sur les groupes de Lie stratifiés. *Bull. Soc. Math. France*, **117**, 1989, 211-232.

[166] LEMARIÉ, P.G. and MEYER, Y. Ondelettes et bases hilbertiennes. *Revista Ibero-Americana*, **2**, 1986, 1-18.

[167] LIÉNARD, J.S. *Speech analysis and reconstruction using short-time, elementary wave-forms.* LIMSI-CNRS, Orsay, France.

[168] LINDENSTRAUSS, J. and PELCZYNSKI, A. Contributions to the theory of classical Banach spaces. *J. Functional Anal.*, **8**, 1971, 225-241.

[169] LINDENSTRAUSS, J. and TZAFIRI, L. *Classical Banach spaces, I.*, Springer-Verlag, New York, 1977.

[170] LITTLEWOOD, J.E. and PALEY, R. Theorems on Fourier series and power series. *J. London Math. Soc.*, **6**, 1931, 230-233.

[171] LITTLEWOOD, J.E. and PALEY, R. Theorems on Fourier series and power series. *Proc. London Math. Soc.*, **42**, 1937, 52-89.

[172] McINTOSH, A. *Square roots of elliptic operators.* Centre for Math. Analysis, Australian National University, GPO Box 4, Canberra, ACT 2601, Australia.

[173] McINTOSH, A. Square roots of elliptic operators. *J. Functional Anal.*, **61**, 1985, 3, 307-327.

[174] McINTOSH, A. Functions and derivations of C^*-algebras. *J. Functional Anal.*, **30**, 1978, 2, 264-275.

[175] McINTOSH, A. Counter-example to a question on commutators. *Proc. Amer. Math. Soc.*, **29**, 1971, 430-434.

[176] McINTOSH, A. On representing closed accretive sesquilinear forms as $(A^{1/2}u, A^{*1/2}v)$. *Macquarie Math. Reports*, August 1981.

[177] McINTOSH, A. Clifford algebras and the higher dimensional Cauchy integral. *Approximation thory and function spaces*, 1986, Banach Centre, Warsaw, Poland.

[178] MALLAT, S. Multiresolution approximation and wavelet orthonormal bases of $L^2(\mathbb{R})$. *Trans. Amer. Math. Soc.*, **315**, 1989, 69-87.

[179] MALLAT, S. A theory for multiresolution signal decomposition: the wavelet representation. *IEEE Trans. on Pattern Analysis and Machine Intelligence, Technical Report MS-CIS-87-22*, 1989, University of Pennsylvania.

[180] MALLAT, S. Dyadic wavelets energy zero-crossings. *IEEE Trans. on Information Theory, Technical Report MS-CIS-88-30*, 1989, University of Pennsylvania.

[181] MALLAT, S. *Multiresolution Representations and Wavelets*, Ph.D. thesis in Electrical Engineering, University of Pennsylvania, Philadelphia, Pennsylvania, 19104, USA.

[182] MANDELBROT, B. *The fractal geometry of nature*, W.H. Freeman, New York, 1983.

[183] MARCINKIEWICZ, J. Sur les multiplicateurs des séries de Fourier. *Studia Math.*, **8**, 1939, 78-91.

[184] MAUREY, B. Isomorphismes entre espaces H^1. *Acta Mathematica*, **145**, 1980, 79-120.

[185] MAUREY, B. Le système de Haar. *Séminaire Maurey-Schwartz*, 1974-75, École, Polytechnique.

[186] MEYER, Y. Wavelets and operators. In *Analysis at Urbana I*, (Berkson et al, eds.), pp. 256-365, London Math. Soc. Lecture Notes, **137**, Cambridge University Press, Cambridge, 1989.

[187] MEYER, Y. Real analysis and operator theory. *Pseudo-differential operators and applications, Proc. Symp. Pure Math.*, **43**, 1985, 219-235.

[188] MEYER, Y. Ondelettes, fonctions splines et analyses graduées. *Cahiers Math. de la Décision*, **8703**, Ceremade.

[189] MEYER, Y. Intégrales singulières, opérateurs multilinéaires, analyse complexe et équations aux dérivées partielles. *Proc. International Congress of Math., Warzawa*, 1983, 1001-1010.

[190] MEYER, Y. Intégrales singulières, opérateurs multilinéaires et équations aux dérivées partielles. *Séminaire Goulaouic-Schwartz*, Ecole Polytechnique.

[191] MEYER, M. Une classe d'espaces de type BMO. Applications aux intégrales singulières. *Arkiv för Mat.*, **27**, 1989, 305-318.

[192] MUCKENHOUPT, B. Weighted norm inequalities for classical operators. *Harmonic analysis in Euclidean spaces, Proc. Symp. Pure Math.*, **35**, I, 69-83.

[193] MURAI, T. *Boundedness of singular integral operators of Calderon type (V and VI)*. Nagoya University preprint series, 1984

[194] MURAI, T. *A real variable method for the Cauchy transform and analytic capacity*. Lect. Notes in Math., **1307**, Springer-Verlag, Berlin, 1988.

[195] NAGEL, A., RIVIÈRE, N. and WAINGER, S. On Hilbert transformations along curves. *Bull. Amer. Math. Soc.*, **80**, 1974, 106-108.

[196] NAGEL, A., RIVIÈRE, N. and WAINGER, S. On Hilbert transforms along curves. *Amer. .J. Math.*, **98**, 1976, 395-403.

[197] NECAS, J. *Les méthodes directes en théorie des équations elliptiques.* Academia, Prague.

[198] O'NEILL, R., SAMPSON, G. and SOARES DE SOUZA, G. Several characterizations of the special atom space with applications. *Revista Matematica Ibero-Americana*, **2**, 1986, 3, 333-355.

[199] PAUL, T. Functions analytic on the half plane as quantum mechanic states. *J. Math. Phys.*, **25**, 1984, 3252-3256.

[200] PAUL, T. Wavelets and path integrals. *Ondelettes, Méthodes temps-fréquence et espaces des phases.* Combes, J.M., Grossman, A. and Tchamitchian, P., eds., C.P.T., CNRS-Luminy, Case 907, 13288 Marseille Cedex 9.

[201] PEETRE, J. On convolution operators leaving $L^{p,\lambda}$ spaces invariant. *Ann. Math. Pura Appl.*, **72**, 1966, 295-304.

[202] PEETRE, J. *New thoughts on Besov spaces.* Duke University Math. Dept., Durham, U.S.A., 1976.

[203] PHONG, D.H. and STEIN, E.M. Hilbert integrals, singular integrals and Radon transforms. *Acta Mathematica*, **157**, 1985, 99-157.

[204] PONCE, G. *Propagation of $L^{q,k}$ smoothness for solutions of the Euler equation.* Preprint, Dept. of Math., University of Chicago, Chicago Il., 60637.

[205] RICCI, F. and WEISS, G. A characterization of $H^1(\Sigma_{n-1})$. *Harmonic analysis in Euclidean spaces, Proc. Symp. Pure Math.*, **35**, 1979, I, 289-294.

[206] RODET, X. Time-Domain Formant-Wave-Function Synthesis. *Computer Music J.*, **8**, 1985, Part 3.

[207] RUBIO DE FRANCIA, J.L. A new technique in the theory of A_p weights. *Topics in modern harmonic analysis*, Proc. Sem. Torino-Milano, **2**, 571-580.

[208] RUBIO DE FRANCIA, J.L. A Littlewood-Paley inequality for arbitrary intervals. *Revista Matematica Ibero-Americana*, **1**, 1985, 2, 1-13.

[209] RUDIN, W. *Fourier analysis on groups*, Interscience, John Wiley, New York, 1962.

[210] SADOSKY, C. *Interpolation of operators and singular integrals*, Marcel Dekker, New York, 1979.

[211] SAKS, C. and ZYGMUND, A. *Analytic functions*, Monografie Matematyczne, Warszawa-Wroclaw, 1952.

[212] SEMMES, S.W. *A criterion for the boundedness of singular integrals on hypersurfaces.* Preprint, Dept. of Math., Yale University, New Haven, Ct, 06520, USA.

[213] SJÖLIN, P. and STRÖMBERG, J.O. Spline systems as bases in Hardy spaces. *Israel J. Math.*, **45**, 1983, 2-3, 147-156.

[214] SJÖLIN, P. and STRÖMBERG, J.O. Basis properties of Hardy spaces. *Arkiv för Mat.*, **21**, 1983, 111-125.

[215] SPANNE, S. Sur l'interpolation entre les espaces..., *Ann. Scuola Norm. Pisa*, **20**, 1966, 625-648.

[216] STEGENGA, D.A. Multipliers of the Dirichlet space. *Illinois J. Math.*, **24**, 1980, 113-139.

[217] STEIN, E.M. *Singular integrals and differentiability properties of functions.* Princeton University Press, Princeton NJ, 1970.

[218] STEIN, E.M. *Topics in harmonic analysis*. Annals of Math. Studies, **63**, Princeton University Press, Princeton NJ, 1970.

[219] STEIN, E.M. On limits of sequences of operators. *Annals of Math.*, **74**, 1961, 140-170.

[220] STEIN, E.M. and WAINGER, S. Problems in harmonic analysis related to curvature. *Bull. Amer. Math. Soc.*, **84**, 1978, 1239-1295.

[221] STEIN, E.M. and WEISS, G. *Introduction to Fourier analysis on Euclidean spaces*. Princeton University Press, Princeton NJ, 1971.

[222] STEIN, E.M. and WEISS, G. On the theory of H^p spaces. *Acta Mathematica*, **103**, 1960, 25-62.

[223] STRÖMBERG, J.O. A modified Franklin system and higher-order spline systems on $\mathbb{R}^n$ as unconditional bases for Hardy spaces. *Conference in Harmonic Analysis in Honor of Antoni Zygmund*, II, 475-493, Wadsworth, Belmont Ca, 1983.

[224] STRÖMBERG, J.O. Bounded mean oscillation with Orlicz norms and duality of Hardy spaces. *Indiana University Math. J.*, **28**, 1979, 511-544.

[225] STRÖMBERG, J.O. and TORCHINSKY, J. Weights, sharp maximal functions and Hardy spaces. *Bull. Amer. Math. Soc.*, **3**, 1980, 1053-1056.

[226] TAKAGI, T. A simple example of a continuous function without derivative. *The collected papers of Teiji Takagi*, 5-6 (III.78), Iwanami Shoten, Tokyo, 1973.

[227] TCHAMITCHIAN, P. Calcul symbolique sur les opérateurs de Calderón-Zygmund et bases inconditionelles de $L^2(\mathbb{R})$. *C. R. Acad. Sci. Paris*, **303**, 1986, 215-218.

[228] TCHAMITCHIAN, P. *Ondelettes adaptées à l'analyse complexe*. Preprint.

[229] TITCHMARSH, E.C. *Introduction to the theory of Fourier integrals*, Oxford at the Clarendon Press, 1937.

[230] TRIEBEL, H. *Theory of function spaces*. Monographs in Mathematics, **78**, Birkhauser, Basel, 1983.

[231] UCHIYAMA, A. A constructive proof of the Fefferman-Stein decomposition for BMO($\mathbb{R}^n$). *Acta Mathematica*, **148**, 1982, 215-241.

[232] VERCHOTA, G. Layer potentials and regularity for the Dirichlet problem for Laplace's equation in Lipschitz domains.*J. Functional Anal.*, **59**, 572-611.

[233] WAINGER, S. On some aspects of differentiation theory. *Topics in modern harmonic analysis*, Proc. Sem. Torino-Milano, May-June 1982, Istituto Nazionale di alta matematica Francesco Severi, 2, 667-700.

[234] WAINGER, S. Averages and singular integrals over lower dimensional sets. *Beijing lectures in harmonic analysis*, Stein, E.M., ed., *Annals of Math. Studies*, **112**, Princeton University Press, Princeton NJ, 1986.

[235] WILSON, J.M. A simple proof of the atomic decomposition for $H^p(\mathbb{R})$. *Studia Math.*, **74**, 1982, 25-33.

[236] WILSON, J.M. *On the atomic decomposition for Hardy spaces*. Preprint.

[237] WOJTASZCZYK, P. The Franklin system is an unconditional basis for H^1. *Arkiv för Mat.*, **20**, 1982, 293-300.

[238] ZINSMEISTER, M. *Domaines de Lavrentiev*. Publications mathématiques d'Orsay, **204**, 1986.

[239] ZYGMUND, A. *Trignometric Series*. Second edition, Cambridge University Press, Cambridge, 1968.

New references on wavelets and their applications

[1] BEYLKIN, G. et al., eds. *Wavelets*. Jones and Bartlett, Boston Ma., 1992.

[2] COIFMAN, R. Adapted multiresolution analysis, computation, signal processing and operator theory. *ICM 90, Kyoto*. Springer-Verlag, Berlin, 1992.

[3] DAVID, G. *Wavelets and singular integrals on curves and surfaces*. Lecture Notes in Mathematics, **1465**, Springer-Verlag, Berlin, 1991.

[4] GASQUET, C. and WITOMSKI, P. *Analyse de Fourier et applications. Filtrage, calcul numérique, ondelettes*. Masson, Paris, 1990.

[5] LEMARIÉ, P.G. *Les ondelettes en 1989*. Lecture Notes in Mathematics, **1438**, Springer-Verlag, Berlin, 1990.

[6] MEYER, Y., ed. *Wavelets and applications, Proceedings of the International Conference, Marseille, France, May 1989*, Masson, Paris & Springer-Verlag, Berlin, 1992.

[7] MEYER, Y. Wavelets and applications. *ICM 90, Kyoto*. Springer-Verlag, Berlin, 1992.

Index

Imprimé en France, Normandie Roto Impression s.a.
61250 Lonrai
Numéro d'impression : I2-2269
Dépôt légal : quatrième trimestre 1992
Hermann, éditeurs des sciences et des arts